Springer Undergraduate Mathematics Series

Springer

New York
Berlin
Heidelberg
Hong Kong
London
Milan
New Delhi
Paris
Tokyo

Other books in this series

A First Course in Discrete Mathematics *I. Anderson*
Analytic Methods for Partial Differential Equations *G. Evans, J. Blackledge, P. Yardley*
Applied Geometry for Computer Graphics and CAD *D. Marsh*
Basic Linear Algebra *T.S. Blyth and E.F. Robertson*
Basic Stochastic Processes *Z. Brzeźniak and T. Zastawniak*
Elementary Differential Geometry *A. Pressley*
Elementary Number Theory *G.A. Jones and J.M. Jones*
Elements of Abstract Analysis *M. Ó Searcóid*
Elements of Logic via Numbers and Sets *D.L. Johnson*
Further Linear Algebra *T.S. Blyth and E.F. Robertson*
Geometry *R. Fenn*
Groups, Rings and Fields *D.A.R. Wallace*
Hyperbolic Geometry *J.W. Anderson*
Information and Coding Theory *G.A. Jones and J.M. Jones*
Introduction to Laplace Transforms and Fourier Series *P.P.G. Dyke*
Introduction to Ring Theory *P.M. Cohn*
Introductory Mathematics: Algebra and Analysis *G. Smith*
Linear Functional Analysis *B.P. Rynne and M.A. Youngson*
Matrix Groups: An Introduction to Lie Group Theory *A. Baker*
Measure, Integral and Probability *M. Capiński and E. Kopp*
Multivariate Calculus and Geometry *S. Dineen*
Numerical Methods for Partial Differential Equations *G. Evans, J. Blackledge, P. Yardley*
Probability Models *J. Haigh*
Real Analysis *J.M. Howie*
Sets, Logic and Categories *P. Cameron*
Special Relativity *N.M.J. Woodhouse*
Symmetries *D.L. Johnson*
Topics in Group Theory *G. Smith and O. Tabachnikova*
Topologies and Uniformities *I.M. James*
Vector Calculus *P.C. Matthews*

D.F. Parker

Fields, Flows and Waves

An Introduction to Continuum Models

With 90 Figures

Springer

David F. Parker, MA, PhD, FRSE
School of Mathematics, University of Edinburgh,
James Clerk Maxwell Building,
The King's Buildings, Mayfield Road,
Edinburgh EH9 3JZ, UK

Parker: Fields, Flows and Waves

First Indian Reprint 2004

ISBN 81-8128-164-0

This edition is published by Springer (India) Private Limited, Registered Office: 906-907, Akash Deep Building, Barakhamba Road, New Delhi – 110 001, India.

Printed in India by Rajkamal Electric Press, Delhi.

Preface

Many phenomena in the physical and biological sciences involve the collective behaviour of (very large) numbers of individual objects. For example, the behaviour of gases ultimately concerns the interacting motions of uncountably many atoms and molecules, but to understand flow in nozzles, around aircraft and in meteorology it is best to treat velocity and density as *continuous* functions of position and time and then to analyse the associated *flows.* Although modern electronics involves ever smaller components, even the semiconductor devices used widely in electronic communications and in digital processing involve collective phenomena, such as *electric currents* and *fields*, which are continuously varying functions of position and time. Diffusion and reaction between various chemical constituents, the growth and spread of biological organisms and the flow of traffic on major highways are all phenomena which may be described and analysed in terms of *fields* and *flows*, while sound, light and various other electromagnetic phenomena involve both fields and *waves.* Treating these using a *continuum model*, which does not attempt to trace the motion and evolution of individual objects, often gives good predictions. The mathematical concepts and techniques which underlie such treatments are the subject of this book.

This book is designed as a first introduction to the *use* of mathematical techniques, within continuum theories. While it presumes some knowledge of several variable calculus and partial derivatives, it does not rely upon vector calculus beyond the concept of gradient of a scalar, despite the fact that we live in a three-dimensional world. The aim is to develop familiarity with the ideas of balance laws and conservation equations, showing how they give rise to ordinary and partial differential equations. Methods for solving and interpreting appropriate solutions are then introduced in context. Each of the ten chapters introduces a new topic and some methods appropriate for its analysis. The physical and practical aspects are used to help in the formulation of the models and in interpreting the mathematical predictions. Many worked examples are included in the text and each chapter includes many student exercises, for which solutions are given at the end of the book.

Chapter 1 deals with conservation and balance for flows in one dimension and with cylindrical and spherical symmetry, showing how relevant ordinary differential equations are derived and then solved subject to appropriate boundary conditions. In Chapter 2, unsteady flow of heat is treated, so illustrating the general concept of balance between supply rate, outflow and rate of change of storage. Partial differential equations are also encountered and solved in terms of separable solutions. Chapter 3 introduces potentials, in the context both of gravitation and of electrostatics. It also motivates Gauss's Law of flux in its integral form. Chapter 4 requires slightly more from vector calculus, in obtaining Laplace's equation, Poisson's equation and the heat equation from the balance laws. It then extends the method of separation of variables both in cartesian variables and to problems having spherical boundaries. Chapter 5 is the first to introduce waves. Both travelling and standing waves on a string are described, before the methods are shown to apply equally to sound waves in tubes and to telegraphy. Fluid flow is the topic of Chapter 6, in which balance and conservation are emphasized, streamlines, velocity potential and stream function are discussed, before pressure and momentum are analysed. While this, necessarily, is only an introduction to the extremely broad subject of fluid mechanics it serves to show the applicability of mathematical treatments introduced earlier. Similarly, Chapter 7 covers only the basics of elasticity, but besides further illustrating the occurrence of balance laws it also shows how the kinematics of deformation borrows from matrix algebra a number of important ideas. While mostly treating static deformations such as shearing and torsion of bars, it also reveals how the wave equation from Chapter 5 reappears. Waves then are the central feature of both Chapters 8 and 9. Using acoustics for motivation, the versatile ideas of the complex exponential and of plane waves are developed. Reflection and refraction at plane surfaces are considered, so leading to an analysis of guided waves in both underwater acoustics and in elasticity. Many of the techniques are then extended within Chapter 9, which concerns electromagnetic waves. This topic, of great importance to modern technology, is the most mathematically demanding in this book, but is included so that the interested reader can see both how techniques already learnt can predict important features and how refinement of the techniques can address recent research issues, such as the design of optical fibres for telecommunications. The final chapter shows how ideas and techniques developed previously apply not just to physical science. Through considering chemical reaction and diffusion, travelling wavefronts are encountered. Very similar mathematics arises in describing the growth and spread of biological organisms. The concepts of self-similar solutions, diffusive instability and pattern formation are introduced. Bio-mathematics is a rapidly growing topic and this chapter serves to indicate how many of the ideas of fields, flows and waves are helpful in its development.

Much of the material in Chapters 1–7 was developed as a 24-lecture course for middle undergraduate years at the University of Edinburgh, for students having some knowledge of Fourier series and of several variable calculus, but not presumed to be familiar with techniques for solving partial differential equations. Chapters 1–5 form the core material, which could be combined with different selections of material from Chapters 6, 7, 8 or 10 to form a course of similar duration. Undeniably, Chapters 8–10 involve some techniques which are more advanced than those in earlier chapters, but it is felt appropriate to show students how core techniques of applied mathematics can indicate methods for analysing and explaining important effects in modern technology and science. Chapter 9 is strongly linked to Chapter 8 and so might be appropriate for a course at more advanced level, or alternatively could be combined with some of the earlier material for students who will study fluid mechanics and elasticity in separate courses. Chapter 10, which is necessarily only an introduction to topics in bio-mathematics, similarly might be appropriate for inclusion with the core material.

Much of the material has been class-tested and so my thanks go to all those students who pointed out errors in earlier versions. My gratitude is due also to the editorial staff (present and past) at Springer-Verlag who were sympathetic about slippage of deadlines and in particular to Karen Borthwick for much helpful advice. Many of the figures were professionally prepared by Aaron Wilson, to whom I am most grateful. I am also much indebted to Margarida Facão for her considerable computational effort in preparing two of the figures in Chapter 10. While all these thanks are heartfelt, they are nothing in comparison to my thanks to Tanya for accepting without demur that so many weekends and evenings should be dominated by my attachment to the laptop computer. Her encouragement has been immense. In this, as in so many other things, my indebtedness to her is unbounded.

Edinburgh
January 2003

Contents

1
The Continuum Description

At the most elementary level, many collective phenomena involve large numbers of individual, but interacting, objects. However, meteorologists do not attempt to trace the motions of individual molecules, but (as a glance at any weather map will show) they deal with averaged properties such as temperature, pressure, humidity and wind speed. Similarly, engineers do not analyse the stretching of bonds between neighbouring molecules in a structural solid, nor do radio antenna designers consider the motions of individual electrons within metals. They use *macroscopic* or *continuum* theories – theories dealing with quantities averaged over very many neighbouring *particles.*

1.1 Densities and Fluxes

The simplest of *continuum models* involve a *density*, a *flux vector* and a *balance law.* To analyse these, a few mathematical preliminaries and definitions are necessary.

Let $\boldsymbol{x} = \overrightarrow{\mathrm{OP}}$ denote the *position vector* of a point P relative to some origin O. In terms of cartesian coordinates x, y and z, it may be resolved into components as

$$\boldsymbol{x} = x\boldsymbol{i} + y\boldsymbol{j} + z\boldsymbol{k} \ ,$$

where $\boldsymbol{i}$, $\boldsymbol{j}$ and $\boldsymbol{k}$ are the *unit vectors* along the orthogonal axes Ox, Oy and Oz. Time is denoted by t. Quantities such as density which occur in the model will typically depend upon t and (one or more of) the coordinates x, y and z, so that *partial derivatives* will arise. For example, if $f(x, y, z, t)$ is a density, then $\partial f/\partial x$ denotes its partial derivative with respect to x, which is the derivative with respect to x when *all* remaining variables (viz. y, z and t) are held fixed. Successive derivatives (e.g. $\partial^2 f/\partial y^2$) may arise and equality of *mixed partial derivatives* (e.g. $\partial^2 f/\partial x \partial t = \partial^2 f/\partial t \partial x$) will be assumed.

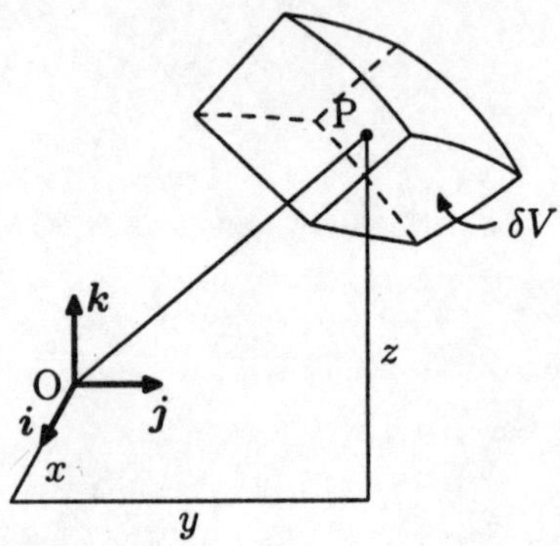

Figure 1.1 Volume element δV around a point P, at $\boldsymbol{x} = x\boldsymbol{i} + y\boldsymbol{j} + z\boldsymbol{k}$.

A *density* $f(\boldsymbol{x}, t)$ measures the amount of some relevant attribute per unit volume surrounding P at time t. The most familiar density is the *mass density* ρ (often abbreviated as *density*). It is obtained by dividing the mass contained within a small volume surrounding P by the volume δV of that *volume element* and then considering that ratio as δV shrinks to sizes containing merely a few thousand atoms (but not so small that the ratio is significantly affected by whether or not a particular atom is inside or outside δV. Since one cubic centimetre of water, oil, etc. contains around 10^{23} molecules, many continuum theories are applicable to phenomena having length scales even as small as 10^{-8} m $= 10^{-2}$ micron).

Other densities which are important are:

charge density in electromagnetism (measuring the charge per unit volume),

energy density, which may measure the amount of energy stored per unit volume as e.g. *thermal energy* (in heat flow), *strain energy* (in elasticity) or *chemical energy*,

population density (for biological organisms) or *species concentration* (measuring the mass of each constituent per unit volume in a mixture).

The attribute (mass, charge, energy, ...) recorded by a density may move. The rate and direction of flow is associated with a *flux vector*, typically $\boldsymbol{q}(\boldsymbol{x}, t)$. This is defined so that, when a small *area element* $\delta\Sigma$ containing P is chosen to have unit normal $\boldsymbol{n}$ parallel to $\boldsymbol{q}$ at time t, the rate at which the quantity crosses the area element is (approximately)

$$|\boldsymbol{q}|\delta\Sigma = \boldsymbol{q}\cdot\boldsymbol{n}\,\delta\Sigma\,.$$

Thus, $\boldsymbol{q}$ points in the direction of flow and has magnitude equal to the rate at which the relevant quantity crosses unit area. More generally, if an element of area $\delta\Sigma$ is centred at P but has arbitrary *unit normal* $\boldsymbol{n}$, the flow rate across that element is (by the rules of scalar products for vectors)

$$\boldsymbol{q}\cdot\boldsymbol{n}\,\delta\Sigma = |\boldsymbol{q}|\cos\alpha\,\delta\Sigma\,, \tag{1.1}$$

where α is the angle between the *flux vector* $\boldsymbol{q}$ and the normal to the surface element, so that $|\boldsymbol{q}|\cos\alpha$ is the component of flux normal to $\delta\Sigma$. The vector $\boldsymbol{n}\,\delta\Sigma$ is known as the *vector element of area.*

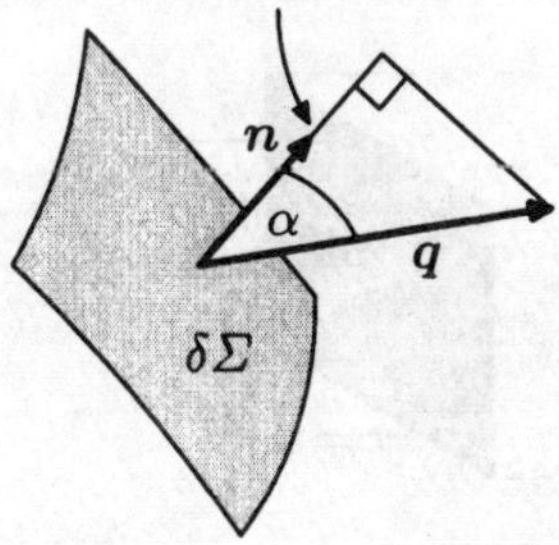

Figure 1.2 A surface element $\delta\Sigma$ having unit normal $\boldsymbol{n}$. The normal component $\boldsymbol{q}\cdot\boldsymbol{n} = |\boldsymbol{q}|\cos\alpha$ of a flux vector $\boldsymbol{q}$ measures the rate of flow per unit area.

1.2 Conservation and Balance Laws in One Dimension

A *conservation law* is a statement that, for any attribute which can be neither created nor destroyed but which (like mass and charge) may merely move, the total rate of outflow from some region must equal the rate of *decrease* of that attribute located within that region. In one-dimensional problems, for example with flow parallel to Ox so that the flux may be treated as a scalar $q(x,t)$ (taken positive along the positive sense of the x-axis), the law is simply found.

Consider a portion of slab having unit area and having faces at $x = x_1$ and $x = x_2$. If the attribute has density $f(x,t)$, the quantity located within the region at time t is

$$\int_{x_1}^{x_2} f(x,t)\,\mathrm{d}x = F(t) \qquad \text{(say)}\,, \tag{1.2}$$

while the flow rate *out of* the region is

$$q(x_2,t) - q(x_1,t) \qquad \text{for} \qquad x_1 < x_2\,.$$

Setting this outflow rate equal to $-\mathrm{d}F/\mathrm{d}t$ yields the *conservation law*:

$$\frac{\mathrm{d}}{\mathrm{d}t}\int_{x_1}^{x_2} f(x,t)\,\mathrm{d}x \;+\; q(x_2,t) - q(x_1,t) = 0\,. \tag{1.3}$$

In words, this may be stated as

Rate of increase stored within a region

+ Rate of outflow across its boundary = 0 .

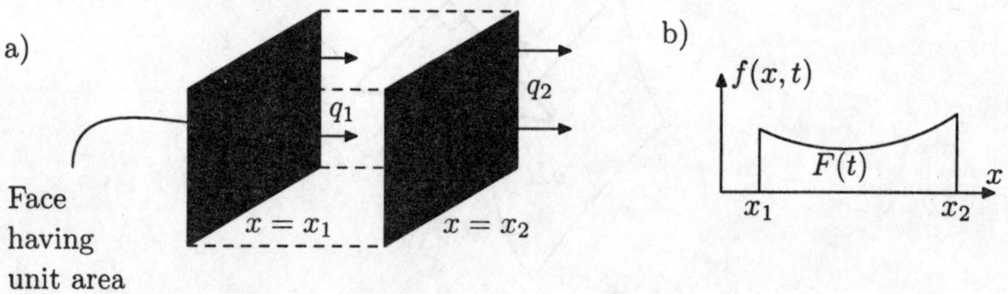

Figure 1.3 a) A rectangular box having unit cross-sectional area and extending over $x_1 < x < x_2$. Total outflow rate is $q_2 - q_1 = q(x_2,t) - q(x_1,t)$. b) At instant t, rate of increase of attribute within the box is $\mathrm{d}F/\mathrm{d}t$ (see (1.2)).

Equation (1.3) expresses the conservation law over an extended region. From it, a useful *point form* (or *local form*) may be deduced by using two fundamental results from calculus:

1. The statement

$$\int_{x_1}^{x_2} \frac{\partial q}{\partial x}\,\mathrm{d}x = [q(x,t)]_{x_1}^{x_2} = q(x_2,t) - q(x_1,t) \equiv q_2(t) - q_1(t) \tag{1.4}$$

follows directly from the *fundamental theorem of calculus*, since t is treated as a constant during both integration and partial differentiation.

2. *Differentiation under an integral sign* gives (since x_1 and x_2 are held fixed)

$$\frac{\mathrm{d}}{\mathrm{d}t}\int_{x_1}^{x_2} f(x,t)\,\mathrm{d}x = \int_{x_1}^{x_2} \frac{\partial f}{\partial t}\,\mathrm{d}x\,. \tag{1.5}$$

(Here, $\partial f/\partial t$ is the time rate of change at each internal point $x \in (x_1, x_2)$.) Applying both these results to (1.3) gives (since x_1 and x_2 are held fixed)

$$0 = \int_{x_1}^{x_2} \frac{\partial f}{\partial t}\,\mathrm{d}x + \int_{x_1}^{x_2} \frac{\partial q}{\partial x}\,\mathrm{d}x$$
$$= \int_{x_1}^{x_2} \left\{ \frac{\partial f}{\partial t} + \frac{\partial q}{\partial x} \right\} \mathrm{d}x\,.$$

However, the endpoints of this interval of integration are arbitrary. For *all* choices, the integral has value zero. Even when x_2 is altered to $x_2 + \delta x_2$, there is no change to the integral. Hence the *integrand* $\partial f/\partial t + \partial q/\partial x$ must equal zero at $x = x_2$ for all t. This result is true for all values of x_2, so yielding the *point form* of the *conservation law* :

$$\frac{\partial f}{\partial t} + \frac{\partial q}{\partial x} = 0\,. \tag{1.6}$$

Example 1.1

Suppose that sediment sinks through water occupying a tank having its bottom at $z = 0$. Suppose also that it is known that at time t (secs) the concentration of sediment (gm/m^3) is $c(z,t) = Cte^{-kzt}$, where C and k are positive constants. (a) Calculate the total mass of sediment within the region $0 \le x \le a$, $0 \le y \le b$, $0 \le z \le h$. (b) Show that the mass of sediment above 1 m^2 of floor area and below height z is $Ck^{-1}(1-\mathrm{e}^{-kzt})$ gm. (c) Assuming that sediment flow is purely vertical and is zero at $z = 0$, find the flux q (in gm/(m^2 sec), parallel to Oz) at height z. (d) Confirm the conservation law in the form $\partial c/\partial t + \partial q/\partial z = 0$.

(a) In this example, the concentration c depends only on z and t (z replaces x in (1.2)–(1.6)). Hence, the sediment mass within the rectangular region is

$$M = ab \int_0^h Cte^{-kzt}\,\mathrm{d}z = abC\left[-k^{-1}\mathrm{e}^{-kzt}\right]_0^h = abCk^{-1}(1 - \mathrm{e}^{-kht})\,.$$

(b) Sediment mass in $0 < z^* < z$ above 1 m^2 of floor is

$$\int_0^z Cte^{-kz^*t}\mathrm{d}z^* = Ct(-kt)^{-1}(\mathrm{e}^{-kzt} - 1) = Ck^{-1}(1 - \mathrm{e}^{-kzt}) \equiv M(z,t) \text{ (say)}.$$

(c) Regard z as fixed. The time derivative of M equals the rate of sediment flow *into* the vertical column of water (since there is no flow through the floor at $z = 0$). Hence

$$-q(z,t) \times 1^2 = \frac{\partial M}{\partial t} = Cze^{-kzt}\quad , \quad \text{i.e. } q = -Cze^{-kzt}\,,$$

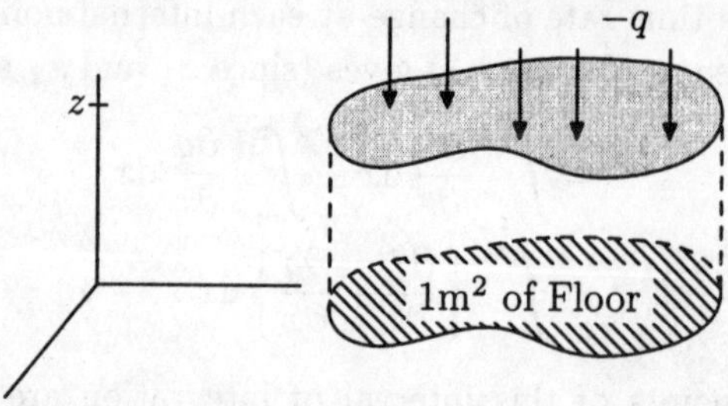

Figure 1.4 Diagram for Example 1.1(c); $-q(x,t)$ = rate of sediment flow into the water column.

with the negative sign implying that flux is downward.

(d) Since $\partial c/\partial t = -kzCte^{-kzt}$ and $\partial q/\partial z = ktCze^{-kzt}$, the law $\partial c/\partial t + \partial q/\partial z = 0$ is satisfied.

Some attributes (e.g. heat, chemical species, ...) may not only flow, but may be *supplied*, *created* or *destroyed*. An additional term must then be included in (1.3), to account for the rate of supply within the region. Let $S(x,t)$ be the *supply rate* (e.g. by electrical heating, chemical reaction, ...) of the attribute per unit volume, so that within the portion of slab considered in (1.3) the attribute is supplied at rate

$$\int_{x_1}^{x_2} S(x,t)\,\mathrm{d}x\,.$$

This must equal the sum of the outflow rate and the rate of increase of the quantity of attribute stored and so should be set equal to the left-hand side of (1.3) to give the *balance law*

$$\frac{\mathrm{d}}{\mathrm{d}t}\int_{x_1}^{x_2} f(x,t)\,\mathrm{d}x + q(x_2,t) - q(x_1,t) = \int_{x_1}^{x_2} S(x,t)\,\mathrm{d}x\,. \tag{1.7}$$

As with the *conservation law*, there is a verbal statement:

Rate of increase stored within a region
+ Rate of outflow across its boundary = Rate of supply.

Also, since the left-hand side of (1.7) is the sum of expressions (1.4) and (1.5), the balance law may be rearranged as

$$\int_{x_1}^{x_2} \left\{ \frac{\partial f}{\partial t} + \frac{\partial q}{\partial x} - S(x,t) \right\} dx = 0\,.$$

Repeating the previous reasoning shows that the integrand must vanish everywhere. This yields the *point form* of the *balance law* as

$$\frac{\partial f}{\partial t} + \frac{\partial q}{\partial x} = S(x,t) \, . \tag{1.8}$$

In a *steady* problem, the density f, the supply rate S and the flux q do not depend on time, so that (1.8) is simplified to

$$\frac{\mathrm{d}q}{\mathrm{d}x} = S(x) \, . \tag{1.9}$$

Good illustrations of these laws arise in considering the *flow of heat.*

1.3 Heat Flow

On a molecular scale, heat is a manifestation of the thermal energy of atomic vibrations. On an everyday scale (in the kitchen, home, engine or living body), *heating* is the input of vibrational energy (often by conversion from chemical, mechanical or electrical energy). This heat then is redistributed by processes such as conduction.

As thermal energy is added to each unit of mass, a change in *temperature* is observable. Temperature is different from heat; it is not a form of energy, but is a measure of how *hot* an object is. As is common experience, *hot* objects tend to give heat to neighbouring *colder* objects. Heat may flow by various mechanisms; in solids it is primarily by *conduction.* In his famous 1822 treatise, Fourier extended the observation that the *rate of heat conduction* from a hot body to a colder one is proportional to the difference in their temperatures and deduced that the flux of heat is proportional to the *temperature gradient.* This follows from an experiment which shows that across a slab of thickness D and having faces at *temperatures* T_1 and T_2, with $T_2 < T_1$, heat flows through area A of the slab at a rate proportional to

$$\frac{A}{D}(T_1 - T_2) \, .$$

In this experiment, the heat flux is normal to the slab and has magnitude $q = K(T_1 - T_2)/D$, where the constant K is the *thermal conductivity* of the material of the slab. The heat flux everywhere within the slab is proportional to the temperature difference divided by the thickness. More generally, if the temperature within a solid is $T = T(x)$, it is postulated that heat will flow parallel to Ox with the *heat flux* (rate of heat flow per unit area) at a typical point being

$$q = -K\frac{\mathrm{d}T}{\mathrm{d}x}\,. \tag{1.10}$$

This is *Fourier's law of heat conduction* in one dimension. It is an empirical law stating that heat flows in the direction in which temperature decreases, with the flow rate at x being proportional to the slope of the graph of $T(x)$ (the local *temperature gradient*). (An empirical law is one which is based on experiment. Fourier's law is not universal. There are some materials in which heat flux obeys more complicated rules.)

Example 1.2

Find the temperature $T(x)$ within a windowpane of thickness D, having thermal conductivity K, and with faces at temperatures T_1 and T_2.

This is a *steady* problem (T is independent of time). Let the pane occupy $0 < x < D$ and let $q = q(x)$ be the *heat flux* (taken positive in the Ox direction). Consider planes at two locations $x = x_1$ and $x = x_2$, where $q(x_1) = q_1$ and $q(x_2) = q_2$. Since there is no *supply* of heat within $x_1 < x < x_2$, Equation (1.9) gives $q_2 = q_1$. Since x_1 and x_2 may be anywhere within the slab, this implies that q is independent of x (i.e. q = constant). Hence

$$\frac{\mathrm{d}T}{\mathrm{d}x} = \frac{-q}{K} = \text{constant}\,.$$

This *ordinary differential equation* (ODE) for $T(x)$ is readily integrated, subject to the boundary condition $T(0) = T_1$, to give

$$T = \frac{-q}{K}x + T_1\,.$$

Setting $T = T_2$ at $x = D$ then gives

$$T_2 = T_1 - \frac{q}{K}D \quad \left(\text{which confirms the result:}\quad \text{Heat flux} = q = K\frac{T_1 - T_2}{D}\right).$$

Some consequences shown by Example 1.2 are:

i) Over an area A, the heat lost per second is $K\left(\dfrac{T_1 - T_2}{D}\right)A$.
Thus, to reduce heat loss, it is best to choose a material with a small conductivity K (i.e. a bad heat conductor – a good thermal insulator).

ii) A double glazing unit has a layer of air (a bad conductor) sandwiched between two sheets of glass. When the air layer is sufficiently thin, convection within the layer is inhibited so that heat transfer is predominantly by conduction. The air may then be treated as a (bad) conductor, with thermal conductivity K_a much smaller than the thermal conductivity of glass K_g.

As in Example 1.2, the heat flux q normal to the pane does not depend on the coordinate x. Hence, the temperature gradient is $\mathrm{d}T/\mathrm{d}x = -q/K_\mathrm{a}$ within the air, while in each layer of glass it is $\mathrm{d}T/\mathrm{d}x = -q/K_\mathrm{g}$. Since $K_\mathrm{a} \ll K_\mathrm{g}$, the temperature gradient is much larger in the air than in the glass:

$$\left|\frac{\mathrm{d}T}{\mathrm{d}x}\right|_\mathrm{air} \gg \left|\frac{\mathrm{d}T}{\mathrm{d}x}\right|_\mathrm{glass} .$$

When the total temperature difference $T_1 - T_2$ across the unit is specified, most of the temperature drop arises across the air layer (see Figure 1.5). The air acts as a jacket (see Exercise 1.4), much reducing the heat flux q below the value for two glass layers without the air layer (when the overall temperature difference $T_1 - T_2$ is fixed).

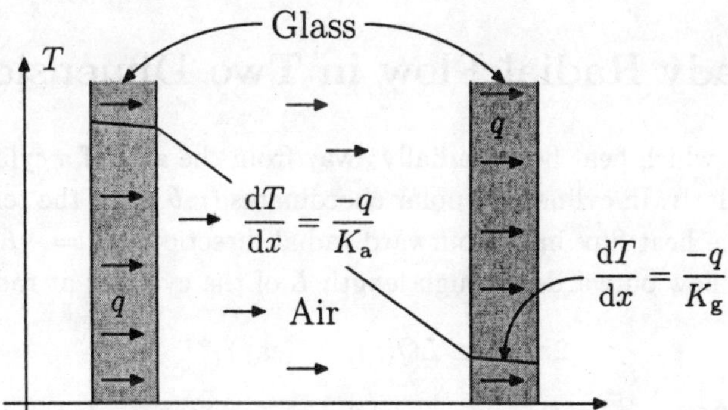

Figure 1.5 Steady temperature $T(x)$ within a double glazing unit for which $K_\mathrm{a} \ll K_\mathrm{g}$ and with uniform heat flux q.

EXERCISES

1.1. Find the total fluid mass in a reservoir $0 \le x \le a$, $0 \le y \le b$, $0 \le z \le h$ filled with fluid having density $\rho = A + B\mathrm{e}^{-\alpha z}$. Confirm that, if $\rho(0) = \rho_0$ and $\rho(h) = \rho_1$, then $A = (\rho_1 - \rho_0 \mathrm{e}^{-\alpha h})/(1 - \mathrm{e}^{-\alpha h})$ and $B = (\rho_0 - \rho_1)/(1 - \mathrm{e}^{-\alpha h})$.

1.2. Heat flows steadily with uniform flux q through two slabs of respective thicknesses h_1 and h_2 and thermal conductivities K_1 and K_2. (a) Find the temperature drops across each individual slab. (b) Show that if the two slabs are placed in contact and an overall temperature drop $\varDelta T$ is imposed then the flux is $K_1K_2\varDelta T/(h_1K_2 + h_2K_1)$.

1.3. Given that the temperature $T(x)$ in one-dimensional, steady heat conduction satisfies $T'(x) = -q/K$, where q is the heat flux and K is the thermal conductivity, (a) find the temperature drop across a slab of thickness h and conductivity K_1; (b) find the total temperature drop $\varDelta T$ across a double-glazing unit composed of two glass sheets of thickness h and conductivity K_1 sandwiching an air layer of thickness H and conductivity K_2 (in each case take the flux to be q).

1.4. For the double-glazing unit of Exercise 1.3(b), suppose that $K_2 = 0.1K_1$ and that $H = h$ and so deduce that $q = K_1\varDelta T/12h$. Compare this flux with the flux q_0 through a single pane of thickness $2h$ with the same temperature drop $\varDelta T$. Is the ratio q/q_0 below 20% ?

1.4 Steady Radial Flow in Two Dimensions

Problems in which heat flows radially away from the axis of a cylinder may be treated similarly. In cylindrical polar coordinates (r, θ, z), if the temperature is $T = T(r)$ the heat flux in the outward radial direction is $q = -K\mathrm{d}T/\mathrm{d}r$. The rate of heat flow outwards through length L of the cylinder at radius r is

$$2\pi rLq \equiv LQ(r) \qquad \text{(say)}\,,$$

where $Q = Q(r) \equiv 2\pi rq$ is the outward flow rate through unit length of cylinder of radius r. If the rate of heat supply per unit volume is $S = S(r)$, then, for steady flow, heat is supplied to the cylindrical shell of radius r, thickness δr and length L at rate $2\pi rL\delta rS(r)$, which must equal the difference between the rates at which heat flows through the two bounding cylinders (i.e. must equal $(Q(r + \delta r) - Q(r))L$, since no heat flows *axially* through the ends):

$$2\pi L\left\{(r+\delta r)q(r+\delta r) - rq(r)\right\} \approx 2\pi rLS(r)\delta r\,. \tag{1.11}$$

Since taking the limit $\delta r \to 0$ gives

$$2\pi\frac{\mathrm{d}}{\mathrm{d}r}(rq) = \frac{\mathrm{d}Q}{\mathrm{d}r} = 2\pi rS(r)\,,$$

two useful deductions are that

i) In *steady radial flow without heat supply*, Q is independent of r so that $rq(r) = \text{constant}$.

ii) In steady radial flow, the heat supply rate S is related to heat flux q by

$$S = S(r) = \frac{1}{r}\frac{\mathrm{d}}{\mathrm{d}r}(rq) . \tag{1.12}$$

Since $T'(r) \equiv \mathrm{d}T/\mathrm{d}r$ is the derivative of temperature with respect to radial distance, *Fourier's Law of heat conduction* gives

$$q = -K\frac{\mathrm{d}T}{\mathrm{d}r} ,$$

which, when combined with (1.12), gives

$$-\frac{K}{r}\frac{\mathrm{d}}{\mathrm{d}r}\left(r\frac{\mathrm{d}T}{\mathrm{d}r}\right) = S(r) . \tag{1.13}$$

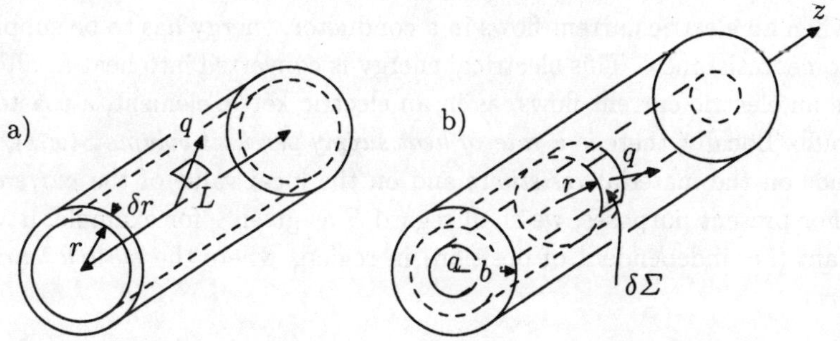

Figure 1.6 a) A cylindrical surface of area $2\pi rL$ surrounded by a shell of volume $2\pi rL\,\delta r$ in which the heat supply rate/volume is $S(r)$. b) Heat flux q at radius r within the pipe wall $a < r < b$ of Example 1.3.

Example 1.3 (Steady heat flow through a pipe wall)

If $T(a) = T_1$ and $T(b) = T_2$ are the temperatures at the inner ($r = a$) and outer ($r = b$) radii of a pipe, find the temperature distribution within the pipe wall $a \le r \le b$.

Under steady conditions the temperature T is independent of axial distance z, azimuthal angle θ and time t. Since $Q = 2\pi rq(r)$ is constant, the temperature satisfies $-2\pi rKT'(r) = Q$. Integrating the differential equation $\mathrm{d}T/\mathrm{d}r = -Q/(2\pi Kr)$ gives

$$T(r) - T(a) = \frac{-Q}{2\pi K}(\ln r - \ln a) .$$

Using both *boundary conditions* gives

$$T_1 - T_2 = T(a) - T(b) = \frac{-Q}{2\pi K}(\ln a - \ln b) ,$$

so that the *heat loss rate per unit length of pipe* is related to the temperature difference $T_1 - T_2$ by

$$Q = 2\pi K \frac{(T_1 - T_2)}{\ln (b/a)} . \tag{1.14}$$

The corresponding dependence of temperature on radius is

$$T(r) = T_1 - (T_1 - T_2)\frac{\ln r - \ln a}{\ln b - \ln a} .$$

N.B. As in one-dimensional heat flow, the rate of heat loss Q is proportional to both K and to the temperature difference.

Heat may be supplied by conversion from other forms of energy. For example, when an electric current flows in a conductor, energy has to be supplied to overcome 'resistance'. This electrical energy is converted into heat at all places where an electric current flows, as in an electric kettle element, a toaster or a lightbulb. Locally, there is a *rate of heat supply per unit volume* $S(\boldsymbol{x}, t)$, which depends on the material *resistivity* and on the local value of the *current density.* For present purposes, we shall regard S as given – for example, it will be constant (i.e. independent of position) in regions where the *electric current* is *uniform.*

Example 1.4

If electric current flows along a cylindrical wire occupying $0 \le r < a$ and supplies heat uniformly at the rate S joules/m^3 sec, find the temperature within the wire when it has thermal conductivity K_0 and is clad by the region $a < r < b$ as in Example 1.3 with thermal conductivity K and outer temperature T_2.

Within $0 \le r < a$ solve the ODE $\mathrm{d}(rq)/\mathrm{d}r = rS$ (i.e. (1.12)) to give $rq(r) = \frac{1}{2}r^2 S$ (N.B. If $Q(0) \neq 0$, then $q(0)$ is infinite). Next, solve $T'(r) = -K^{-1}q = -\frac{1}{2}SK_0^{-1}r$ to give

$$T(r) = T(a) - \frac{1}{4}SK_0^{-1}(r^2 - a^2) .$$

The outward rate of heat flow into the cladding is $Q = 2\pi a q(a) = \pi a^2 S$. Using results from Example 1.3 gives

$$T(a) - T(b) = \frac{\pi a^2 S}{2\pi K}(\ln b - \ln a) .$$

Hence, within the wire the temperature is

$$T(r) = T_2 + \frac{S}{4}\left\{K_0^{-1}(a^2 - r^2) + 2K^{-1}a^2(\ln b - \ln a)\right\}.$$

The temperature is maximum on the axis and varies parabolically across the wire (and logarithmically within the cladding).

1.5 Steady Radial Flow in Three Dimensions

Let r be the radial distance OP of spherical polars (i.e. $r^2 = x^2 + y^2 + z^2$). Steady, radially symmetric temperature distributions $T = T(r)$ give rise to radial heat flux $q = q(r) = -KT'(r)$. Hence, the *outward* rate of heat flow through the sphere Σ of radius r and centred at O is

$$Q(r) = 4\pi r^2 q(r) = -4\pi r^2 K \frac{\mathrm{d}T}{\mathrm{d}r} \tag{1.15}$$

(since the sphere has surface area $4\pi r^2$ and heat flows radially).

In the absence of heat supply, $r^2 q(r)$ is constant (i.e. independent of r), so that the temperature gradient is proportional to r^{-2}.

Example 1.5

Find the steady temperature $T(r)$ when $T(r_1) = T_1$ and $T(r_2) = T_2$ and the heat supply rate is zero.

Since Q is constant in the absence of heat supply, integrating the ODE

$$\frac{\mathrm{d}T}{\mathrm{d}r} = \frac{-Q}{4\pi K}\frac{1}{r^2} \qquad \text{with} \quad T(r_1) = T_1$$

gives

$$T(r) - T(r_1) = T(r) - T_1 = \frac{Q}{4\pi K}\left\{\frac{1}{r} - \frac{1}{r_1}\right\}.$$

Hence, Q is found from $4\pi K(T_1 - T_2) = Q\{(r_1)^{-1} - (r_2)^{-1}\}$. This yields

$$T(r) = \left\{T_1\{r^{-1} - (r_2)^{-1}\} - T_2\{r^{-1} - (r_1)^{-1}\}\right\} / \{(r_1)^{-1} - (r_2)^{-1}\}.$$

Within planets, stars, chemical reactors, compost heaps and living bodies, there is heat generation, or supply, which causes heat to flow. In steady, spherically symmetric cases with *heat supply rate* $S(r)$ and with radial heat flux $q(r)$, the *heat balance law* is found by considering the flow rate $Q = 4\pi r^2 q$ at the internal and external radii r and $r+\delta r$ of a thin spherical shell. Since this shell has volume $\approx 4\pi r^2\delta r$, it receives heat at rate $\approx 4\pi r^2 \delta r S(r)$, so giving

$$Q(r+\delta r) - Q(r) \approx 4\pi r^2 S(r)\delta r\ .$$

This yields the relation $\mathrm{d}Q/\mathrm{d}r = 4\pi r^2 S(r)$, which may be combined with (1.15) to yield

$$r^2 S = \frac{1}{4\pi}Q'(r) = -\frac{\mathrm{d}}{\mathrm{d}r}\left(Kr^2\frac{\mathrm{d}T}{\mathrm{d}r}\right), \tag{1.16}$$

which is the *heat balance equation* for spherical symmetry (analogous to (1.13)).

Example 1.6

Relate the temperature $T(r)$ within a sphere of radius $3a$ and thermal conductivity K to the surface temperature $T(3a)$, when the heat supply rate is

$$S(r) = \begin{cases} A(a-r) & 0 \le r \le a\ , \\ 0 & a \le r \le 3a\ . \end{cases}$$

The outward heat flow $Q(r)$ is found, in the central core $r < a$ from

$$\frac{Q(r)}{4\pi} = \int r^2 S(r)\,\mathrm{d}r = \int A(ar^2 - r^3)\,\mathrm{d}r = A(\tfrac{1}{3}ar^3 - \tfrac{1}{4}r^4)$$

(since $Q = 0$ at $r = 0$ for q to remain finite). In the outer shell $a \le r \le 3a$, there is zero heat supply so that $Q(r) = Q(a) = 4\pi a(\frac{1}{3}a^4 - \frac{1}{4}a^4) = \frac{1}{3}\pi a^4 A$. Then, by integrating the expression for the temperature gradient

$$\frac{\mathrm{d}T}{\mathrm{d}r} = \frac{-Q(r)}{4\pi K r^2} = \begin{cases} K^{-1}A(\frac{1}{4}r^2 - \frac{1}{3}ar) & 0 \le r \le a\ , \\ -K^{-1}A\frac{1}{12}a^4 r^{-2} & a \le r \le 3a \end{cases}$$

while ensuring that T is continuous at $r = a$, the temperature at arbitrary radius r is found in terms of the outer temperature $T(3a)$ as

$$T = \begin{cases} \dfrac{A}{12K}(r^3 - 2ar^2 + \frac{5}{3}a^3) + T(3a) & 0 \le r \le a\ , \\ \dfrac{A}{12K}a^3\left(\dfrac{a}{r} - \frac{1}{3}\right) + T(3a) & a \le r \le 3a\ . \end{cases}$$

[The temperature at the centre of the sphere is $T_{\mathrm{C}} = T(3a) + \frac{5}{36}a^3K^{-1}A$.]

EXERCISES

1.5. If heat flows radially through a pipe wall $a < r < b$ of thermal conductivity K, with inner and outer temperatures $T(a) = T_0$ and $T(b) = T_1$ (with $T_0 > T_1$), derive the expression $T_0 - T_1 = Q(2\pi K)^{-1}\ln(b/a)$ (where Q is the outward heat flow rate per unit length).

By writing $b = a + h$, show that, for $h \ll a$, $Q \approx 2\pi a K(T_0 - T_1)/h$.

1.6. Explain how the result of Exercise 1.5 relates to the heat flow rate $K(T_0 - T_1)/h$ through unit area of a slab of thickness h. Show, by treating h/a as small, that an improved approximation arising from Exercise 1.5 is $Q \approx \pi(a + b)K(T_0 - T_1)/h$ and that this may be interpreted as the flow rate through a slab of thickness h and having the same area as unit length of a cylinder with the mean radius $\frac{1}{2}(a + b)$.

1.7. If the pipe of Exercise 1.5 is enclosed in thermal cladding with thermal conductivity K_2 and occupying $b < r < c$, let Q_2 be the outward heat flow rate when $T(a) = T_0$ and $T(c) = T_1$ with $T(r)$ continuous.

Derive two expressions involving $T(b)$ and so find the ratio Q_2/Q.

1.8. If a spherical shell $a < r < b$ has inner temperature $T(a) = T_1$ and outer temperature $T(b) = T_2$, find the total outward heat flow rate $4\pi r^2 q(r)$ at each radius when there is no heat supply ($S = 0$ in $a < r < b$). Find also $T(r)$ in $a < r < b$.

1.9. Suppose that the radial heat flow in Exercise 1.8 is entirely caused by a uniform heat supply rate $S(r) = S_0$ throughout $r < a$, determine $4\pi a^2 q(a)$. Hence, determine the temperature difference $T_1 - T_2$.

1.10. In a simple model of star structure, it is assumed that nuclear reactions supply heat within an inner core $0 \le r \le a$, so that the temperature there is $T(r) = T_c e^{-\alpha r^2}$. Calculate the steady radial heat flux $-KT'(r)$ in $r < a$, where $K = K_c$ is constant.

Calculate the rate of heat supply $S(r) = r^{-2}\mathrm{d}\left(-r^2KT'(r)\right)/\mathrm{d}r$ in $r < a$. Find also the total rate of heat flow out of the core.

Supposing that the star has an outer shell $a < r < b$ of conductivity $0.5K_c$ and within which $S = 0$, determine the surface temperature $T(b)$.

2
Unsteady Heat Flow

2.1 Thermal Energy

In *unsteady* heat flow problems, in which the temperature $T(x,t)$ depends upon both position x and time t, the balance between heat supply, heat flow and the rate of heat storage must be considered.

Since the temperature of an element δm of mass can change only if *thermal energy* is either supplied or removed, each material has an associated *specific internal energy density* which is a function $U = U(T)$. This function varies from material to material, e.g. it is very different for water, air or steel. It is a *constitutive property*, which is determined either by experiment or from a more fundamental physical theory. It is defined so that $U(T+\delta T) - U(T)$ is the heat required to raise *unit mass* from temperature T to $T + \delta T$. Hence,

$(U(T+\delta T) - U(T))\,\delta m =$ heat required to raise mass δm from T to $T + \delta T$.

For small changes δT in temperature, this heat requirement is approximately $U'(T)\delta T\,\delta m$.

Definition The *specific heat* of a material is

$$c = U'(T) = \frac{\mathrm{d}U}{\mathrm{d}T}\,.$$

Hence, if a *mass element* δm has temperature $T(t)$, it has *internal energy* $U(T(t))\delta m$ and the rate at which it receives heat must be

$$\frac{\mathrm{d}}{\mathrm{d}t}U(T(t))\,\delta m = U'(T)\frac{\mathrm{d}T}{\mathrm{d}t}\delta m = cT'(t)\delta m\,.$$

For many materials, c is effectively constant over significant ranges of temperature, so simplifying the relation between temperature change and energy input.

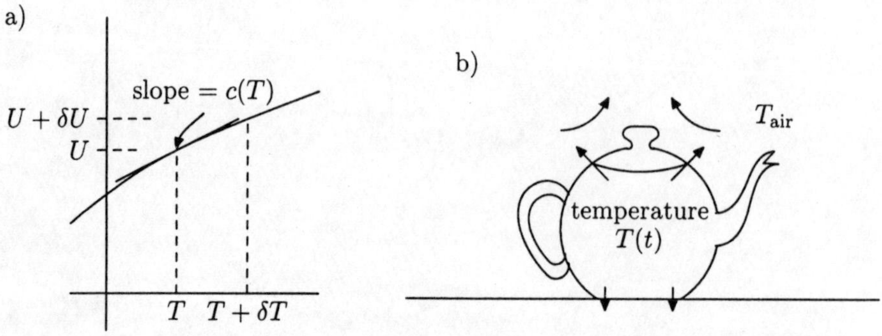

Figure 2.1 a) Specific internal energy $U(T)$ and specific heat $c \equiv U'(T)$. b) Heat flow from teapot by conduction to the table and to currents in the surrounding air.

Example 2.1

If a teapot contains M kg of water, initially at temperature 98°C, how much heat does the water lose as it cools to 36°C? (Assume that the tea is well-stirred, so allowing the approximation $T = T(t)$; i.e. T is independent of position.)

Let $U(T)$ be the specific internal energy. Then, the heat lost during cooling is $M\,(U(98^o) - U(36^o))$.
But, over the temperature range 0–100°C, water has virtually constant specific heat c_w = 4185 joules/(kg °C) (= 1 cal/gm °C). Hence, the heat lost to the environment is $62\,Mc_w \approx 26 \times 10^4\,M$ joules.

(N.B. Heat is lost both by *conduction* through the teapot wall to the table and by *convection* (through air currents). Each of these occurs at rates depending on the difference between T and the room temperature T_{air}. If each rate is assumed proportional to $T - T_{air}$, this predicts (see Exercise 2.5) that temperature decays exponentially with time.)

2.1.1 Heat Balance in One-dimensional Problems

The connection between heat flow, heat supply and temperature change is provided by the heat balance law, which is an example of (1.7).

Taking $T = T(x,t)$ and inserting into the law $U = U(T)$ gives the internal energy per unit mass as $U = U(T(x,t)) \equiv \hat{U}(x,t)$, so that $\rho\hat{U}(x,t)$ is the *internal energy density*. Expressing the rate of heat supply per unit volume in the usual form as $S = S(x,t)$ and inserting into (1.7) gives

$$\frac{\mathrm{d}}{\mathrm{d}t}\int_{x_1}^{x_2} \rho\hat{U}(x,t)\,\mathrm{d}x + q(x_2,t) - q(x_1,t) = \int_{x_1}^{x_2} S(x,t)\,\mathrm{d}x\,.$$

The equivalent point form (1.8) becomes

$$\rho\frac{\partial\hat{U}}{\partial t} + \frac{\partial q}{\partial x} = S(x,t)\,, \tag{2.1}$$

when the *density* ρ is assumed to be constant (i.e. thermal expansion is neglected). As earlier, this *heat balance law* has a verbal statement:

Rate of increase of stored heat + Net outflow rate = Rate of heat supply.

After use of the relation

$$\frac{\partial\hat{U}}{\partial t} = \frac{\mathrm{d}U}{\mathrm{d}T}\frac{\partial T}{\partial t} = c\frac{\partial T}{\partial t}$$

(where c is the specific heat) and *Fourier's law of heat conduction* $q = -\partial T/\partial x$, this equation becomes

$$\rho c\frac{\partial T}{\partial t} = K\frac{\partial^2 T}{\partial x^2} + S(x,t)\,. \tag{2.2}$$

Equation (2.2) may be interpreted for a typical thin slice occupying $(x, x+\delta x)$

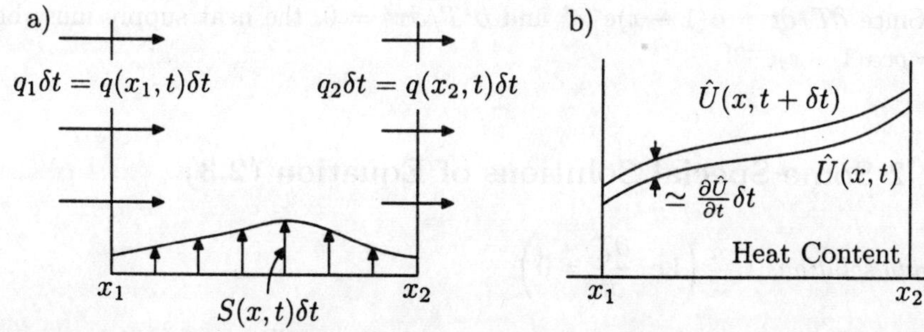

Figure 2.2 a) Heat supply $\int_{x_1}^{x_2} S(x,t)\,\delta t\,\mathrm{d}x$ and heat inflow $(q_1 - q_2)\delta t$ to $x_1 < x < x_2$ over the time interval $(t, t+\delta t)$. b) The corresponding change $\int_{x_1}^{x_2} \rho\{\hat{U}(x,t+\delta t) - \hat{U}(x,t)\}\mathrm{d}x$ in thermal energy.

as a statement of the *warming* during a time interval $(t, t+\delta t)$ due to heat conduction (heat 'in' at x and 'out' at $x+\delta x$) and to the *heat supply rate* $S(x,t)$. Henceforward, the parameters K, ρ and c will be taken to be constants.

Some special cases of (2.2) are significant:

i) *No heat supply:* ($S \equiv 0$)
$$\frac{\partial T}{\partial t} = \nu \frac{\partial^2 T}{\partial x^2}\,. \tag{2.3}$$
This important *partial differential equation* (PDE) is known as the *heat equation* (or *heat conduction equation*) in one dimension, in which the coefficient $\nu \equiv K/(\rho c)$ is the *thermal diffusivity* of the material.

ii) *Perfect insulation:* ($K = 0$) In this (idealized) limit, the heat supplied at each location warms *only* that point, so that the PDE (2.3) may be treated as an ordinary differential equation (ODE) for the change of T at *each* fixed value x (i.e. x may be treated merely as a parameter), namely
$$\rho c \frac{\partial T}{\partial t} = S(x,t)\,.$$

iii) *Perfect conductor:* For K very large, the heat spreads very rapidly, so that the temperature is virtually independent of position ($T(x,t) \approx \hat{T}(t)$). The (mathematical) limit is described by $T = \hat{T}(t)$ (i.e. $K \to \infty$ implies that $|\partial T/\partial x| \to 0$), with heat supply, loss and storage averaged over the heated body (see Example 2.1).

Example 2.2

Find the distribution $S(x,t)$ of heat supply rate required if, in $0 < x < 1$, the temperature distribution $T = (1-x)(1-\mathrm{e}^{-\alpha t})$ should satisfy (2.2).

Since $\partial T/\partial t = \alpha(1-x)\mathrm{e}^{-\alpha t}$ and $\partial^2 T/\partial x^2 = 0$, the heat supply must be $S = \rho c \alpha (1-x)\mathrm{e}^{-\alpha t}$.

2.1.2 Some Special Solutions of Equation (2.3)

Steady solutions $\left(\text{i.e. } \dfrac{\partial T}{\partial t} = 0\right)$

Since this gives $T = T(x)$, Equation (2.3) reduces to $T''(x) = 0$, which may be integrated twice to give
$$T = ax + b \qquad \text{(for some constants } a \text{ and } b\text{)}.$$

In this case, the temperature gradient must be uniform ($T' = a$), which corresponds to uniform heat flux $q = -Ka$, as in Example 1.2 and in Exercise 1.2 and Exercise 1.3.

Solutions polynomial in x and t

An illustrative example is to assume that $T = ax^2 + bx + c + pt$. This gives the expressions

$$\frac{\partial^2 T}{\partial x^2} = 2a + 0 \quad , \qquad \frac{\partial T}{\partial t} = p ,$$

which are consistent with (2.3) whenever $p = \nu 2a$. No restrictions on b or c arise. Hence, the most general solution of the assumed type is

$$T = a(x^2 + 2\nu t) + bx + c , \tag{2.4}$$

with a, b and c *arbitrary.*

It may be observed that the contribution $bx+c$ is just a steady temperature with uniform gradient, as treated earlier. The remaining contribution describes temperature varying linearly with time at rate $2a\nu$ at each fixed x (warming, for $a > 0$; cooling, for $a < 0$), with T quadratic in x. The temperature gradient (hence also the heat flux q) is independent of t and linear in x; the difference $2aL$ in its values at $x = L$ and at $x = 0$ (see Figure 2.3) accounts for the rate of heat inflow and hence the rate of change of temperature. The general solution (2.4) is just the sum of these two contributions.

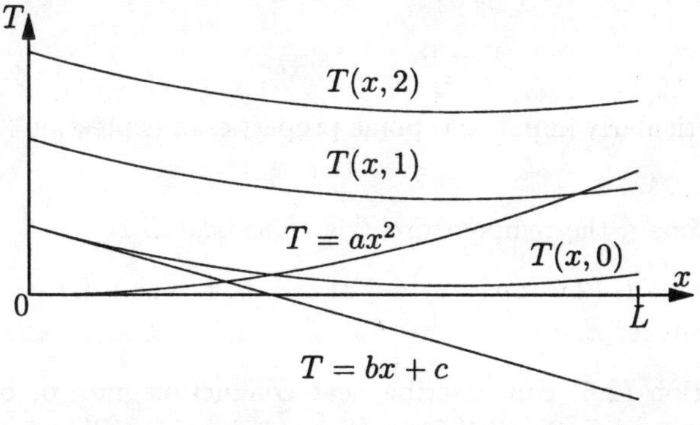

Figure 2.3 Contributions $T = bx + c$, $T = ax^2$, $T(x, 0) = ax^2 + bx + c$ and $T(x, 1)$, $T(x, 2)$ for (2.4).

Although polynomial solutions are quite instructive and useful, exponential solutions are even more interesting and applicable:

Solutions exponential in t (i.e. with $T(x,t) \sim \mathrm{e}^{-\alpha t}$)

In equation (2.3), try $T = \mathrm{e}^{-\alpha t} f(x)$, so that

$$\frac{\partial T}{\partial t} = -\alpha \mathrm{e}^{-\alpha t} f(x) \quad , \qquad \frac{\partial^2 T}{\partial x^2} = \mathrm{e}^{-\alpha t} f''(x) .$$

When these are substituted into (2.3) to give

$$-\alpha e^{-\alpha t} f(x) = \nu e^{-\alpha t} f''(x) \,,$$

a term in $e^{-\alpha t}$ may be cancelled so leaving an ordinary differential equation (ODE)

$$\frac{d^2 f}{dx^2} + \frac{\alpha}{\nu} f = 0 \,. \tag{2.5}$$

This may be recognized as very similar to the SHM equation $u''(t) + p^2 u = 0$ governing *simple harmonic motion*, for which the general solution is a *linear combination* of $\cos pt$ and $\sin pt$. Consequently, the general solution to (2.5) may be written as

$$f = A \cos px + B \sin px \qquad \text{for some constants } A, B \,,$$

where $p = \sqrt{\alpha/\nu}$ is the positive root of $p^2 = \alpha/\nu$ (or, equivalently of $\alpha = \nu p^2$).

The special case of this solution

$$f = B \sin px$$

leads to the unsteady temperature distributions

$$T = B e^{-\nu p^2 t} \sin px \tag{2.6}$$

which are particularly important. Some properties of expression (2.6) are listed below:

i) At each time t, the temperature T is *sinusoidal* in x.

ii) At each time t, (2.6) gives $T = 0$ at $x = 0, \pi/p$ and indeed at $x = n\pi/p$, for each integer n.

iii) The solution (2.6) can describe heat conduction into, or out of, a slab occupying $0 < x < h$, with faces at $x = 0, h$ held at $T = 0$, provided that p is *chosen* to satisfy $p = \pi/h$ (or, indeed $p = n\pi/h$). Thus, the solution

$$T = B \exp\left(\frac{-\nu \pi^2 t}{h^2}\right) \sin \frac{\pi x}{h} \tag{2.7}$$

to (2.3) describes cooling of a slab with maximum *initial* (i.e. at $t = 0$) temperature $T = B$ at $x = h/2$, with the faces $x = 0, h$ held at $T = 0$.

iv) At *each* position x, the temperature T tends exponentially to 0 (the *wall temperature*) as time increases.

v) The *decay rate* $\nu\pi^2/h^2$ is proportional to (thickness)$^{-2}$. Hence, if the thickness is *doubled*, the *cooling time* is multiplied by 4. (This explains how, even in the 18th century, ice could be stored through the summer in an *icehouse* – a chamber under a thick, thermally insulating mound of earth.)

vi) Various different solutions to (2.3) may be added together (i.e. *superposed*) to yield more solutions:

(a) By adding to (2.7) a constant temperature T_0 to give

$$T = T_0 + Be^{-\nu p^2 t}\sin px\,.$$

This corresponds to constant wall temperatures T_0, with $T(x,t) \to T_0$ as $t \to \infty$.

(b) By adding to (2.7) a steady, linearly varying temperature

$$T = T_0 + (T_1 - T_0)\frac{x}{h} + Be^{-\nu\pi^2 t/h^2}\sin \pi x/h\,.$$

This solution has constant, but different, temperatures T_0 and T_1 at the two faces $x = 0, h$ of the slab.
Again, as $t \to \infty$, the temperature *decays* towards a *steady state.*

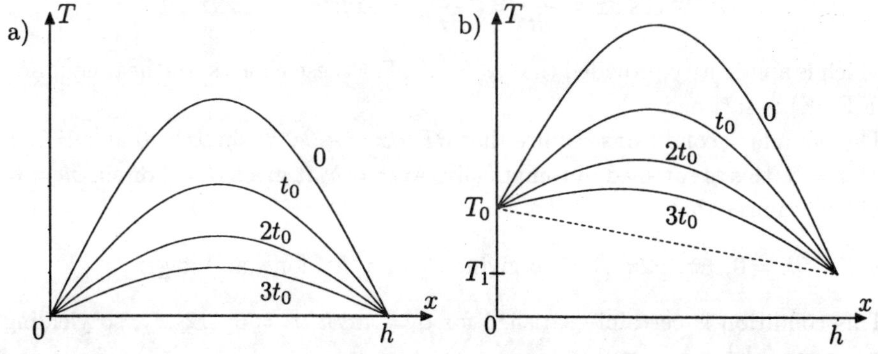

Figure 2.4 a) The temperature $T(x,t)$ of (2.7) at times $t = 0$, t_0, $2t_0$, $3t_0$ (for $t_0 = 0.5h^2/\nu\pi^2$). b) Similar temperature distributions with $T(0,t) = T_0$ and $T(h,t) = T_1$, as in vi) (b) above.

Example 2.3

Find restrictions on the constants A, B, C and D so that $T = Ax^6 + Bx^4t + Cx^2t^2 + Dt^3$ satisfies the heat equation (2.3). Obtain the most general corresponding expression for T.

Since

$$\frac{\partial T}{\partial t} = Bx^4 + 2Cx^2t + 3Dt^2\,, \quad \frac{\partial^2 T}{\partial x^2} = 30Ax^4 + 12Bx^2t + 2Ct^2\,,$$

equating coefficients in $\partial T/\partial t = \nu\partial^2 T/\partial x^2$ gives

$$B = 30\nu A\,, \quad 2C = 12\nu B = 360\nu^2 A\,, \quad 3D = 2\nu C = 360\nu^3 A\,.$$

Hence, allowable temperature distributions are

$$T = A\left(x^6 + 30\nu x^4 t + 180\nu^2 x^2 t^2 + 120\nu^3 t^3\right).$$

Example 2.4

If a slab occupying $0 < x < h$ has both faces insulated, the temperature $T(x,t)$ satisfies the heat equation $\partial T/\partial t = \nu\partial^2 T/\partial x^2$, together with the *boundary conditions* $\partial T/\partial x = 0$ at both $x = 0$ and $x = h$.
Show that the PDE is satisfied for some functions of the form $T = \mathrm{e}^{-\alpha t}\cos\beta x$ and that the boundary conditions are satisfied when $\beta = n\pi/h$, $n = 0, 1, 2, \ldots$.
Obtain the corresponding values α_n for α.

For functions $T = \mathrm{e}^{-\alpha t}\cos\beta x$, it is found that

$$-\alpha\mathrm{e}^{-\alpha t}\cos\beta x = \frac{\partial T}{\partial t} = \nu\frac{\partial^2 T}{\partial x^2} = -\nu\beta^2\mathrm{e}^{-\alpha t}\cos\beta x\,,$$

which is an identity provided that $\alpha = \nu\beta^2$. For these choices, the heat equation (a PDE) is satisfied.
The boundary conditions require that $\partial T/\partial x = -\beta\mathrm{e}^{-\alpha t}\sin\beta x = 0$ at $x = 0, h$.
At $x = 0$, this is satisfied automatically. At $x = h$, it gives $\beta = 0$ or $\sin\beta h = 0$, which requires that

$$\beta h = 0, \pm\pi, \pm 2\pi, \ldots \quad \text{so giving } \beta = n\pi/h \quad \text{for } n \text{ an integer.}$$

This condition is certainly satisfied for $\beta = n\pi/h$, $n = 0, 1, 2, \ldots$, so yielding $\alpha = \nu n^2\pi^2/h^2 \equiv \alpha_n$ (say).

Note that the case $n = 0$ gives $T(x,t) = 1$ (a uniform, constant temperature), while solutions $T(x,t) = \mathrm{e}^{-\alpha_n t}\cos n\pi x/h \equiv T_n(x,t)$ for n negative are identical to those for n positive. All these solutions may be *superposed* to give a more general solution of (2.3) satisfying the boundary conditions describing insulation at $x = 0, h$, namely

$$T(x,t) = \sum_{n=0}^{\infty} a_n\mathrm{e}^{-\alpha_n t}\cos\frac{n\pi x}{h}\,, \qquad \text{where} \quad \alpha_n = \nu n^2\pi^2/h^2\,. \tag{2.8}$$

Here, the coefficients a_n, $n = 0, 1, \ldots$ are arbitrary.

Example 2.5

Suppose that at time $t = 0$ the temperature within a slab $0 < x < h$ is $T(x,0) = A + B\cos\pi x/h + C\cos^2\pi x/h$. Express this function in the form $\sum A_n \cos n\pi x/h$ and so identify a solution (2.8) of (2.3) describing the temperature in an insulated slab with this initial distribution.
Comment on the temperature distribution as $t \to \infty$.

Use $\cos 2\theta = 2\cos^2\theta - 1$ to give $\cos^2 \pi x/h = (1 + \cos 2\pi x/h)/2$, so that

$$T(x,0) = A + \tfrac{1}{2}C + B\cos\pi x/h + \tfrac{1}{2}C\cos 2\pi x/h\,.$$

Comparing coefficients in this with those in $T(x,0) = \sum_{n=0}^{\infty} a_n \cos n\pi x/h$ gives

$$a_0 = A + \tfrac{1}{2}C\,,\quad a_1 = B\,,\quad a_2 = \tfrac{1}{2}C\,,\quad a_n = 0 \text{ for } n \geq 3\,.$$

Hence, for $t \geq 0$ the temperature is

$$T = A + \tfrac{1}{2}C + B\mathrm{e}^{-\nu\pi^2 t/h^2}\cos\frac{\pi x}{h} + \tfrac{1}{2}C\mathrm{e}^{-4\nu\pi^2 t/h^2}\cos\frac{2\pi x}{h}\,.$$

As $t \to \infty$, this temperature tends to $A + \frac{1}{2}C$ everywhere within the slab. This is the average over $0 < x < h$ of the initial temperature distribution (since no heat can escape through the insulated faces).

EXERCISES

2.1. In the absence of heat supply ($S \equiv 0$), find and solve the conditions restricting A, B, C, D, etc. so that the following functions $T(x,t)$ satisfy the heat equation $\partial T/\partial t = \nu\partial^2 T/\partial x^2$:
(a) $T = Axt + Bt + Cx^3 + Dx^2 + Ex + F$,
(b) $T = \mathrm{e}^{At+Bx+C}$,
(c) $T = \mathrm{e}^{At}\cos(Bx + C)$,
(d) $T = A\mathrm{e}^{Bt}\cos Cx + D\mathrm{e}^{Et}\sin Fx$,
(e) $T = At^B \exp\left(\dfrac{x^2}{Ct}\right)$.

2.2. Show that, for all choices of the constants $b_1, b_2, \ldots, b_N$, the expression

$$T(x,t) = \sum_{n=1}^{N} b_n \exp\left(\frac{-\nu n^2\pi^2 t}{h^2}\right)\sin\frac{n\pi x}{h}$$

satisfies the heat equation (2.3) and has $T = 0$ at both $x = 0$ and $x = h$ for all t.

For $b_1 = 32/\pi^3$, $b_3 = 32/(27\pi^3)$ with $b_n = 0$ otherwise:

(a) Evaluate $T(h/2, 0)$ and sketch the graph of $T(x, 0)$ in $0 \le x \le h$. Compare this graph with that for $f(x) \equiv 4x(x-h)/h^2$ (If you use Maple or Mathematica, write $x/h \equiv X$, so that $0 \le X \le 1$).

(b) Simplify the expression for $T(x,t)$ when $\nu t = h^2\pi^{-2}$. What do you observe about the *relative importance* of the terms involving b_1 and b_3?

(c) Does the observation in (b) suggest that a simple approximation for $T(x,t)$ exists as $t \to \infty$ (e.g. for $t > h^2(\nu\pi^2)^{-1}$) for general values of $b_1, b_2, \ldots, b_N$ (with $b_1 \neq 0$)?

2.3. Using the fact that $T = \mathrm{e}^{-\nu p^2 t} \sin px$ satisfies (2.3) for all values of p, find all the (positive) values of p which give solutions having $T = 0$ at $x = 0$ and having $\partial T/\partial x = 0$ at $x = h$ (corresponding to insulation at $x = h$). By superposing these solutions, obtain an expression for $T(x,t)$ involving an (infinite) set of arbitrary coefficients.

2.4. Show that the heat equation $\partial T/\partial t = \nu \partial^2 T/\partial x^2$ has solutions of the form $T = u(x)\cos\omega t + v(x)\sin\omega t$ provided that $u(x)$ and $v(x)$ satisfy the ODEs $u''(x) = \omega\nu^{-1}v\,,\quad v''(x) = -\omega\nu^{-1}u$. By seeking solutions of the form $u \propto \mathrm{e}^{kx}$ (allowing k to be complex), find solutions which are bounded as $x \to -\infty$. Select from these the solution for which $T = T_0 \cos\omega t$ at $x = 0$.

This is a model for heat conduction into soil due to diurnal fluctuations in temperature of amplitude T_0 around the mean seasonal temperature. Do the peaks and troughs of heat flux (into $x < 0$) at $x = 0$ precede or follow those of temperature? By how much (N.B. $2\pi/\omega = 1$ day)?

2.2 Effects of Heat Supply

If there is a steady rate of heat supply $S(x)$ distributed throughout the slab in $0 < x < h$, it is possible to determine an associated *steady* temperature $T = \hat{T}(x)$ satisfying the balance law (2.2) and conditions $\hat{T}(0) = 0 = \hat{T}(h)$. Thus, since $\partial\hat{T}/\partial t = 0$, this reduces to

$$0 = K\hat{T}''(x) + S\,,$$

which is an ODE for $\hat{T}(x)$ of the form $\hat{T}''(x) = -K^{-1}S(x)$. Once $S(x)$ is specified, two integrations with respect to x yield $\hat{T}(x)$.

The simplest case is $S =$ constant, which yields

$$\hat{T} = -\left(\frac{S}{2K}\right)x^2 + ax + b\,.$$

The condition $\hat{T}(0) = 0$ gives $b = 0$, while the condition $\hat{T}(h) = 0$ requires that $-(Sh^2/2K) + ah = 0$, so that $a = Sh/2K$. Thus, the steady temperature arising from uniform heat supply rate is parabolic in x:

$$\hat{T}(x) = \frac{-S}{2K}x(x-h)\,.$$

Other solutions to (2.2) corresponding to *the same* heat supply $S =$ const., are obtained by adding to this any solution to (2.3) (which corresponds to zero heat supply rate). In particular, adding to the *steady solution* $\hat{T}$ the decaying sinusoidal contribution (2.7) yields the solution (inserting $\nu = K/\rho c$)

$$T(x,t) = \frac{-S}{2K}x(x-h) + B\exp\left(\frac{-K\pi^2 t}{\rho c h^2}\right)\sin\frac{\pi x}{h}\,. \tag{2.9}$$

Note that this solution also has $T = 0$ at both $x = 0$ and $x = h$.

Generally, solutions to (2.2) (or to (2.3)) may be built up from

i) steady solutions accounting for the heat supply and agreeing with the boundary conditions,

ii) one (or many) contributions decaying exponentially with time (as in (2.8)).

This is a combination of two standard procedures for treating linear PDEs – the *superposition principle* and the *method of separation of variables.*

Example 2.6

A slab occupying $0 < x < h$ is held at temperature $T = T_0$ for times $t \leq 0$ (i.e. $T(x,0) = T_0$ in $0 < x < h$). Find $T(x,t)$ for $t > 0$ when the face $x = 0$ remains at temperature $T(0,t) = T_0$ but the temperature at the face $x = h$ is abruptly changed to give $T(h,t) = T_0 + T_1$ for $t > 0$.
This problem involves no heat sources. Hence, T satisfies the heat equation (2.3):

$$\frac{\partial T}{\partial t} = \nu\frac{\partial^2 T}{\partial x^2}\,.$$

Construct T as the sum $T = \hat{T}(x) + u(x,t)$ of

i) a steady part $\hat{T}(x)$ satisfying the *boundary conditions* at $x = 0$ and $x = h$:

$$\nu\hat{T}''(x) = 0 \quad \text{in} \quad 0 < x < h\,, \quad \text{with} \quad \hat{T}(0) = T_0\,, \quad \hat{T}(h) = T_0 + T_1\,,$$

ii) and an unsteady part $u(x,t)$ which must satisfy the heat equation (2.3) with $u = 0$ at $x = 0, h$:

$$\frac{\partial u}{\partial t} = \nu \frac{\partial^2 u}{\partial x^2} \quad \text{in} \quad 0 < x < h\,, \quad \text{with} \quad u(0,t) = 0\,, \quad u(h,t) = 0\,. \tag{2.10}$$

The defining properties of these problems are summarized in Figure 2.5. The solution to i) is just a temperature distribution $\hat{T} = T_0 + T_1 x/h$ which has uniform temperature gradient (and constant heat flux).

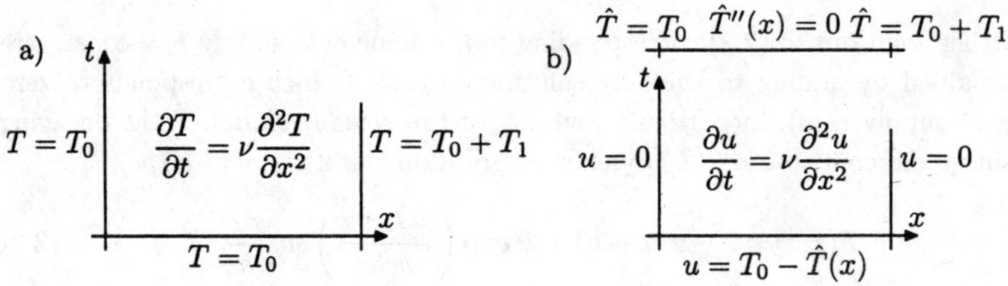

Figure 2.5 a) Schematic diagram of initial condition $T(x,0) = T_0$ and boundary conditions $T(0,t) = T_0$, $T(h,t) = T_0 + T_1$ for Example 2.6. b) Equivalent schematic diagram for the steady part $\hat{T}(x)$ and unsteady part $u(x,t)$ of $T(x,t) = \hat{T} + u$.

The function $u(x,t)$ may be built up from many solutions like (2.7) as

$$u(x,t) = \sum_{n=1}^{\infty} b_n \exp\left(\frac{-\nu n^2 \pi^2 t}{h^2}\right) \sin\frac{n\pi x}{h} \tag{2.11}$$

(see also Exercise 2.2). Then, superposing these as $T(x,t) = \hat{T}(x) + u(x,t)$ gives a function T satisfying (2.3) and the required conditions at $x = 0$, h. The remaining condition is the *initial condition*, namely that, at $t = 0$,

$$\hat{T}(x) + u(x,0) = T(x,0) = T_0\,.$$

This, together with expression (2.11) evaluated at $t = 0$, yields the requirement

$$u(x,0) = T_0 - \hat{T}(x) = \frac{-T_1}{h} x = \sum_{n=1}^{\infty} b_n \sin\frac{n\pi x}{h}\,. \tag{2.12}$$

The right-hand side is just a *Fourier sine series*. It represents an odd periodic function of x, having period $2h$. Its *coefficients* b_n are related to the function $u(x,0)$ over the half-period $0 < x < h$ by the usual *half-range Euler formula*

$$b_n = \frac{2}{h} \int_0^h u(x,0) \sin\frac{n\pi x}{h}\, \mathrm{d}x\,.$$

This gives (after integration by parts)

$$b_n = \frac{2}{h}\int_0^h \frac{-T_1}{h}x\sin\frac{n\pi x}{h}\,dx = \frac{-2T_1}{h^2}\frac{-h}{n\pi}\left[x\cos\frac{n\pi x}{h}\right]_0^h - \frac{2T_1}{hn\pi}\int_0^h \cos\frac{n\pi x}{h}\,dx$$
$$= \frac{2T_1}{\pi}\frac{(-1)^n}{n}, \qquad \text{so that } b_1 = \frac{-2T_1}{\pi}, \quad b_2 = \frac{2T_1}{2\pi}, \quad b_3 = \frac{-2T_1}{3\pi}, \text{ etc.}$$

With these coefficients inserted into (2.11), the function $T(x,t) = \hat{T}(x) + u(x,t)$ describes how the temperature adjusts from the *uniform state* $T = T_0$ towards the *steady state* $T = \hat{T}(x)$ after the abrupt change of temperature at $x = h$, $t = 0$.

Example 2.7

Verify that, for $\hat{T}(x) \equiv (-S/2\rho c)x(x-h)$ and $u(x,t)$ as in (2.11), the *superposition* $T(x,t) = \hat{T}(x) + u(x,t)$ describes heat conduction in a slab $0 < x < h$ with uniform, constant heat supply rate S and with both faces $x = 0$ and $x = h$ held at temperature $T = 0$. Determine the coefficients b_n so that $T(x,0) = 0$.

Since

$$\frac{\partial^2 T}{\partial x^2} = \frac{-S}{\rho c} + \sum_{n=1}^{\infty} b_n\left(\frac{-n^2\pi^2}{h^2}\right)\exp\left(\frac{-\nu n^2\pi^2 t}{h^2}\right)\sin\frac{n\pi x}{h},$$
$$\frac{\partial T}{\partial t} = \sum_{n=1}^{\infty}\left(\frac{-\nu n^2\pi^2}{h^2}\right) b_n \exp\left(\frac{-\nu n^2\pi^2 t}{h^2}\right)\sin\frac{n\pi x}{h},$$

then $\rho c\,\partial T/\partial t - K\partial^2 T/\partial x^2 = S$, as in (2.2). Also, $\hat{T}(0) = 0 = \hat{T}(h)$ and $\sin n\pi x/h = 0$ at $x = 0, h$ for each n, so giving $T(0,t) = 0 = T(h,t)$.

At $t = 0$, the initial condition gives

$$0 = T(x,0) = \frac{-S}{\rho c}x(x-h) + \sum_{n=1}^{\infty} b_n \sin\frac{n\pi x}{h}.$$

Thus, finding the Fourier sine coefficients b_n in

$$\sum_{n=1}^{\infty} b_n \sin\frac{n\pi x}{h} = \frac{S}{\rho c}x(x-h) \qquad \text{in} \quad 0 < x < h$$

gives

$$b_n = \frac{2}{h}\int_0^h \frac{S}{\rho c}x(x-h)\sin\frac{n\pi x}{h}\,dx$$
$$= \frac{2S}{\rho ch}\left[\frac{-h}{n\pi}x(x-h)\cos\frac{n\pi x}{h}\right]_0^h - \frac{2S}{\rho ch}\left(\frac{-h}{n\pi}\right)\left[(2x-h)\frac{h}{n\pi}\sin\frac{n\pi x}{h}\right]_0^h$$
$$- \frac{2S}{\rho ch}\left(\frac{h}{n\pi}\right)^3\left[2\cos\frac{n\pi x}{h}\right]_0^h = \frac{4Sh^2}{\rho c(n\pi)^3}[1 - (-1)^n].$$

Hence, the temperature within $0 \le x \le h$ is given by

$$T = \frac{S}{\rho c}\left\{x(h-x) + \frac{8h^2}{\pi^3}\left[e^{-\nu\pi^2 t/h^2}\sin\frac{\pi x}{h} + \frac{1}{27}e^{-9\nu\pi^2 t/h^2}\sin\frac{3\pi x}{h} + \ldots\right]\right\}.$$

2.3 Unsteady, Spherically Symmetric Heat Flow

The unsteady version of the spherically symmetric treatment in §1.5 has $T = T(r,t)$ with radial heat flux $q(r,t)$ and heat supply rate $S(r,t)$, giving rise to

$$\text{Rate of change of heat in a shell } (r, r+\delta r) \;=\; 4\pi r^2 \rho c \frac{\partial T}{\partial t}\,.$$

Equating this to the sum of the heat supply rate to the shell and the difference of the heat inflow and outflow rates through its surfaces gives

$$r^2 \rho c \frac{\partial T}{\partial t} = K \frac{\partial}{\partial r}\left(r^2 \frac{\partial T}{\partial r}\right) + r^2 S(r,t)\,. \tag{2.13}$$

For analysis and construction of solutions, it turns out that this has a more convenient form, because there is an identity

$$\frac{\partial}{\partial r}\left(r^2 \frac{\partial T}{\partial r}\right) = r^2 \frac{\partial^2 T}{\partial r^2} + 2r\frac{\partial T}{\partial r} = r\,\frac{\partial^2 (rT)}{\partial r^2}\,.$$

Hence, Equation (2.13) is equivalent to

$$\rho c \frac{\partial (rT)}{\partial t} = K \frac{\partial^2 (rT)}{\partial r^2} + rS(r,t)\,, \tag{2.14}$$

which is mathematically equivalent to (2.2) after x is replaced by r and T is replaced by rT (with $rS(r,t)$ acting as the supply rate). Hence, methods used for constructing steady and unsteady solutions in one dimension are readily adapted to spherical symmetry.

Example 2.8 (Cooking a plum pudding or a haggis)

Assuming spherical symmetry, find the temperature $T(r,t)$ within a spherical body occupying $r \le a$, when $T = T_0$ for $t \le 0$ but for $t > 0$ the outer boundary $r = a$ is raised to temperature $T = T_0 + T_1$.

It is necessary to solve

$$\frac{\partial (rT)}{\partial t} = \nu \frac{\partial^2 (rT)}{\partial r^2} \quad \text{in} \quad 0 \le r < a\,,$$
$$T(r,0) = T_0\,, \quad 0 \le r < a\,; \qquad T(a,t) = T_0 + T_1\,, \quad t > 0\,.$$

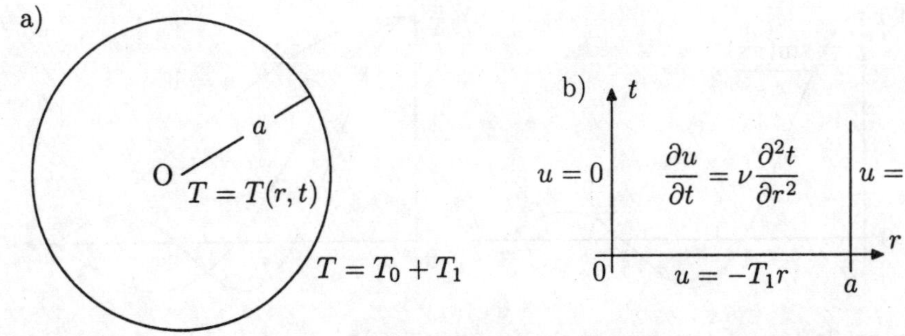

Figure 2.6 a) A spherically symmetric temperature distribution $T(r,t)$. b) Boundary and initial conditions for $u(r,t) \equiv r(T - T_0 - T_1)$, appropriate for Example 2.8.

Write $T = T_0 + T_1 + \bar{T}(r,t)$, since $T = T_0 + T_1$ is the *steady* temperature distribution which satisfies the condition at $r = a$. Then, after writing $r\bar{T}(r,t) \equiv u(r,t)$, it is necessary to solve

$$\frac{\partial u}{\partial t} = \nu \frac{\partial^2 u}{\partial t^2}\,; \quad u(0,t) = 0 \quad \text{since } \bar{T} \text{ is finite as } r \to 0\,; \quad u(a,t) = 0$$

with initial condition $u(r,0) = -rT_1$ in $0 \le r < a$. As in Example 2.6, $u(r,t)$ may be built up by superposing solutions exponential in t and sinusoidal in r, viz.

$$u = \sum_{n=1}^{\infty} b_n \exp\left(\frac{-\nu n^2 \pi^2 t}{a^2}\right) \sin \frac{n\pi r}{a}\,.$$

Moreover, since $u(r,0)$ is proportional to r in $0 < r < a$, the coefficients b_n are readily adapted from those in (2.11) as $b_n = -2T_1 a(-1)^{n-1}/(\pi n)$. Thus,

$$T = T_0 + T_1 + \frac{2T_1 a}{\pi} \sum_{n=1}^{\infty} \frac{(-1)^n}{n} r^{-1} \exp\left(\frac{-\nu n^2 \pi^2 t}{a^2}\right) \sin \frac{n\pi r}{a}\,.$$

Conclusions:

i) $T \to T_0 + T_1$ (the oven temperature) as $t \to +\infty$.

ii) The temperature deficit ($T_0 + T_1 - T = -\bar{T}$) is a sum of *decaying modes* which have respective radial dependences $\propto r^{-1} \sin(n\pi r/a)$ and corresponding *decay rates* $\nu n^2 \pi^2/a^2$ which increase rapidly with n. Note also that, since $\lim_{x \to 0} x^{-1} \sin x = 1$, each function $(rn)^{-1} \sin n\pi r/a$ has the same finite limit π/a as $r \to 0$.

iii) The smallest decay rate (for $n = 1$) is $\nu\pi^2/a^2$. Since each mode has decay rate $\propto a^{-2}$, the cooking time will be $\propto a^2 \propto$ ('weight')$^{2/3}$ (which accounts for instructions to increase the cooking time for larger dishes.)

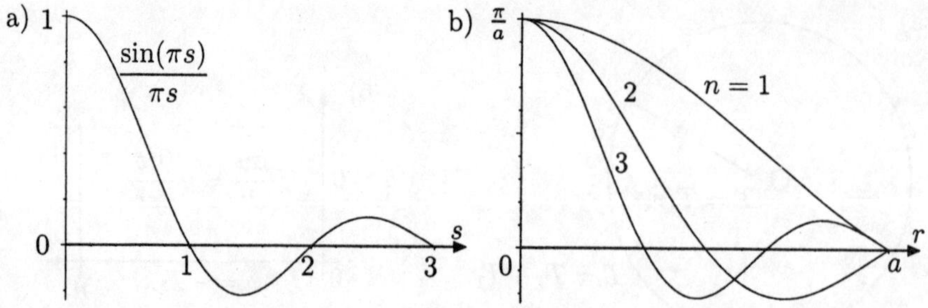

Figure 2.7 Graphs a) of the function $(\pi s)^{-1} \sin \pi s$ and b) of the modes $(nr)^{-1} \sin(n\pi r/a)$ arising in the solution to Example 2.8.

iv) Cooling of hot objects obeys a similar rule; i.e. the cooling time is proportional to ('weight')$^{2/3}$ (which is why small babies should be well wrapped in winter, to avoid rapid cooling).

Cooking by microwave heating is substantially different. Throughout the region $0 \le r < a$ there is a distributed source of heat supply, due to conversion from electromagnetic radiation.

Example 2.9

Suppose that the spherical body of Example 2.8 is again at $T = T_0$ for $t \le 0$, but for $t > 0$ there is a uniform source of heating $S = S_0$ throughout $r < a$. Find an expression for the temperature $T(r, t)$ when it is assumed (for simplicity) that $T = T_0$ at $r = a$.

Write $T = \hat{T}(r) + \bar{T}(r, t)$. Then, the steady contribution $\hat{T}$ satisfies $-K(r\hat{T})'' = rS_0$ with $\hat{T}(a) = T_0$. Integrating twice gives

$$Kr\hat{T}(r) = -\tfrac{1}{6}r^3 S_0 + Ar \qquad \text{(since } \hat{T} \text{ must be finite at } r = 0\text{)}.$$

Then, the boundary condition at $r = a$ yields $A = \frac{1}{6}a^2 S_0$, so that the *steady contribution* is

$$\hat{T}(r) = T_0 + \tfrac{1}{6}S_0 K^{-1}(a^2 - r^2)\,.$$

To find $\bar{T}$, write $u(r, t) \equiv r\bar{T}(r, t)$ which satisfies

$$\frac{\partial u}{\partial t} = \nu \frac{\partial^2 u}{\partial t^2}\,; \quad u(0, t) = 0 \ \text{ to keep } \bar{T} \text{ finite}\,; \quad u(a, t) = 0\,.$$

As in Example 2.7, the expression for $u(r, t)$ may be written as

$$u = \sum_{n=1}^{\infty} b_n \exp\left(\frac{-\nu n^2 \pi^2 t}{a^2}\right) \sin \frac{n\pi r}{a}\,,$$

involving a set of coefficients $b_0, b_1, b_2, \ldots$. These coefficients are determined from the *initial condition*, rewritten as $u(r,0) + r\hat{T}(r) = rT_0$. Thus,

$$\sum_{n=1}^{\infty} b_n \sin\frac{n\pi r}{a} = u(r,0) = r(T_0 - \hat{T}(r)) = \tfrac{1}{6} S_0 K^{-1}(r^3 - a^2 r)\,.$$

Applying the *half-range Euler formula* in the form

$$b_n = \frac{2}{a}\int_0^a u(r,0)\sin\frac{n\pi r}{a}\,dr$$

gives

$$\begin{aligned}
\frac{3aK}{S_0} b_n &= \int_0^a (r^3 - a^2 r)\sin\frac{n\pi r}{a}\,dr \\
&= \left[\frac{-a}{n\pi}(r^3 - a^2 r)\cos\frac{n\pi r}{a}\right]_0^a + \frac{a}{n\pi}\int_0^a (3r^2 - a^2)\cos\frac{n\pi r}{a}\,dr \\
&= 0 + \left(\frac{a}{n\pi}\right)^2\left[(3r^2 - a^2)\sin\frac{n\pi r}{a}\right]_0^a - \left(\frac{a}{n\pi}\right)^2\int_0^a 6r\sin\frac{n\pi r}{a}\,dr \\
&= 0 + 6\left(\frac{a}{n\pi}\right)^3\left[r\cos\frac{n\pi r}{a}\right]_0^a - 6\left(\frac{a}{n\pi}\right)^3\int_0^a \cos\frac{n\pi r}{a}\,dr \\
&= 6\frac{a^4}{(n\pi)^3}\cos n\pi - 0 = \frac{6a^4}{n^3\pi^3}(-1)^n\,.
\end{aligned}$$

Combining the resulting expression for $u(r,t)$ with that for $\hat{T}$ gives

$$T(r,t) = T_0 + \frac{S_0}{6K}(a^2 - r^2) + \frac{2S_0 a^3}{K\pi^3}\sum_{n=1}^{\infty}\frac{(-1)^n}{n^3}\exp\left(\frac{-\nu n^2\pi^2 t}{a^2}\right)\sin\frac{n\pi r}{a}\,.$$

When compared with the prediction from Example 2.8, the factor n^{-3} shows an even stronger decay of coefficients with n. Thus, for $t > a^2/(\nu\pi^2)$, the departure $T - T$ from the steady state temperature $\hat{T}(r)$ is well approximated by the single decaying mode $-2S_0 a^3/(K\pi^3)\exp(-\nu\pi^2 t/a^2)\sin(\pi r/a)$. More crucially, the steady state is ***hotter in the centre*** than near the outside, so that microwave power (rate of energy supply) must be limited (i.e. applied with care) if excessively hot regions (***hotspots***) are to be avoided.

EXERCISES

2.5. If the teapot of Example 2.1 loses heat to its surroundings at rate $\beta(T - 20^\circ)$ joules/min, determine the differential equation governing $T(t)$. Solve it, given that $T = 98^\circ$C at time $t = 0$. How long does the teapot take to cool to 36°C?

2.6. A slab having thermal conductivity K extends from $x = 0$ to $x = 4a$. It is perfectly thermally insulated at $x = 0$ and is held at temperature $T = T_1$ at $x = 4a$. Find the steady temperature $T = T(x)$ when heat is supplied at uniform rate $S = S_0$ in $0 < x < a$, with no heat supply in $a < x < 4a$.

2.7. A flat heating element extending over $0 < x < h$ is initially at uniform temperature T_0. At time $t = 0$, a uniform heat supply rate $S = S_0$ is switched on, while the surfaces are maintained at temperature T_0. For the governing equations

$$\rho c \frac{\partial T}{\partial t} = K \frac{\partial^2 T}{\partial x^2} + S_0 \,, \quad 0 < x < h \,, \, t > 0 \,; \quad T(0,t) = T(h,t) = T_0 \,,$$

(a) determine the steady state solution $\hat{T}(x)$,

(b) find a representation for $u(x,t) \equiv T - \hat{T}(x)$ as an infinite series,

(c) obtain the coefficients b_n in that series.

Hence, obtain $T(x,t)$.

2.8. By relating the equation governing spherically symmetric heat conduction $\partial(rT)/\partial t = \nu \partial^2(rT)/\partial r^2$ to the one-dimensional heat conduction equation, or otherwise, show that each of the following is a possible temperature distribution:

$$\text{(a) } r^{-1}\exp(Ar + \nu A^2 t) \,, \quad \text{(b) } r^{-1} \mathrm{e}^{-\nu p^2 t} \sin pr \,,$$
$$\text{(c) } A(r + 2\nu t/r) + Br^{-1} + C \,.$$

2.9. Suppose that, in the spherically symmetric body of Example 2.9, the boundary condition at $r = a$ is replaced by one in which the heat flux is $q = B(T(a,t) - T_0)$, but the problem is otherwise unchanged. Determine and solve the differential equation and boundary conditions defining the steady solution $T = \hat{T}(r)$. Obtain the boundary condition at $r = a$ for the unsteady contribution $\bar{T}(r,t) \equiv T - \hat{T}(r)$. Show that this condition is compatible with separable solutions $\bar{T} = r^{-1}\mathrm{e}^{-\nu p^2 t} \sin pr$ for some choices of p. Obtain, but do not solve, the equation defining possibilities for p.
(N.B. This is more physically reasonable than the condition used in Example 2.9. It states that the rate of conduction through the surface is proportional to the difference in temperature between the body and its surroundings.)

3

Fields and Potentials

3.1 Gradient of a Scalar

Fourier's law, that heat flux is proportional to the *gradient* of temperature $T(\boldsymbol{x}, t)$, is a statement about both the direction and the rate of heat transport. It is a statement about vectors. In one-dimensional examples and in cases of cylindrical or spherical symmetry, the flow direction is evident. More generally, the *gradient* of the scalar function $T(\boldsymbol{x}, t)$ is found using a basic concept from *vector calculus* (see e.g. Matthews (1998)).

Consider first the values of a scalar quantity $f(\boldsymbol{x})$ at points on some curve Γ. If (x, y, z) are cartesian coordinates of the point $\boldsymbol{x} = x\boldsymbol{i} + y\boldsymbol{j} + z\boldsymbol{k}$ of Γ, the difference in the values of f at $\boldsymbol{x}$ and at a neighbouring point $\boldsymbol{x} + \delta\boldsymbol{x}$ is found (using Taylor series in several variables) as

$$\delta f \equiv f(x + \delta x, y + \delta y, z + \delta z) - f(x, y, z) \approx \frac{\partial f}{\partial x}\delta x + \frac{\partial f}{\partial y}\delta y + \frac{\partial f}{\partial z}\delta z\,.$$

Let s measure the *distance* along Γ from some chosen (*reference*) point, so that the coordinates are given parametrically by $x = x(s)$, $y = y(s)$, $z = z(s)$. Then, taking the limit $\delta s \to 0$ (in which $\delta x \to 0$, etc.) yields the *chain rule*

$$\frac{df}{ds} = \frac{\partial f}{\partial x}\frac{dx}{ds} + \frac{\partial f}{\partial y}\frac{dy}{ds} + \frac{\partial f}{\partial z}\frac{dz}{ds} = \frac{\partial f}{\partial x}x'(s) + \frac{\partial f}{\partial y}y'(s) + \frac{\partial f}{\partial z}z'(s)\,. \tag{3.1}$$

This has a vectorial interpretation. The quantities $x'(s) \equiv \mathrm{d}x/\mathrm{d}s$, etc. are the components of the unit tangent $\boldsymbol{t} = \dfrac{d\boldsymbol{x}}{ds} = \boldsymbol{i}x'(s) + \boldsymbol{j}y'(s) + \boldsymbol{k}z'(s)$ to Γ at

$\boldsymbol{x} = \boldsymbol{x}(s)$. Hence, the right-hand side of (3.1) is the *scalar product* of $\boldsymbol{t}$ with the vector

$$\text{grad } f = \boldsymbol{i}\frac{\partial f}{\partial x} + \boldsymbol{j}\frac{\partial f}{\partial y} + \boldsymbol{k}\frac{\partial f}{\partial z} \equiv \boldsymbol{\nabla} f\,, \tag{3.2}$$

namely

$$\frac{df}{ds} = \boldsymbol{t} \cdot \text{grad } f\,.$$

Definition: The symbol $\boldsymbol{\nabla}$, pronounced as 'del' or 'nabla', is the *vector operator*

$$\boldsymbol{\nabla} \equiv \boldsymbol{i}\frac{\partial}{\partial x} + \boldsymbol{j}\frac{\partial}{\partial y} + \boldsymbol{k}\frac{\partial}{\partial z}\,. \tag{3.3}$$

Definition: The *gradient* of a scalar function $f = f(\boldsymbol{x})$, is $\boldsymbol{\nabla} f = \text{grad } f = \boldsymbol{\nabla} f(\boldsymbol{x})$.

For *any* unit vector $\boldsymbol{t}$, the quantity $\boldsymbol{t} \cdot \boldsymbol{\nabla} f$ measures the *rate of increase of f with respect to distance* in the direction of $\boldsymbol{t}$, measured at $\boldsymbol{x}$.

Thus, associated with *any* scalar function $f(\boldsymbol{x})$ of position, the vector function of position (i.e. *vector field*) $\boldsymbol{\nabla} f(\boldsymbol{x})$ defined by (3.2) has the property that $\boldsymbol{t} \cdot \boldsymbol{\nabla} f$ is the *directional derivative* of f at $\boldsymbol{x}$ in the direction of $\boldsymbol{t}$. Note that

i) $\boldsymbol{\nabla} f = \text{grad}\, f$ depends on position $\boldsymbol{x}$ but not upon the choice of Γ.

ii) The directional derivative at $\boldsymbol{x}$ does not depend upon the choice of curve Γ through $\boldsymbol{x}$, but merely upon the direction of the unit vector $\boldsymbol{t}$.

iii) At any point $\boldsymbol{x}$, the vector grad f is normal to the *level surface* (i.e. the surface $f =$ constant) which passes through $\boldsymbol{x}$.

3.1.1 Some Applications

i) For any temperature field $T(\boldsymbol{x}, t)$ (steady, or unsteady), the *temperature gradient* is the vector field grad $T = \boldsymbol{\nabla} T$.

The general statement of Fourier's law of heat conduction in three dimensions is

$$\boldsymbol{q} = -K \text{grad } T\,.$$

This states that the *heat flux vector* $\boldsymbol{q}$

(a) is perpendicular to the *isothermal surfaces* $T =$ constant;

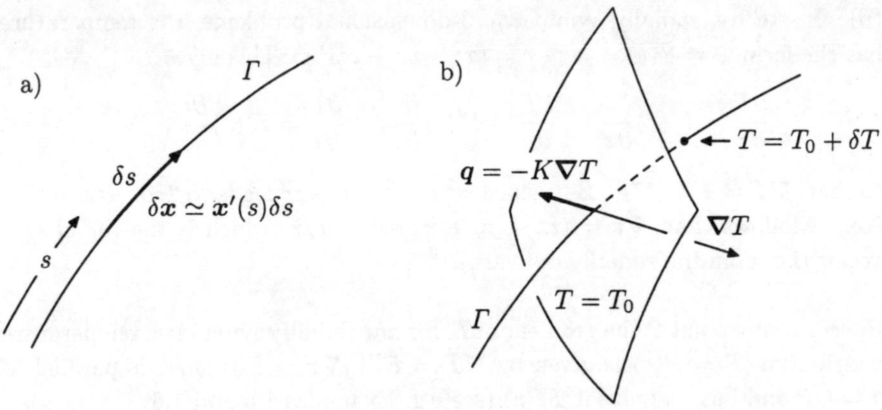

Figure 3.1 a) A small displacement $\delta \boldsymbol{x} \approx \boldsymbol{x}'(s)\,\delta s = \delta \boldsymbol{s}$ along a curve Γ. b) The associated temperature increment $\delta T \approx \boldsymbol{\nabla}T \cdot \delta \boldsymbol{x}$, where $\boldsymbol{\nabla}T$ is orthogonal to the *isothermal surface* $T = T_0$. The *heat flux* vector $\boldsymbol{q} = -K\boldsymbol{\nabla}T$ is also normal to the *isothermal* $T = T_0$.

(b) has magnitude proportional to the derivative of temperature with respect to distance;

(c) points in the direction of most rapid *decrease* of temperature.

ii) In a liquid or gas, the *pressure* varies with position, viz. $p = p(\boldsymbol{x}, t)$. At each time t, the surfaces $p =$ constant are the *isobaric surfaces* – a typical weather map shows the *isobars*, their intersections with the ground surface.

In fluid dynamics and meteorology, the *pressure gradient* $\boldsymbol{\nabla}p$ (a vector field) is important in determining the fluid motion.

iii) In diffusive processes, a chemical species moves so that differences in concentration c tend to be smoothed out. By analogy with Fourier's law for the flux of heat, the flux $\boldsymbol{q}$ of the species is taken, in *Fick's law of diffusion*, to be

$$\boldsymbol{q} = -K_c\boldsymbol{\nabla}c\,, \qquad \text{where } \boldsymbol{\nabla}c \text{ is the } \textit{concentration gradient.}$$

N.B. Diffusion drives material from regions of high concentration c to regions of lower c. K_c is the *diffusion coefficient* of the species.

Example 3.1

(a) When the temperature is $T = ax+b$, the temperature gradient is $\boldsymbol{\nabla}T = a\boldsymbol{i}$; i.e. it is *uniform* and is parallel to Ox, as in one-dimensional examples within Chapter 1.

(b) In steady, radially symmetric 3-dimensional problems, the temperature has the form $T = F(r)$, where $r = (x^2 + y^2 + z^2)^{1/2}$. This gives

$$\frac{\partial T}{\partial x} = F'(r)\frac{\partial r}{\partial x}, \quad \frac{\partial T}{\partial y} = F'(r)\frac{\partial r}{\partial y}, \quad \frac{\partial T}{\partial z} = F'(r)\frac{\partial r}{\partial z},$$

so that $\nabla T = F'(r)\nabla r$. But, since $r^2 = x^2+y^2+z^2$ leads to $2r\partial r/\partial x = 2x$, etc., it follows that $\nabla r = (\boldsymbol{i}x + \boldsymbol{j}y + \boldsymbol{k}z)/r = \boldsymbol{x}/r$, which is the *radial unit vector* (i.e. pointing radially outward).

Hence, at any point P the gradient of T, for *any* radially symmetric temperature distribution $T = F(r)$ is given by $\nabla T = F'(r)\nabla r = F'(r)\boldsymbol{x}/r$, is parallel to $\boldsymbol{x} = \overrightarrow{\mathrm{OP}}$ and has magnitude $F'(r)$ (see e.g. Examples 1.5 and 1.6).

Example 3.2

If $T = Ar^{-1} + B$, find ∇T and show that the heat flux given by Fourier's law has $|\boldsymbol{q}| \propto r^{-2}$.

Since $\nabla r^{-1} = -r^{-2}\nabla r = -r^{-2}\boldsymbol{x}/r$, then $\nabla T = -a\boldsymbol{x}/r^3$. The corresponding heat flux is $\boldsymbol{q} = -K\nabla T = KA\boldsymbol{x}/r^3$, so that $|\boldsymbol{q}| = KAr^{-2}$.

As was seen in Example 1.5, heat flux with $|\boldsymbol{q}| \propto r^{-2}$ arises in $r_1 < r < r_2$ if the heat supply rate S is zero in $r_1 < r < r_2$. Similar radially symmetric *vector fields*, with modulus $\propto r^{-2}$ are important in both gravitational theory and electrostatics.

3.2 Gravitational Potential

Centuries of astronomical observation led to the deduction that a mass m_0 regarded as concentrated at a single point O exerts on a mass m at a distant point P a force

$$\boldsymbol{F}_m = -Gm_0m\frac{\boldsymbol{x}}{r^3}, \quad \text{where } \boldsymbol{x} = \overrightarrow{\mathrm{OP}}, \quad r = |\boldsymbol{x}|,$$

and where G is the *gravitational constant.* Thus, the force acts radially *towards* O, with strength $\propto r^{-2}$ and proportional also to each of the masses. This is *Newton's law of gravitation.* If other masses influence the motion of m, the total (or *resultant*) force at P is found by vector addition. It is convenient to regard each mass as giving a contribution to a *gravitational field* at P.
Definition: The *gravitational field* at P due to *any* system of masses is found as the ratio $\boldsymbol{F}_m/m$, where $\boldsymbol{F}_m$ is the force acting on a (suitably small) test

mass m placed at P (m must remain small in order that the test mass does not disturb the location of the other masses).

Example 3.3

The gravitational field at P due to a point mass M placed at O is $\boldsymbol{F} \equiv \dfrac{-GM\boldsymbol{x}}{r^3}$.

The field in Example 3.3 is the gradient of the scalar GMr^{-1} (see Example 3.2). This motivates the concept of gravitational potential.
Definition: The *gravitational potential* due to a point mass M at $\boldsymbol{x} = \boldsymbol{0}$ is $\phi(\boldsymbol{x}) \equiv \dfrac{-GM}{|\boldsymbol{x}|}$. It has the property that $\boldsymbol{F} = -\boldsymbol{\nabla}\phi = -\text{grad}\,\phi$.

Example 3.4

Show that the *gravitational potential* due to M at $\boldsymbol{x} = \boldsymbol{x}_0$ is $\phi = \dfrac{-GM}{|\boldsymbol{x} - \boldsymbol{x}_0|}$.

The gradient of ϕ is found by noting that $|\boldsymbol{x} - \boldsymbol{x}_0|^2 = (x - x_0)^2 + (y - y_0)^2 + (z - z_0)^2$, so that (cf. Example 3.1) $\partial|\boldsymbol{x} - \boldsymbol{x}_0|/\partial x = (x - x_0)/|\boldsymbol{x} - \boldsymbol{x}_0|$, etc. Then,

$$\boldsymbol{\nabla}\left(\frac{GM}{|\boldsymbol{x} - \boldsymbol{x}_0|}\right) = -GM\frac{(x - x_0)\boldsymbol{i} + (y - y_0)\boldsymbol{j} + (z - z_0)\boldsymbol{k}}{|\boldsymbol{x} - \boldsymbol{x}_0|^3} = -GM\frac{\boldsymbol{x} - \boldsymbol{x}_0}{|\boldsymbol{x} - \boldsymbol{x}_0|^3}$$

which is a vector acting at P along $\overrightarrow{\text{PP}_0}$, with the correct magnitude $GM|\text{PP}_0|^{-2}$.

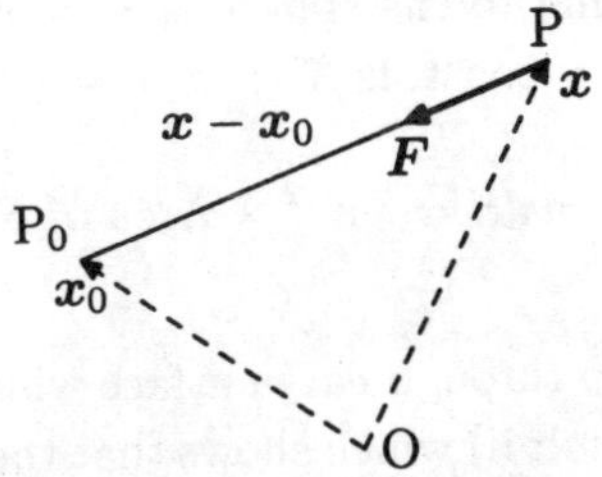

Figure 3.2 The gravitational field at P, due to mass M at P_0, is directed along $\overrightarrow{\text{PP}_0} = \boldsymbol{x}_0 - \boldsymbol{x}$.

Since forces add as vectors, a mass m at a typical point $\boldsymbol{x}$ and influenced by a set of masses $m_1, \ldots, m_N$ at points $\boldsymbol{x}_1, \ldots, \boldsymbol{x}_N$ will experience a *resultant force*

$$\boldsymbol{F}_m = \sum_{n=1}^{N} -Gmm_n \frac{\boldsymbol{x} - \boldsymbol{x}_n}{|\boldsymbol{x} - \boldsymbol{x}_n|^3} = m\,\text{grad}\left\{\sum_{n=1}^{N} \frac{Gm_n}{|\boldsymbol{x} - \boldsymbol{x}_n|}\right\}. \tag{3.4}$$

Thus, the *gravitational field* (i.e. $\boldsymbol{F}_m/m$) may be calculated at *any* $\boldsymbol{x}$ as the gradient of a *gravitational potential*

$$\phi = -\sum_{n=1}^{N} \frac{Gm_n}{|\boldsymbol{x} - \boldsymbol{x}_n|}, \tag{3.5}$$

namely $\boldsymbol{F} = -\boldsymbol{\nabla}\phi = -\text{grad}\,\phi$. In problems where mass is distributed continuously (not just at discrete points), introduction of a potential allows much simpler manipulations, since only a scalar need be considered.

3.2.1 Special Properties of the Function $\phi = r^{-1}$

i) The function $\phi = r^{-1} = (x^2 + y^2 + z^2)^{-1/2}$ satisfies *Laplace's equation:*

$$\nabla^2\,\phi \equiv \frac{\partial^2\phi}{\partial x^2} + \frac{\partial^2\phi}{\partial y^2} + \frac{\partial^2\phi}{\partial z^2} = 0\,, \tag{3.6}$$

except at the point $\boldsymbol{x} = \boldsymbol{0}$ (i.e. at $r = 0$). [Readers familiar with *divergence* (see Chapter 6) will recognize (3.6) as $\text{div}(\text{grad}\ \phi) = \boldsymbol{\nabla}\cdot(\boldsymbol{\nabla}\,\phi) = 0$.]

ii) The gradient $-\boldsymbol{x}/r^3$ of r^{-1} is *radial* and gives the same *total flux* -4π through *each* sphere $r =$ constant centred at O; since $\text{grad}\,\phi = -\boldsymbol{x}/r^3$ and the outward unit normal to the sphere is $\boldsymbol{n} = \boldsymbol{x}/r$, then the flux at each point of a sphere $\mathcal{S}$: $r =$ const. is $\boldsymbol{\nabla}\,\phi\cdot\boldsymbol{x}/r = -r^{-2}$, then the total flux is

$$\iint_{\mathcal{S}} \boldsymbol{\nabla}\,\phi\cdot\boldsymbol{n}\,dS = -r^{-2} \times \text{Area of } \mathcal{S} = -4\pi\,.$$

iii) The total flux of $\text{grad}\,\phi$ through *each* surface which encloses O equals -4π (this generalizes the result ii) which shows that the total flux is independent of radius, when the surface is a sphere centred at O).

Example 3.5 (*Outline proof of* i))

Show that $\nabla^2 r^{-1} = 0$. $\qquad (r \neq 0)$

Since $\partial r^{-1}/\partial x = -r^{-2}(x/r)$ (see Example 3.1), then

$$\frac{\partial^2 r^{-1}}{\partial x^2} = \frac{\partial}{\partial x}\left(\frac{-x}{r^3}\right) = \frac{-1}{r^3} - x\left(\frac{-3}{r^4}\right)\frac{\partial r}{\partial x} = -r^{-3} + \frac{3x}{r^4}\frac{x}{r} = 3x^2 r^{-5} - r^{-3}$$

(for $r \neq 0$). Similarly, $\partial^2 r^{-1}/\partial y^2 = 3y^2 r^{-5} - r^{-3}$, etc. so that

$$\nabla^2 r^{-1} = 3r^{-5}(x^2 + y^2 + z^2) - r^{-3}(1 + 1 + 1) = 3r^{-3} - 3r^{-3} = 0\,.$$

N.B. Alternatively, treating $\boldsymbol{\nabla}$ as a *vector operator* and using $\boldsymbol{\nabla} r^{-1} = -r^{-2}\boldsymbol{\nabla} r$ $= -\boldsymbol{x}/r^3 = -\boldsymbol{i}(x/r^3) - \boldsymbol{j}(y/r^3) - \boldsymbol{k}(z/r^3)$ gives

$$\nabla^2 r^{-1} = \boldsymbol{\nabla}\cdot(\boldsymbol{\nabla} r^{-1}) = \left(\boldsymbol{i}\frac{\partial}{\partial x} + \boldsymbol{j}\frac{\partial}{\partial y} + \boldsymbol{k}\frac{\partial}{\partial z}\right)\cdot\left(\boldsymbol{i}\frac{-x}{r^3} + \boldsymbol{j}\frac{-y}{r^3} + \boldsymbol{k}\frac{-z}{r^3}\right)$$

$$= \frac{\partial}{\partial x}\left(\frac{-x}{r^3}\right) + \frac{\partial}{\partial y}\left(\frac{-y}{r^3}\right) + \frac{\partial}{\partial z}\left(\frac{-z}{r^3}\right) \qquad \text{etc.}$$

Also, by using a shift of origin $\boldsymbol{0} \to \boldsymbol{x}_0$, it is found that $\nabla^2(|\boldsymbol{x} - \boldsymbol{x}_0|^{-1}) = 0$, except at $\boldsymbol{x} = \boldsymbol{x}_0$.

The result iii) is of great importance for *potential theory*, which applies to gravitation, to electrostatics and elsewhere in applied mathematics. An *outline proof* is as follows:
Suppose that $\mathcal{S}$ is *any* surface enclosing O. Around a point P of $\mathcal{S}$, where the unit normal is $\boldsymbol{n}$, consider a surface element having area δS. The set of straight lines (rays) from O to the boundary of δS forms an elementary cone with vertex O. Denote this cone by Π. The first objective is to relate the flux $\boldsymbol{\nabla} r^{-1}\cdot\boldsymbol{n}\,\delta S$ through this surface element to the equivalent flux through the elementary area cut out by Π from the sphere of unit radius, centred at O. On this *unit sphere*, let δA_0 be the elementary area. This area is related to the area δA cut by Π from the sphere through P, of radius r and centre O, by $\delta A = r^2\delta A_0$ (since the elementary areas have the same shape but have dimensions scaled by the factor r = OP).

Now, since *any* sphere centred at O has *unit normal* $\boldsymbol{x}/r$ and since $\boldsymbol{\nabla} r^{-1} = -\boldsymbol{x}/r^3$, the flux of $\boldsymbol{\nabla} r^{-1}$ through the elements cut out from the two spheres are equal:

$$\frac{-\boldsymbol{x}}{r^3}\cdot\frac{\boldsymbol{x}}{r}\,\delta A = -\frac{\boldsymbol{x}\cdot\boldsymbol{x}}{r^4}r^2\,\delta A_0 = -\delta A_0 \qquad \text{(independently of radius).}$$

However, if α measures the angle at P between the unit normal $\boldsymbol{n}$ and the unit radial vector $\boldsymbol{x}/r$, then $\cos\alpha = \boldsymbol{n}\cdot\boldsymbol{x}/r$. Moreover, the projection of the surface element δS at P onto the tangent plane to the sphere through P has area $\delta S\,\cos\alpha$, so that

$$\frac{\boldsymbol{x}}{r^3}\cdot\boldsymbol{n}\,\delta S \approx \frac{1}{r^2}\delta A = \delta A_0\,.$$

Hence the flux of ∇r^{-1} through the surface element δS at P is approximately $-\delta A_0$. This statement, which improves in accuracy as the dimensions of δS become negligible with respect to the radii of curvature at P, is independent of

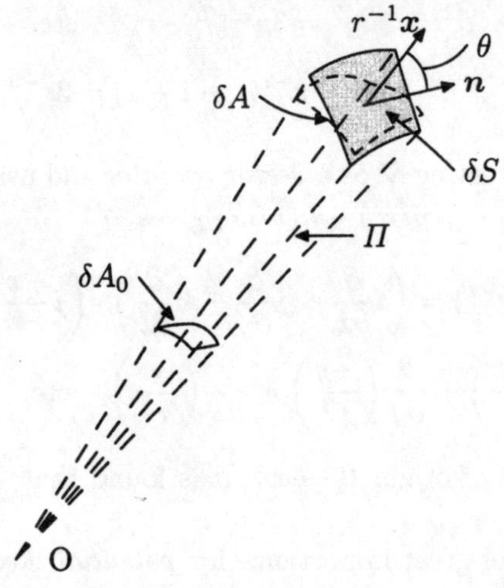

Figure 3.3 Diagram illustrating why ∇r^{-1} has equal flux through all portions of surface intersecting an elementary cone Π.

the shape of δS and of the orientation relative to $\overrightarrow{\mathrm{OP}} = \boldsymbol{x}$. Then, if $\mathcal{S}$ encloses P in such a way that each ray (i.e. half-line) from O meets $\mathcal{S}$ exactly once, there is a one-to-one mapping between points of $\mathcal{S}$ and points of the unit sphere. The *total flux* of ∇r^{-1} through $\mathcal{S}$ is obtained by integration. But $\iint \delta A_0 = 4\pi$, so that

$$\iint_{\mathcal{S}} \frac{-\boldsymbol{x}}{r^3} \cdot \boldsymbol{n} \mathrm{d}S = -\iint \delta A_0 = -4\pi \,.$$

This result is *independent* of the shape of $\mathcal{S}$, provided that O is within $\mathcal{S}$ (each ray through O must meet a closed surface in an *odd* number of points, with each point where $\boldsymbol{n}$ points inwards being cancelled by one where $\boldsymbol{n}$ points outward. Exceptional cases, where the ray is tangential to $\mathcal{S}$, give zero contribution to the flux, since $\boldsymbol{x} \cdot \boldsymbol{n} = 0$).

If O is *outside* $\mathcal{S}$, then each non-tangential ray either does not intersect $\mathcal{S}$ or meets $\mathcal{S}$ in as many inward-facing as outward-facing points. In such cases $\iint_{\mathcal{S}} (-\boldsymbol{x}/r^3) \cdot \boldsymbol{n}\, \mathrm{d}S = 0$. Combining these two cases gives the important result

$$\iint_{\mathcal{S}} \nabla r^{-1} \cdot \boldsymbol{n}\, \mathrm{d}S = \begin{cases} -4\pi & \text{for } all \text{ surfaces } \mathcal{S} \text{ enclosing O} \\ 0 & \text{for } all \text{ surfaces } \mathcal{S} \text{ not enclosing O.} \end{cases} \tag{3.7}$$

The function $\phi = -1/(4\pi r)$ is often regarded as the *fundamental solution of Laplace's equation.* It satisfies $\nabla^2 \phi = 0$ everywhere except at $\boldsymbol{r} = \boldsymbol{0}$ and has the property that the total outward flux of $\nabla \phi$ through *any* surface $\mathcal{S}$ equals unity if the surface $\mathcal{S}$ encloses $\boldsymbol{r} = \boldsymbol{0}$, but vanishes otherwise. It is sometimes

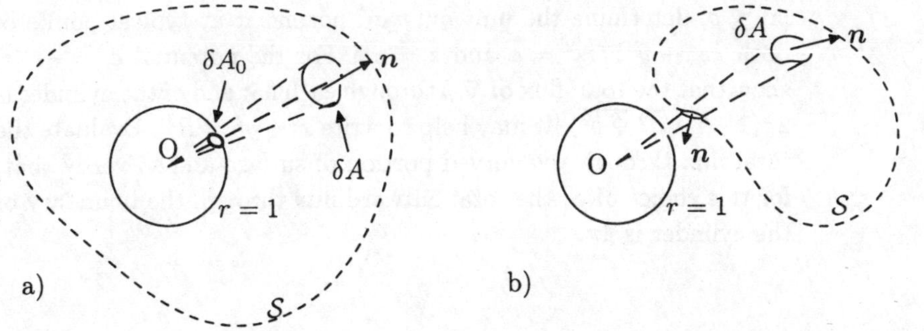

Figure 3.4 The total flux of the field ∇r^{-1} through a surface $\mathcal{S}$ a) equals -4π if $\mathcal{S}$ encloses O, but b) equals 0 if O is outside $\mathcal{S}$.

regarded as the solution due to a *unit source* at $\boldsymbol{r} = \mathbf{0}$. A simple change of origin shows that $\phi \equiv -1/(4\pi|\boldsymbol{r} - \boldsymbol{r}_0|)$ corresponds to a unit source at $\boldsymbol{r} = \boldsymbol{r}_0$.

EXERCISES

3.1. If $T = A(x^2 + y^2) + Bz$, calculate ∇T. Show that the heat flux vector $\boldsymbol{q} = -K\nabla T$ is the sum of a uniform flux parallel to Oz and (for $A \geq 0$) a flux radially towards the Oz axis.

3.2. Using $\Phi(\boldsymbol{x}) = -G|\boldsymbol{x}|^{-1}$ to denote the potential of a unit mass at $\boldsymbol{x} = \mathbf{0}$, show that each of the following describes a gravitational potential. Find the corresponding gravitational fields $\boldsymbol{F} = -\nabla\phi$ and identify the locations and magnitudes of the point masses creating the fields:
(a) $\phi = M\Phi(\boldsymbol{x})$, (b) $\phi = M_1\Phi(\boldsymbol{x} - \boldsymbol{x}_1) + M_2\Phi(\boldsymbol{x} - \boldsymbol{x}_2)$,
(c) $\phi = m\{\Phi(\boldsymbol{x} - a\boldsymbol{i}) + \Phi(\boldsymbol{x} + a\boldsymbol{i})\} + M\{\Phi(\boldsymbol{x} - b\boldsymbol{j}) + \Phi(\boldsymbol{x} + b\boldsymbol{j})\}$,
where $\boldsymbol{i}$ and $\boldsymbol{j}$ are the usual unit vectors. In case (c), evaluate $\boldsymbol{F}$ at $\boldsymbol{x} = \mathbf{0}$ and explain your answer.

3.3. Suppose that a binary star has mass m_1 at $\boldsymbol{x} = a\boldsymbol{i}$ and mass m_2 at $\boldsymbol{x} = -a\boldsymbol{i}$. Write down the gravitational potential ϕ at general position $\boldsymbol{x} = x\boldsymbol{i} + y\boldsymbol{j} + z\boldsymbol{k}$.
By writing $|\boldsymbol{x}| = r$, express the potential ϕ in terms of r and x. For $r \gg a$, show that

$$\phi \approx -\frac{(m_1 + m_2)G}{r} - \frac{(m_1 - m_2)Gax}{r^3} .$$

3.4. If $\mathcal{S}$ is the bounding surface to the cylindrical region $x^2 + y^2 \leq a^2$,

$|z| \le b$, determine the unit outward normal $\boldsymbol{n}$ at typical points of both $(x^2+y^2)^{-1/2} = a$ and $z = \pm b$. For the potential $\phi = -r^{-1}$, show that the total flux of $\boldsymbol{\nabla}\phi$ through each flat end of the cylinder is $2\pi(1-b/\sqrt{a^2+b^2})$ [It may help to write $x^2+y^2 = R^2$]. Evaluate the total flux through the curved portion of surface and so verify that, for this choice of $\mathcal{S}$, the total outward flux through the boundary of the cylinder is 4π.

3.3 Continuous Distributions of Mass

The *gravitational field* $\boldsymbol{F}$ due to *any* distribution of mass may be derived from a potential $\phi(\boldsymbol{x})$ as $\boldsymbol{F} = -\text{grad}\ \phi = -\boldsymbol{\nabla}\phi$. Moreover, the property (3.7) leads to the general result for any closed surface $\mathcal{S}$ having outward unit normal $\boldsymbol{n}$

$$\iint_{\mathcal{S}} \boldsymbol{F}\cdot\boldsymbol{n}\,\mathrm{d}S = -4\pi G \times \text{mass enclosed within } \mathcal{S} \tag{3.8}$$

(since a mass element $\delta m = \rho\delta V$ occupying an elementary volume δV will contribute to the total flux out of $\mathcal{S}$ if, and only if, the volume element is contained within $\mathcal{S}$).

The result (3.8) readily allows the calculation of gravitational fields due to spherically symmetric distributions of mass. In spherical polar coordinates, the potential may be written as $\phi = \phi(r)$, where $r = (x^2+y^2+z^2)^{-1/2}$. Then $\boldsymbol{F} = -\phi'(r)\boldsymbol{x}/r$ is purely radial. Choosing surfaces $\mathcal{S}$ as r = constant then gives

$$4\pi r^2\phi'(r) = 4\pi G \times \text{mass enclosed within the sphere of radius } r \equiv M(r) \quad \text{(say)}\,.$$

Example 3.6

Suppose that a planet having outer radius a is spherically symmetric (i.e. it has density $\rho = \rho(r)$). Find an expression for the mass $M(r)$ within the sphere centred at O and having radius r. Show that, for all radii $r \ge a$, the potential may be taken as $\phi = -GM(a)/r$.

The spherical symmetry implies that $\phi = \phi(r)$ and that $\boldsymbol{\nabla}\phi = \phi'(r)\boldsymbol{x}/r$. The mass $M(r)$ within a *test surface* $\mathcal{S}$ which is the sphere of radius r centred at O is

$$M(r) = \int_0^r 4\pi\bar{r}^2\rho(\bar{r})\,\mathrm{d}\bar{r}\,.$$

Also, from (3.8) it follows that $4\pi GM(r) = -4\pi r^2 \boldsymbol{F} \cdot \boldsymbol{n} = 4\pi r^2 \phi'(r)$, so that $\phi'(r) = Gr^{-2}M(r)$. Now, within $r > a$, the (mass) density is zero, so that $M(r) = M(a)$ for $r > a$. Hence,

$$\phi'(r) = \frac{GM(a)}{r^2} \qquad \text{throughout } r \geq a\,.$$

Integrating this gives

$$\phi(r) = -\frac{GM(a)}{r} + C\,;$$

while, without loss of generality, the constant of integration C may be set equal to zero, since it has no mechanical significance.

[Notice that, outside the mass, the potential is exactly as if *all* the mass were concentrated at O (this is the justification for treating stars and planets as point masses, when considering their influence on distant objects).]

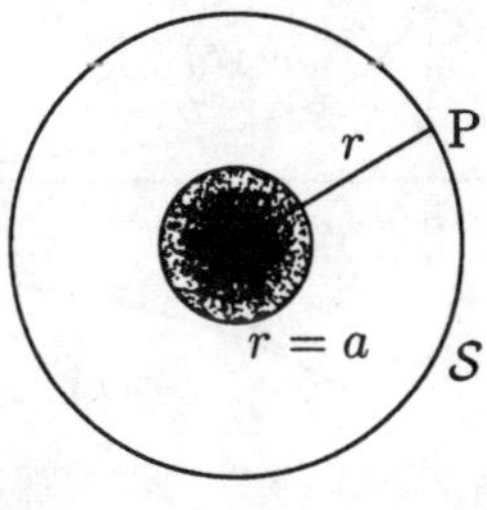

Figure 3.5 Diagram for Example 3.6, showing that for a *test surface* $r =$ constant, with $r \geq a$, the total mass M is enclosed.

Example 3.7

Suppose that the planet of Example 3.6 has uniform density $\rho = \rho_0$ in $r < a$. Then its total mass is $M = \frac{4}{3}\pi a^3 \rho_0 = M(a)$. Show that at radii $r < a$, then $\phi' \propto r$. Find the gravitational potential at all radii.

Take $\mathcal{S}$ as a sphere of radius r centred at O. Then, if $r < a$, the enclosed mass equals $\frac{4}{3}\pi r^3 \rho_0$. A slight modification of the reasoning in Example 3.6 gives

$$4\pi r^2 \phi'(r) = 4\pi G \tfrac{4}{3}\pi r^3 \rho_0\,.$$

Integrating the equation $\phi'(r) = \frac{4}{3}\pi \rho_0 G r$ gives

$$\phi = \frac{2\pi\rho_0 G}{3} r^2 + k \qquad \text{in } r \leq a\,,$$

while $\phi(r) = -GM/r$ for $r > a$ (see Example 3.6). Now, the potential must be continuous (its derivative $\nabla\phi = -\boldsymbol{F}$ must be finite), so that equating values at $r = a$ gives

$$\frac{-MG}{a} = \frac{2\pi\rho_0 Ga^2}{3} + k \qquad \Rightarrow k = -\frac{3}{2}\frac{MG}{a}\,.$$

Thus
$$\phi(r) = \begin{cases} \dfrac{MG}{a}\left(\dfrac{-3}{2} + \dfrac{r^2}{2a^2}\right) & \text{for} \quad r \geq a\,, \\ \dfrac{-MG}{r} & \text{for} \quad r \leq a\,. \end{cases}$$

This predicts that, below the Earth's surface, the gravitational field decreases linearly with depth – a phenomenon which is detectable when measurements are taken down deep mine shafts. Also, unsurprisingly, $\boldsymbol{F}$ is continuous at $r = a$.

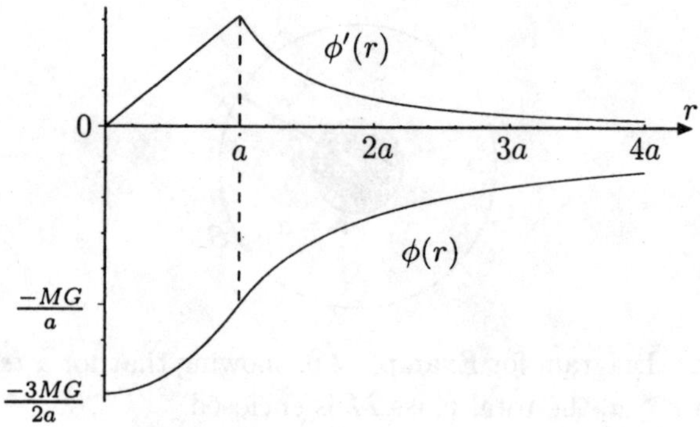

Figure 3.6 The potential $\phi(r)$ and associated outward radial field strength $\phi'(r)$ for the mass distribution of Example 3.7.

3.4 Electrostatics

Observations show that electric charges at rest sometimes *repel* each other, but sometimes *attract* each other. However, the force always acts along their line of centres and has strength $\propto r^{-2}$. This is summarized in *Coulomb's law of electrostatic force*

$$\boldsymbol{F}_{21} = KQ_1Q_2\frac{\boldsymbol{x}_2 - \boldsymbol{x}_1}{|\boldsymbol{x}_2 - \boldsymbol{x}_1|^3} \tag{3.9}$$

for the force $\boldsymbol{F}_{21}$ exerted on charge Q_2 at $\boldsymbol{x} = \boldsymbol{x}_2$ by Q_1 at $\boldsymbol{x}_1$, where K is a physical constant (in the S.I. system of units, in which force is measured in newtons, length is measured in metres and charge is measured in coulombs, $K = (4\pi\epsilon_0)^{-1}$ where ϵ_0 is the *permittivity of free space*). Notice that $\boldsymbol{F}_{12} = -\boldsymbol{F}_{21}$ and that Equation (3.9) has the same rule for decay with distance as the Newtonian law of gravitation, namely $|\boldsymbol{F}| \propto (\text{distance})^{-2}$. The sole difference is that the charges Q_1 and Q_2 may be either positive or negative. The force is *repulsive* if the charges have the same sign, but *attractive* if they have opposite signs. This is often summarized in the statement

Like charges repel, but *unlike charges attract.*

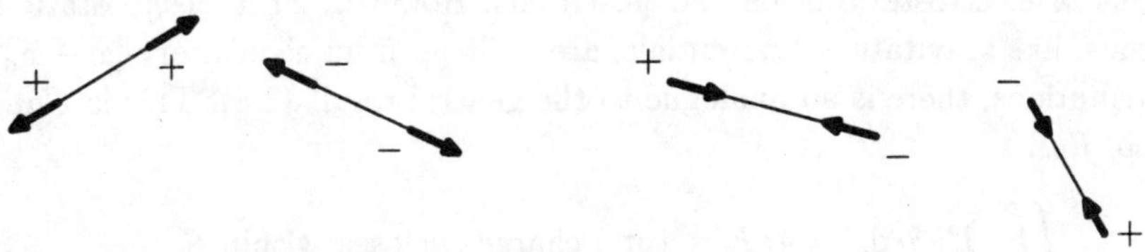

Figure 3.7 Repulsive forces between charges of like sign and forces of attraction between charges of unlike sign.

As with Newtonian gravitation, the force law (3.9) may be interpreted as stating that a point charge Q at location $\boldsymbol{x}_0$ creates an *electrostatic field*

$$\boldsymbol{E} = KQ\frac{\boldsymbol{x} - \boldsymbol{x}_0}{|\boldsymbol{x} - \boldsymbol{x}_0|^3} \tag{3.10}$$

which is directed radially away from $\boldsymbol{x}_0$ if Q is positive (and towards $\boldsymbol{x}_0$ if Q is negative). Moreover, that field may be written as $\boldsymbol{E} = -\boldsymbol{\nabla}\phi$, in terms of the *electrostatic potential* $\phi = KQ|\boldsymbol{x} - \boldsymbol{x}_0|^{-1}$.
Definition: For *any* configuration of charges, static or moving, the *electric field* $\boldsymbol{E}$ is such that a *small* (test) charge q at $\boldsymbol{x}$ experiences a force $q\boldsymbol{E}$.

Thus, a set of point charges q_n static at $\boldsymbol{x} = \boldsymbol{x}_n$ $(n = 1, \ldots, N)$ produces electrostatic field

$$\boldsymbol{E} = K\sum_{n=1}^{N} q_n \frac{\boldsymbol{x} - \boldsymbol{x}_n}{|\boldsymbol{x} - \boldsymbol{x}_n|^3} \tag{3.11}$$

which is readily derivable from the *electrostatic potential*

$$\phi = K\sum_{n=1}^{N} \frac{q_n}{|\boldsymbol{x} - \boldsymbol{x}_n|}, \tag{3.12}$$

as $\boldsymbol{E} = -\boldsymbol{\nabla}\phi$. Note the appearance of the $-$ sign in the law $\boldsymbol{E} = -\boldsymbol{\nabla}\phi$. This ensures that force on a charge of magnitude $+1$ at $\boldsymbol{x}$ is $+\boldsymbol{E}$ and that the physical significance of $\phi(\boldsymbol{x})$ is that it equals the work required to bring that *unit charge* 'from infinity' to $\boldsymbol{x}$:

$$\int_{\infty}^{\boldsymbol{x}} -\boldsymbol{E}(\bar{\boldsymbol{x}}) \cdot \mathrm{d}\bar{\boldsymbol{x}} = \int_{\infty}^{\boldsymbol{x}} \boldsymbol{\nabla}\phi(\bar{\boldsymbol{x}}) \cdot \mathrm{d}\bar{\boldsymbol{x}} = \phi(\boldsymbol{x}) \,.$$

3.4.1 Gauss's Law of Flux

The summation formulae (3.11) and (3.12) are inconvenient for most calculations of electrostatic fields and potentials. However, since electrostatic potentials, like gravitational potentials, are built up from elementary $|\boldsymbol{x} - \boldsymbol{x}_n|^{-1}$ contributions, there is an analogue to the general result (3.8). This is *Gauss's law of flux*:

$$\iint_S \boldsymbol{E} \cdot \boldsymbol{n}\, \mathrm{d}S = 4\pi K \times \text{ total charge enclosed within } S \,.$$

In S.I. units, so that $(4\pi K)^{-1}$ equals the physical constant ϵ_0, Gauss's law becomes

$$\iint_S \epsilon_0 \boldsymbol{E} \cdot \boldsymbol{n}\, \mathrm{d}S = \text{total charge within } S \,. \tag{3.13}$$

As for gravitational problems, the calculation of electrostatic fields due to *any* spherically symmetric distribution of charge is then easily performed. Examples in which the charge is distributed with a *charge density* $\rho(\boldsymbol{x})$ follow.

3.4.2 Charge-free Regions

In gases, liquids and metallic conductors, electrons may move very readily. As a *static* field becomes established, the (positive and negative) charge redistributes itself so that in many regions the *charge density* is zero. Such a region is said to be *charge-free.* It is a consequence of Gauss's law that, in any charge-free region, ϕ satisfies Laplace's equation $\nabla^2\phi = 0$.
[Actually, within a metallic conductor, electrostatic charge resides only in (very) thin layers at the material surface. Moreover, the field $\boldsymbol{E}$ is zero elsewhere within the conducting material.]

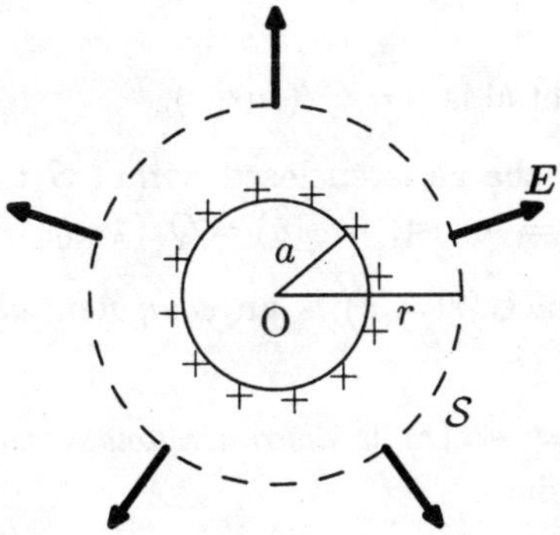

Figure 3.8 A layer of positive charge over the surface $r = a$ of a conductor, as in Example 3.8. The total flux of $\boldsymbol{E}$ is calculated through a sphere $\mathcal{S}$ of large radius.

Example 3.8

Find the electrostatic field outside a spherical metallic conductor of radius a, carrying total charge Q, distributed evenly over the surface $r = a$.

By symmetry, the potential has the form $\phi = \phi(r)$, where $r^2 = x^2+y^2+z^2$. For the choice of *test surface* $\mathcal{S}$: $r =$ const., the unit normal is $\boldsymbol{n} = \boldsymbol{x}/r$ while $\boldsymbol{\nabla}\phi = \phi'(r)\boldsymbol{x}/r$, so that the (outward) radial component of the electric field $\boldsymbol{E}$ is

$$E \equiv -\boldsymbol{\nabla}\phi \cdot \boldsymbol{n} = -\boldsymbol{n} \cdot \boldsymbol{\nabla}\phi = -\phi'(r) .$$

Applying Gauss's law (3.13) then gives

$$\begin{aligned} Q &= \iint_{\mathcal{S}} \epsilon_0 \boldsymbol{E} \cdot \boldsymbol{n}\, \mathrm{d}S = -\epsilon_0 \iint_{\mathcal{S}} \phi'(r)\, \mathrm{d}S \\ &= -\epsilon_0 \phi'(r) \times \text{Area of}\, \mathcal{S} = -4\pi r^2 \epsilon_0 \phi'(r) . \end{aligned}$$

Hence, the field strength at radius r is

$$E = -\phi'(r) = \frac{Q}{4\pi\epsilon_0 r^2}$$

so that the (radially symmetric) field $\boldsymbol{E}$ is

$$\boldsymbol{E} = \frac{Q}{4\pi\epsilon_0} \frac{\boldsymbol{x}}{r^3} . \tag{3.14}$$

Notice that this formula is just like that for a charge of magnitude Q, concentrated at O (a *point charge*), even though the charge is distributed as a thin layer, idealized mathematically as a sheet having *surface charge density* of magnitude $\sigma \equiv Q/(4\pi a^2)$ at radius $r = a$ (i.e. a layer of zero thickness).

Observe that

i) For $r > a$, the potential is $\phi = Q/(4\pi\epsilon_0 r)$.

ii) In $r < a$, since no charge is enclosed within $\mathcal{S}$ then $\phi'(r) = -E = 0$. Hence, for $r < a$, $\phi =$ const. $= \phi(a) = Q/(4\pi\epsilon_0 a)$.

iii) The conductor surface ($r = a$) is an *equipotential surface*, i.e. a surface $\phi =$ constant.

iv) The radial field $E = -\phi'(r)$ is *discontinuous* at the surface $r = a$ where the charge resides, since
On the exterior (i.e. at $r = a + 0$) , $E = Q/(4\pi\epsilon_0 a^2)$,
On the interior (i.e. at $r = a - 0$) , $E = 0$.
Taking the difference gives

$$E_+ - E_- \equiv E(a+0) - E(a-0) = Q/(4\pi\epsilon_0 a^2) = (\epsilon_0)^{-1}\sigma \,.$$

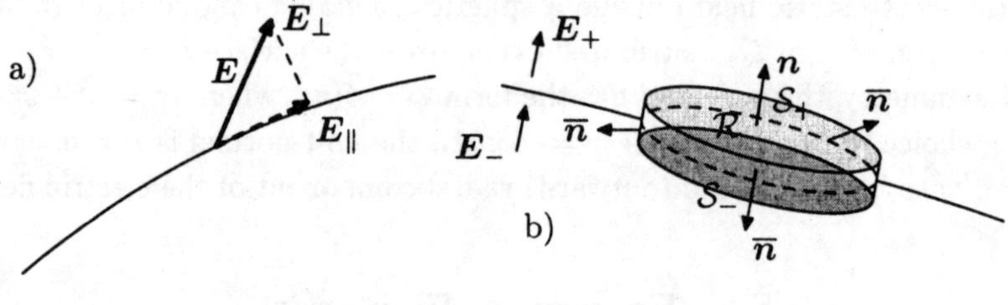

Figure 3.9 a) The field $\boldsymbol{E}$ resolved into components $\boldsymbol{E}_\perp$ and $\boldsymbol{E}_\parallel$, normal and tangential to a conducting surface. b) Illustration of the surface of a coin-shaped region $\mathcal{R}$ used as test surface in Gauss's law.

3.4.3 Surface Charge Density

Properties iii) and iv) are examples of *general* properties of conductors in electrostatics:

- At a conducting surface the field $\boldsymbol{E}$ is perpendicular to the conductor.
- If the surface bears a sheet of charge having *surface charge density* $\sigma = \sigma(\boldsymbol{x})$, the normal component of $\epsilon_0 \boldsymbol{E}$ has a jump equal to σ.

Otherwise, if the tangential component $\boldsymbol{E}_\parallel$ of the field $\boldsymbol{E}$ were nonzero, this component would cause charge to move along the conducting surface until a new equilibrium was established. [An equivalent, and convenient, statement is that the conducting surface is an *equipotential surface* (ϕ = constant).]

The discontinuity in $\epsilon_0 \boldsymbol{E}$ is analysed by applying Gauss's law (3.13) to a coin-shaped region $\mathcal{R}$ bounded by two flat surfaces $\mathcal{S}_\pm$ parallel to and adjacent to an element ΔS of the conducting surface $\mathcal{S}$ and a rim which is everywhere orthogonal to $\mathcal{S}$. Let $\boldsymbol{n}$ be the unit normal to $\mathcal{S}$, pointing from the negative side of $\mathcal{S}$ to the positive side and denote the outward unit normal to $\mathcal{R}$ by $\bar{\boldsymbol{n}}$. Then, on the rim $\bar{\boldsymbol{n}} \cdot \boldsymbol{n} = 0$ everywhere, so that $\boldsymbol{E} \cdot \boldsymbol{n} = 0$ also. The rim thus gives no contribution to the outward flux of $\epsilon_0 \boldsymbol{E}$ on the left-hand side of (3.13). On $\mathcal{S}_+$, the outward normal is $\bar{\boldsymbol{n}} = \boldsymbol{n}$, while on $\mathcal{S}_-$ the outward normal is $\bar{\boldsymbol{n}} = -\boldsymbol{n}$. Then, since the region $\mathcal{R}$ contains charge $\sigma \Delta S$, Gauss's law (3.13) gives

$$\sigma \, \Delta S = \iint_{\mathcal{S}_+ + \mathcal{S}_-} \epsilon_0 \boldsymbol{E} \cdot \bar{\boldsymbol{n}} \, \mathrm{d}S \approx \epsilon_0 (\boldsymbol{E}_+ \cdot \boldsymbol{n} - \boldsymbol{E}_- \cdot \boldsymbol{n}) \, \Delta S \, .$$

Taking the limit $\Delta S \to 0$ gives at each point of $\mathcal{S}$ the result

$$\epsilon_0 (\boldsymbol{E}_+ - \boldsymbol{E}_-) \cdot \boldsymbol{n} = \sigma \, , \tag{3.15}$$

where $\boldsymbol{E}_+$ and $\boldsymbol{E}_-$ denote the limiting values of $\boldsymbol{E}$ as $\mathcal{S}$ is approached from the positive and negative sides.

Equivalently, the conditions at the conducting surface may be simply expressed as

$$\boldsymbol{n} \cdot [\boldsymbol{\nabla} \phi]_-^+ = \left[\frac{\partial \phi}{\partial n} \right]_-^+ = -\sigma / \epsilon_0 \quad \text{and} \quad \phi_+ = \phi_- \, , \tag{3.16}$$

i.e. ϕ is continuous and the jump in $\partial \phi / \partial n$ is proportional to the surface charge density (here $\partial / \partial n \equiv \boldsymbol{n} \cdot \boldsymbol{\nabla}$ is a shorthand notation for the *normal derivative* at a point of $\mathcal{S}$).

Example 3.9 (A cylindrical capacitor)

An electric capacitor is a device for storing charge, typically by transferring charge from one conductor to another, so creating a *potential difference.* Suppose that two coaxial, metallic cylinders of length L have radii a and b. When 'end effects' are neglected, the electrostatic field may be treated as two-dimensional with potential $\phi = \phi(r)$ for $r = (x^2 + y^2)^{1/2}$. If the inner cylinder carries total charge Q, show that $\phi = -Q/(2\pi\epsilon_0 L) \ln r$ + constant in $a \le r \le b$. Determine the *capacitance* $C = Q/V$ of the device, where $V = \phi(a) - \phi(b)$. Find also the electric field within $a < r < b$.

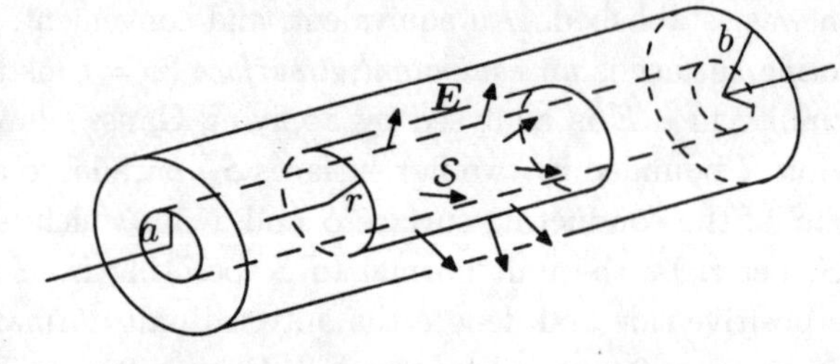

Figure 3.10 Radial field $\boldsymbol{E}$ within a cylindrical capacitor and a convenient test cylinder S of radius r, with $a < r < b$.

Charge resides solely at radii $r = a$ and $r = b$. Unit length of the inner cylinder carries charge Q/L. Then, by Gauss's law, the outward flux of $\epsilon_0 \boldsymbol{E}$ from a cylinder of length r is

$$-\epsilon_0 2\pi r \frac{\partial \phi}{\partial r} = \iint_S \epsilon_0 \boldsymbol{E} \cdot \boldsymbol{n} \, \mathrm{d}S = Q/L \qquad \text{for } a < r < b .$$

Then, integrating the ODE $\phi'(r) = -Q/(2\pi\epsilon_0 L) r^{-1}$, gives the result $\phi = -Q/(2\pi\epsilon_0 L) \ln r +$ constant.
The potential difference between the conductors is $V = \phi(a) - \phi(b) = Q/(2\pi\epsilon_0 L) \ln(b/a)$, so that the capacitance is $C = 2\pi\epsilon_0 L[\ln(b/a)]^{-1}$.
In $a < r < b$, the field is $\boldsymbol{E} = -\boldsymbol{\nabla}\phi = Q/(2\pi\epsilon_0 L)(x\boldsymbol{i} + y\boldsymbol{j})r^{-2}$ i.e. a field radially away from the axis, with strength $\propto r^{-1}$.

EXERCISES

3.5. For a spherically symmetric function $\phi = \phi(r)$, where $r^2 = x^2 + y^2 + z^2$, calculate $\partial\phi/\partial x$, $\partial\phi/\partial y$, $\partial\phi/\partial z$ and $\partial^2\phi/\partial x^2$. Hence show that

$$\nabla^2 \phi \equiv \frac{\partial^2 \phi}{\partial x^2} + \frac{\partial^2 \phi}{\partial y^2} + \frac{\partial^2 \phi}{\partial z^2} = \frac{\mathrm{d}^2 \phi}{\mathrm{d}r^2} + \frac{2}{r}\frac{\mathrm{d}\phi}{\mathrm{d}r} .$$

3.6. If the potential $\phi(r)$ of Exercise 3.5 is the gravitational potential due to a spherically symmetric density distribution $\rho(r)$, show that the flux law (3.8) implies that

$$4\pi r^2 G\rho(r) = \frac{\mathrm{d}}{\mathrm{d}r}\left(r^2 \frac{\mathrm{d}\phi}{\mathrm{d}r}\right) .$$

Confirm that such a spherically symmetric potential satisfies $\nabla^2\phi = 4\pi G\rho(r)$.

3.7. For a spherically symmetric star of outer radius R and density $\rho = \rho(r)$ in $0 \le r < R$, express the mass $m(r)$ within the sphere of radius r centred at O in terms of an integral. Find the strength of the gravitational field at radius r if the density is assumed to be $\rho = \rho_0 R^{-3}\{R^3 - 2Rr^2 + r^3\}$. Confirm that the total mass is $M = 0.4\pi\rho_0 R^3$ and that, at radius R, the gravitational field strength is MG/R^2.

3.8. For each of the following potentials ϕ, determine the electric field $\boldsymbol{E} = -\boldsymbol{\nabla}\phi$ and confirm that $\nabla^2\phi = 0$:
(a) $\phi = x^2+4xy+2y^2-3z^2$, (b) $\phi = \{(x-a)^2+y^2+z^2\}^{-1/2}$,
(c) $\phi = y/r^3$, where $r^2 = x^2 + y^2 + z^2$.

3.9. In the region $r_1 < r < r_2$ between a spherical conductor of radius r_1 carrying evenly distibuted charge Q and a concentric spherical conductor of radius r_2 carrying total charge $-Q$, show that the potential $\phi = \phi(r)$ must have the form $\phi = Ar^{-1} + B$. Determine the field $\boldsymbol{E} = -\boldsymbol{\nabla}\phi$, the potential difference $V \equiv \phi(r_1) - \phi(r_2)$ and the *capacitance* $C \equiv Q/V$.

3.10. Confirm that the function

$$\phi = A\left(\frac{1}{|\boldsymbol{x} - a\boldsymbol{i}|} - \frac{1}{|\boldsymbol{x} + a\boldsymbol{i}|}\right)$$

satisfies Laplace's equation everywhere except at $\boldsymbol{x} = \pm a\boldsymbol{i}$. If ϕ is an electrostatic potential, show that the equipotential $\phi = 0$ is the infinite plane $x = 0$ and determine the flux $\iint_S \boldsymbol{\nabla}\phi \cdot \boldsymbol{n}\,\mathrm{d}S$ through any closed surface $\mathcal{S}$ which encloses the point $\boldsymbol{x} = a\boldsymbol{i}$ and lies in the region $x \ge 0$. Choose the value A so that ϕ is the potential due to a point charge q located at $\boldsymbol{x} = a\boldsymbol{i}$ adjacent to a conducting surface at $x = 0$.

In this physical problem there is no electric field in $x < 0$, so that the potential equals zero in $x < 0$. By considering the jump in $\partial\phi/\partial x$ at $x = 0$, determine the surface charge density σ induced over $x = 0$.

4

Laplace's Equation and Poisson's Equation

4.1 The Ubiquitous Laplacian

In Chapter 3, it was noted that both gravitational and electrostatic potentials satisfy Laplace's equation $\nabla^2\phi = 0$ in any region where the mass and charge densities vanish. In fact, the Laplacian operator ∇^2 occurs naturally in many applications of mathematics, since it is intimately related to the total outflow rate from a region in which a flux vector is proportional to the gradient of a scalar. Derivation of the important result

$$\iint_{\partial\mathcal{R}} \boldsymbol{\nabla}\phi \cdot \boldsymbol{n}\,\mathrm{d}S = \iiint_{\mathcal{R}} \nabla^2\phi\,\mathrm{d}V \tag{4.1}$$

equating the total flux of the field $\boldsymbol{\nabla}\phi$ through the bounding surface $\partial\mathcal{R}$ of a general three-dimensional region $\mathcal{R}$ to the volume integral of $\nabla^2\phi$ over $\mathcal{R}$ is a central result in *vector calculus* (see e.g. Matthews (1998)). Its truth is readily demonstrated, as indicated below, by first considering a general rectangular block, then any combination of adjoining blocks and, finally, by combining this with a result for an infinitesimal tetrahedron.

Let $\mathcal{R}$ be the rectangular box occupying $x_1 \le x \le x_2, y_1 \le y \le y_2, z_1 \le z \le z_2$. Suppose that a vector field $\boldsymbol{F}$ has the component form $\boldsymbol{F}(x, y, z) = F_1\boldsymbol{i} + F_2\boldsymbol{j} + F_3\boldsymbol{k}$, so that its total *outward* flux through the plane face at $x = x_2$

(where $\boldsymbol{i}$ is the outward unit normal) is

$$\iint_{y_1<y<y_2,\, z_1<z<z_2} F_1(x_2, y, z)\, \mathrm{d}y\mathrm{d}z\,.$$

Over the face at $x = x_1$, the outward normal is $\boldsymbol{n} = -\boldsymbol{i}$, so that the integrated outward flux is the negative of a similar integral evaluated at $x = x_1$. Adding together these two contributions to the flux out of $\mathcal{R}$ and using the fundamental theorem of calculus gives

$$\begin{aligned}\iint_{y_1<y<y_2,\, z_1<z<z_2} & F_1(x_2, y, z)\, \mathrm{d}y\mathrm{d}z + \iint_{y_1<y<y_2,\, z_1<z<z_2} -F_1(x_1, y, z)\, \mathrm{d}y\mathrm{d}z \\ &= \iint_{y_1<y<y_2,\, z_1<z<z_2} \{F_1(x_2, y, z) - F_1(x_1, y, z)\}\, \mathrm{d}y\mathrm{d}z \\ &= \iiint_{\mathcal{R}} \frac{\partial F_1}{\partial x}\, \mathrm{d}x\mathrm{d}y\mathrm{d}z = \iiint_{\mathcal{R}} \frac{\partial F_1}{\partial x}\, \mathrm{d}V\,.\end{aligned}$$

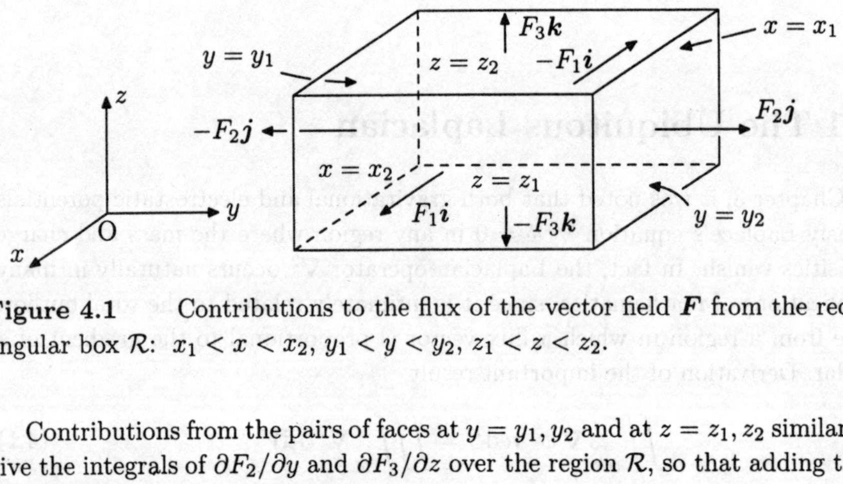

Figure 4.1 Contributions to the flux of the vector field $\boldsymbol{F}$ from the rectangular box $\mathcal{R}$: $x_1 < x < x_2$, $y_1 < y < y_2$, $z_1 < z < z_2$.

Contributions from the pairs of faces at $y = y_1, y_2$ and at $z = z_1, z_2$ similarly give the integrals of $\partial F_2/\partial y$ and $\partial F_3/\partial z$ over the region $\mathcal{R}$, so that adding the contributions from *all six* faces gives

$$\iint_{\partial\mathcal{R}} \boldsymbol{F}\cdot\boldsymbol{n}\mathrm{d}S = \iiint_{\mathcal{R}} \left\{\frac{\partial F_1}{\partial x} + \frac{\partial F_2}{\partial y} + \frac{\partial F_3}{\partial z}\right\} \mathrm{d}V\,. \tag{4.2}$$

[N.B. For *any* vector field $\boldsymbol{F}$, the integrand on the right-hand side of (4.2) is known as the *divergence* of $\boldsymbol{F}$; it may be calculated as

$$\begin{aligned}\operatorname{div}\boldsymbol{F} &= \frac{\partial F_1}{\partial x} + \frac{\partial F_2}{\partial y} + \frac{\partial F_3}{\partial z} = \left(\boldsymbol{i}\frac{\partial}{\partial x} + \boldsymbol{j}\frac{\partial}{\partial y} + \boldsymbol{k}\frac{\partial}{\partial z}\right)\cdot(\boldsymbol{i}F_1 + \boldsymbol{j}F_2 + \boldsymbol{k}F_3) \\ &= \boldsymbol{\nabla}\cdot\boldsymbol{F} \end{aligned} \tag{4.3}$$

involving the same vector operator $\boldsymbol{\nabla}$ as appears in the gradient of a scalar.]

When the result (4.2) is applied to two adjacent blocks which share part of one face, the outflow from one block is an inflow to its neighbour. The integral of $\operatorname{div} \boldsymbol{F}$ over the combined region again equals the integral of $\boldsymbol{F} \cdot \boldsymbol{n}$ over the outer boundary (the contributions over the common portion of the interface cancelling). By combining the result for many blocks assembled as in a masonry wall, the result may be shown to apply to regions which are very close in shape to an arbitrary region $\mathcal{R}$, but with surface composed of (arbitrarily small) rectangular faces parallel to one of the coordinate planes.

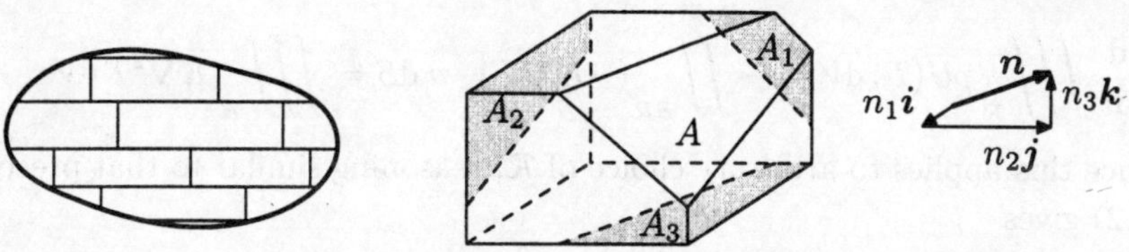

Figure 4.2 Division of a region $\mathcal{R}$ into an assemblage of blocks and truncated blocks, showing sloping portions of the surface of $\mathcal{R}$ and opposite portions of the plane faces.

The true region $\mathcal{R}$ differs from the assemblage of blocks by a large number of truncated blocks each having a sloping, but approximately plane, face $\boldsymbol{n}$, as shown in Figure 4.2. If this sloping face has area A, then its projections onto the three opposite faces with respective unit normals $-\boldsymbol{i}$, $-\boldsymbol{j}$ and $-\boldsymbol{k}$ are A_1, A_2 and A_3, where

$$A_1 = A\boldsymbol{n} \cdot \boldsymbol{i} \quad , \qquad A_2 = A\boldsymbol{n} \cdot \boldsymbol{j} \quad , \qquad A_3 = A\boldsymbol{n} \cdot \boldsymbol{k} \, .$$

Thus, since

$$\boldsymbol{F} \cdot \boldsymbol{n} \, A = F_1 A_1 + F_2 A_2 + F_3 A_3 \, ,$$

each contribution to the outward flux is the same as to flux through a portion of a plane parallel to one of the axes and located at a representative point of the sloping face. Hence, the total outflow through the collection of faces parallel to the coordinate planes becomes (in the limit of ever finer subdivision into rectangular blocks) equal to the outflow from the general region $\mathcal{R}$ (it can be shown that the integrals of $\boldsymbol{\nabla} \cdot \boldsymbol{F}$ over the union of all the truncated blocks may be made arbitrarily small, as may the errors due to the dependence of $\boldsymbol{F}$ upon position over elements of the bounding surface).

The result (together with appropriate conditions on the smoothness of the vector field $\boldsymbol{F}(\boldsymbol{x})$ and on the bounding surface $\partial\mathcal{R}$ to the three-dimensional region $\mathcal{R}$)

$$\iint_{\partial\mathcal{R}} \boldsymbol{F} \cdot \boldsymbol{n} \, \mathrm{d}S = \iiint_{\mathcal{R}} \boldsymbol{\nabla} \cdot \boldsymbol{F} \, \mathrm{d}V = \iiint_{\mathcal{R}} \operatorname{div} \boldsymbol{F} \, \mathrm{d}V \qquad (4.4)$$

is known as the *divergence theorem.* For the special choice $\boldsymbol{F} = \boldsymbol{\nabla}\phi$ it becomes the identity (4.1), since

$$\boldsymbol{\nabla}\cdot(\boldsymbol{\nabla}\phi) \equiv \left(\boldsymbol{i}\frac{\partial}{\partial x} + \boldsymbol{j}\frac{\partial}{\partial y} + \boldsymbol{k}\frac{\partial}{\partial z}\right)\cdot\left(\boldsymbol{i}\frac{\partial\phi}{\partial x} + \boldsymbol{j}\frac{\partial\phi}{\partial y} + \boldsymbol{k}\frac{\partial\phi}{\partial z}\right)$$
$$= \frac{\partial^2\phi}{\partial x^2} + \frac{\partial^2\phi}{\partial y^2} + \frac{\partial^2\phi}{\partial z^2} = \nabla^2\phi .$$

Applying the result (4.4) to the heat flux vector $\boldsymbol{q} = -K\boldsymbol{\nabla}T$ in a three-dimensional region $\mathcal{R}$ in which there are no heat sources, gives

$$\frac{\mathrm{d}}{\mathrm{d}t}\iiint_{\mathcal{R}} \rho U(T)\,\mathrm{d}V = -\iint_{\partial\mathcal{R}} (-K\boldsymbol{\nabla}T)\cdot\boldsymbol{n}\,\mathrm{d}S = \iiint_{\mathcal{R}} K\nabla^2 T\,\mathrm{d}V .$$

Since this applies to arbitrary choice of $\mathcal{R}$, reasoning similar to that preceding (2.2) gives

$$\rho c\frac{\partial T}{\partial t} = K\nabla^2 T = K\left(\frac{\partial^2 T}{\partial x^2} + \frac{\partial^2 T}{\partial y^2} + \frac{\partial^2 T}{\partial z^2}\right), \tag{4.5}$$

which generalizes the *heat equation* (2.3) to three dimensions. In the special case when the temperature is *steady,* it reduces to Laplace's equation for $T(x, y, z)$:

$$\nabla^2 T = 0 .$$

These examples are typical of the many in which the Laplacian ∇^2 and Laplace's equation arise. They show the importance of methods for constructing solutions in useful and representative regions.

4.2 Separable Solutions

Many solutions to Laplace's equation $\nabla^2\phi = 0$ may be built up by using special *separable solutions* of the form $\phi = X(x)Y(y)Z(z)$. This procedure is well suited to problems in which ϕ (or its normal derivative) is specified at each point of the surface of a rectangular region. Since solutions to Laplace's equation may be added together to create other solutions, a great variety of solutions may be constructed *by superposition.* For the function $\phi = X(x)Y(y)Z(z)$, the required derivatives are

$$\frac{\partial^2\phi}{\partial x^2} = X''(x)Y(y)Z(z)\,, \quad \frac{\partial^2\phi}{\partial y^2} = X(x)Y''(y)Z(z)\,, \quad \frac{\partial^2\phi}{\partial z^2} = X(x)Y(y)Z''(z)\,,$$

where primes denote differentiation. Substitution of these into $\nabla^2\phi = 0$ gives

$$X''(x)Y(y)Z(z) + X(x)Y''(y)Z(z) + X(x)Y(y)Z''(z) = 0 .$$

The next step is to obtain an equation in which the terms depending on x, y and z appear separately. This is achieved by dividing through by the product XYZ to give

$$\frac{X''(x)}{X(x)} + \frac{Y''(y)}{Y(y)} + \frac{Z''(z)}{Z(z)} = 0\,. \tag{4.6}$$

As x varies, the last two terms remain constant. Hence, the ratio $X''(x)/X(x)$ must equal a constant. Similarly, the ratios $Y''(y)/Y(y)$ and $Z''(z)/Z(z)$ *each* are constants. For each of $X(x), Y(y)$ and $Z(z)$, three types of behaviour are possible, as is illustrated by first considering $X(x)$.

The condition $X''(x)/X(x) = \alpha$ (where α is a *constant of separation*) is equivalent to the ODE

$$\frac{\mathrm{d}^2X}{\mathrm{d}x^2} = \alpha X\,,$$

for which the cases $\alpha > 0$, $\alpha = 0$ and $\alpha < 0$ give three distinct possibilities:

(i) For $\alpha > 0$, the function $X(x)$ behaves exponentially, with

$$X = A\mathrm{e}^{\sqrt{\alpha}x} + B\mathrm{e}^{-\sqrt{\alpha}x}\,.$$

(ii) For $\alpha = 0$, X is linear in x, of the form $X = A + Bx$.

(iii) For $\alpha < 0$, the behaviour of $X(x)$ is sinusoidal. Writing $\alpha = -p^2$ leads to a general solution of the form $X = A\cos px + B\sin px$.

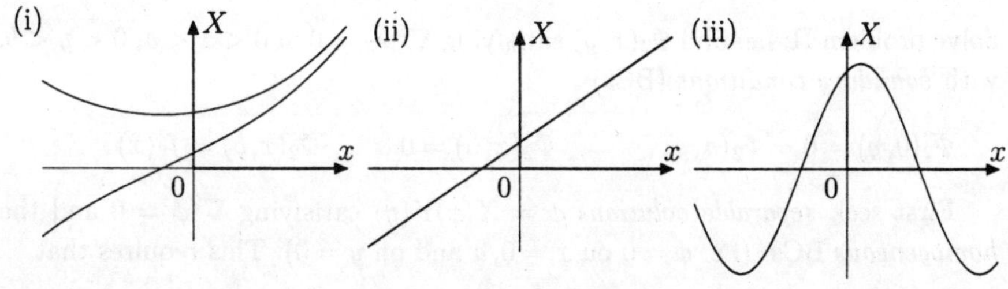

Figure 4.3 Typical behaviours for $X(x)$ with (i) $\alpha > 0$, (ii) $\alpha = 0$ and (iii) $\alpha < 0$.

In each of (i), (ii) and (iii), A and B are constants. Analogous possibilities arise for $Y(y)$ and $Z(z)$. Relevant choices are dictated by the boundary conditions on the faces of the box, as is illustrated for a typical *boundary value problem* (BVP) in two dimensions, namely:
Solve $\nabla^2\phi = 0$ within a rectangle (e.g. $0 < x < a\,,\;\; 0 < y < b$), with the

potential ϕ specified at all points of the boundary.
This may be split into four simpler problems, in each of which ϕ has non-zero values on only one side of the rectangle, as shown diagramatically in Figure 4.4. The sum $\phi(x,y) = \Phi_1 + \Phi_2 + \Phi_3 + \Phi_4$ of the solutions to the problems I–IV is the required solution to the original problem. The procedure is best illustrated by examples.

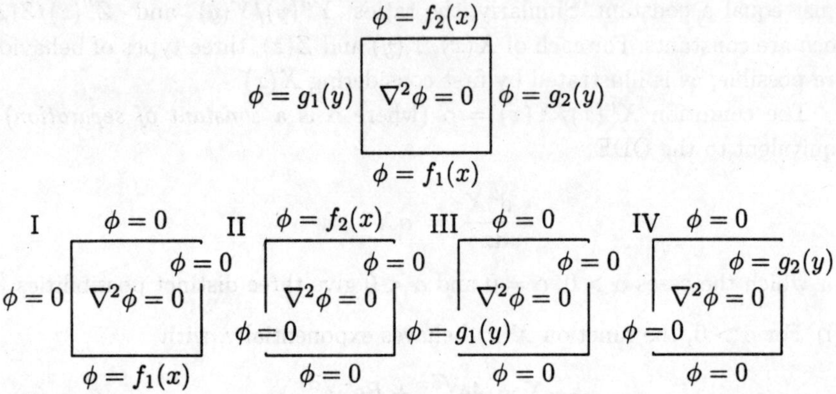

Figure 4.4 Diagram illustrating the splitting of a BVP for Laplace's equation within a rectangle into four simpler problems for $\Phi_i(x,y)$, $i = 1,\ldots,4$.

Example 4.1

Solve problem II; i.e. find $\Phi_2(x,y)$ satisfying $\nabla^2\Phi_2 = 0$ in $0 < x < a$, $0 < y < b$, with *boundary conditions* (BCs)

$$\Phi_2(0,y) = 0 = \Phi_2(a,y)\,, \qquad \Phi_2(x,0) = 0\,, \qquad \Phi_2(x,b) = f_2(x)\,.$$

First seek *separable solutions* $\phi = X(x)Y(y)$ satisfying $\nabla^2\phi = 0$ and the *homogeneous* BCs (i.e. $\phi = 0$ on $x = 0, a$ and on $y = 0$). This requires that

$$X(0) = 0 = X(a) \quad , \qquad Y(0) = 0\,.$$

Dividing the equation $\nabla^2\phi = X''(x)Y(y) + X(x)Y''(y) = 0$ by XY leads to

$$\frac{X''(x)}{X(x)} + \frac{Y''(y)}{Y(y)} = 0\,,$$

so that both X''/X and Y''/Y must be constants. Focus first on $X(x)$, since it must satisfy *two* boundary conditions. Only case (iii) of $X''(x) = \alpha X(x)$ allows $X(x)$ to take the value 0 at two distinct values of x (for (ii), $X \equiv 0$

is the only linear function taking the value 0 at two distinct points, while for (i), setting $X(0) = 0$ gives $B = -A$ so that $0 = X(a) = 2A \sinh \sqrt{\alpha} a$ with either possibility $\alpha = 0$ or $A = 0$ giving only the *trivial solution* $X \equiv 0$). In case (iii), the solution $X = A \cos px + B \sin px$ satisfies $X(0) = 0$ only if $A = 0$; it simultaneously satisfies $X(a) = 0$ only if $pa = n\pi$, for some integer n (the *trivial case* $B = 0$ is omitted since it gives $\phi \equiv 0$). Hence, $X(x)$ must be some constant multiple of $\sin(n\pi x/a)$, for $n = \pm 1, \pm 2, \ldots$ The values $\alpha = -(n\pi/a)^2 \equiv \alpha_n$ are known as the *eigenvalues* for solution of the *two-point boundary value problem* $X''(x) = \alpha X(x)$, $X(0) = 0 = X(a)$; the solutions $X = X_n(x) \equiv \sin(n\pi x/a)$ are the corresponding *eigenfunctions.*

To determine related functions $Y(y)$, observe that

$$\frac{Y''(y)}{Y(y)} = -\frac{X''(x)}{X(x)} = -\alpha_n = \left(\frac{n\pi}{a}\right)^2 > 0 .$$

The corresponding solutions $Y(y)$ must have the form of case (i), namely

$$Y = A\mathrm{e}^{n\pi y/a} + B\mathrm{e}^{-n\pi y/a} .$$

The condition $Y(0) = 0$ gives $A + B = 0$ (or $B = -A$), so that

$$Y(y) = A\left\{\mathrm{e}^{n\pi y/a} - \mathrm{e}^{-n\pi y/a}\right\} = 2A \sinh\left(n\pi y/a\right) .$$

Multiplying $X(x)$ by $Y(y)$ gives the result that *any* constant multiple of *each* function

$$\phi_n(x, y) \equiv \sin\frac{n\pi x}{a} \sinh\frac{n\pi y}{a}, \qquad n = 1, 2, \ldots \tag{4.7}$$

satisfies Laplace's equation $\nabla^2 \phi = 0$ and the homogeneous BCs on $x = 0, a$

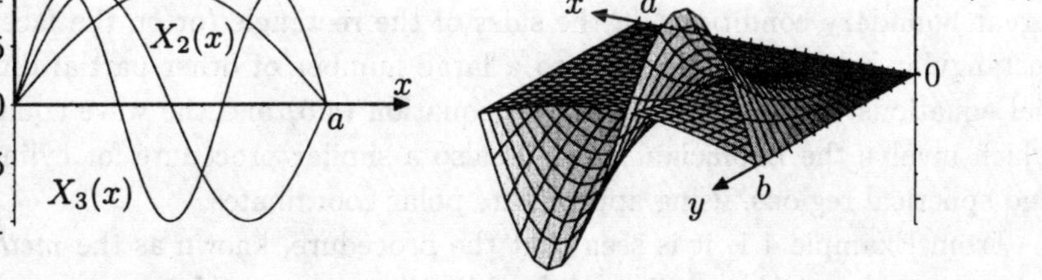

Figure 4.5 The functions $X_n(x) = \sin(n\pi x/a)$ for $n = 1, 2, 3$ and associated eigenfunctions $\phi_n(x, y)$ for $n = 1, 2$ with $b = 0.5a$, for Example 4.1.

and on $y = 0$:

$$\nabla^2 \phi_n = 0 \quad , \qquad \phi_n(0, y) = 0 = \phi_n(a, y), \qquad \phi_n(x, 0) = 0 .$$

Clearly, any *linear combination* $\sum_n A_n \phi_n(x,y)$ also satisfies Laplace's equation and the homogeneous BCs, i.e. by *superposing* the separable solutions as

$$\phi = \sum_{n=1}^{\infty} A_n \, \phi_n(x,y)\,, \tag{4.8}$$

a very broad class of solutions satisfying every requirement other than the BC $\Phi_2(x,b) = f_2(x)$ is constructed. (N.B. There is no need to include ϕ_n for negative n, since $\phi_{-n}(x,y) = \phi_n(x,y)$. Also, the term with $n = 0$ is omitted, since $\phi_0(x,y) \equiv 0$.)

The final requirement (the *inhomogeneous* BC on $y = b$ namely $\phi(x,b) = f_2(x)$) is satisfied by setting $y = b$ in (4.8), to give

$$\sum_{n=1}^{\infty} A_n \sinh\frac{n\pi b}{a} \, \sin\frac{n\pi x}{a} = \Phi_2(x,b) = \phi(x,b) = f_2(x)\,. \tag{4.9}$$

In Equation (4.9), the function $f_2(x)$ should be regarded as given. The task is to identify the coefficients A_n. However, Equation (4.9) may be regarded as a *Fourier sine series* for $f_2(x)$, with *coefficients* $A_n \sinh(n\pi b/a)$. These coefficients are then found from the standard *half-range formulae* (see e.g. Brown and Churchill (1996), Evans, Blackledge & Yardley (1999)) as

$$A_n \sinh\frac{n\pi b}{a} = \frac{2}{a}\int_0^a f_2(x)\, \sin\frac{n\pi x}{a}\, dx\,.$$

Each coefficient A_n is then known (in terms of an integral over the whole extent $0 < x < a$ of the side $y = b$) and may be inserted into the infinite sum (4.8).

The method illustrated in Example 4.1 is versatile, if tedious. It may be extended to three dimensions and to unsteady problems, for a variety of different boundary conditions on the sides of the rectangle (or on the faces of a rectangular box). It applies also to a large number of other partial differential equations (in particular, the heat equation (4.5) and the wave equation), which involve the Laplacian. There is also a similar procedure for cylindrical and spherical regions, using appropriate polar coordinates.

From Example 4.1, it is seen that the procedure, known as the *method of separation of variables* consists of the following sequence of steps:

- Break the problem into sub-problems each having an *inhomogeneous* BC on only one face of the region of solution. For each sub-problem, undertake the following steps:
- Propose *separable solutions*, i.e. products of functions of the individual coordinates (and time).

- Substitute the separable solution into the PDE and rearrange so as to generate ODEs for functions of each individual coordinate.
- For each ODE with associated homogeneous BCs at two points, construct the solutions. This process typically selects certain discrete values (*eigenvalues*) for the *constants of separation.*
- Now solve the ODE which has only one associated homogeneous BC.
- *Superpose* the resulting separable solutions as an infinite sum (summing over the set of eigenvalues and eigenfunctions associated with *each* coordinate direction in which two homogeneous BCs are imposed).
- Impose the single *inhomogeneous* BC upon the solution expressed as the infinite sum. Identify the coefficients in the sum (typically as (generalized) Fourier coefficients). Insert these computed coefficients into the sum representing the solution.
- Finally, if necessary, *superpose* solutions to the various sub-problems to yield the solution to the original problem.

Example 4.2

Construct the solution to $\nabla^2\phi = 0$ in $0 < x < a$, $0 < y < b$ satisfying $\phi(x,b) = \sin(\pi x/2a) \equiv f_2(x)$, $\phi(a,y) = \sin(\pi y/2b) \equiv g_2(y)$, and $\phi = 0$ on the portions $x = 0$ and $y = 0$ of the boundary (i.e. $f_1(x) \equiv 0$, $g_1(y) \equiv 0$).
Write $\phi(x,y) = \Phi_2(x,y) + \Phi_4(x,y)$, where $\Phi_2 = f_2(x)$ on $y = b$, while $\Phi_4 = g_2(y)$ on $x = a$, with $\Phi_2 = 0$ and $\Phi_4 = 0$ on all other edges of the rectangle (i.e. $\Phi_2(x,y)$ is the solution to sub-problem II and $\Phi_4(x,y)$ is the solution to sub-problem IV).

Thus, Φ_2 is found as in Example 4.1. Then, since

$$\begin{aligned}\int_0^a f_2(x)\sin\frac{n\pi x}{a}\,\mathrm{d}x &= \int_0^a \sin\frac{\pi x}{2a}\sin\frac{n\pi x}{a}\,\mathrm{d}x \\ &= \frac{1}{2}\int_0^a \left\{\cos\frac{(2n-1)\pi x}{2a} - \cos\frac{(2n+1)\pi x}{2a}\right\}\mathrm{d}x \\ &= \frac{a}{(2n-1)\pi}\sin(n-\tfrac{1}{2})\pi - \frac{a}{(2n+1)\pi}\sin(n+\tfrac{1}{2})\pi \\ &= \frac{a}{\pi}\left[\frac{-(-1)^n}{2n-1} - \frac{(-1)^n}{2n+1}\right] = \frac{4an(-1)^{n-1}}{\pi(4n^2-1)},\end{aligned}$$

the coefficients A_n in the representation of the function Φ_2 are

$$A_n = \frac{2}{a\sinh n\pi b/a}\cdot\frac{4an(-1)^{n-1}}{\pi(4n^2-1)} = \frac{8n(-1)^{n-1}}{\pi(4n^2-1)\sinh n\pi b/a}.$$

This gives

$$\Phi_2(x,y) = \sum_{n=1}^{\infty} \frac{8n(-1)^{n-1}}{\pi(4n^2-1)} \cdot \frac{\sinh n\pi y/a}{\sinh n\pi b/a} \sin\frac{n\pi x}{a}. \tag{4.10}$$

Construction of the function $\Phi_4(x,y)$ solving problem IV is very closely analogous, with

$$\Phi_4(x,y) = \sum_{n=1}^{\infty} B_n \sinh\frac{n\pi x}{b}\sin\frac{n\pi y}{b}$$

and with

$$\frac{b}{2}\sinh\frac{n\pi a}{b}B_n = \int_0^b g_2(y)\sin\frac{n\pi y}{b}\,\mathrm{d}y = \frac{4bn(-1)^{n-1}}{\pi(4n^2-1)}.$$

Adding the resulting expression for $\Phi_4(x,y)$ to that for $\Phi_2(x,y)$ in (4.10) gives

$$\phi(x,y) = \sum_{n=1}^{\infty} \frac{8n(-1)^{n-1}}{\pi(4n^2-1)} \cdot \frac{\sinh n\pi y/a}{\sinh n\pi b/a} \sin\frac{n\pi x}{a} + \sum_{n=1}^{\infty} \frac{8n(-1)^{n-1}}{\pi(4n^2-1)} \cdot \frac{\sinh n\pi x/b}{\sinh n\pi a/b} \sin\frac{n\pi y}{b}.$$

[In fact, the symmetry of the original problem with respect to interchanges of x and a with y and b yields a simple means for identifying the function $\Phi_4(x,y)$.]

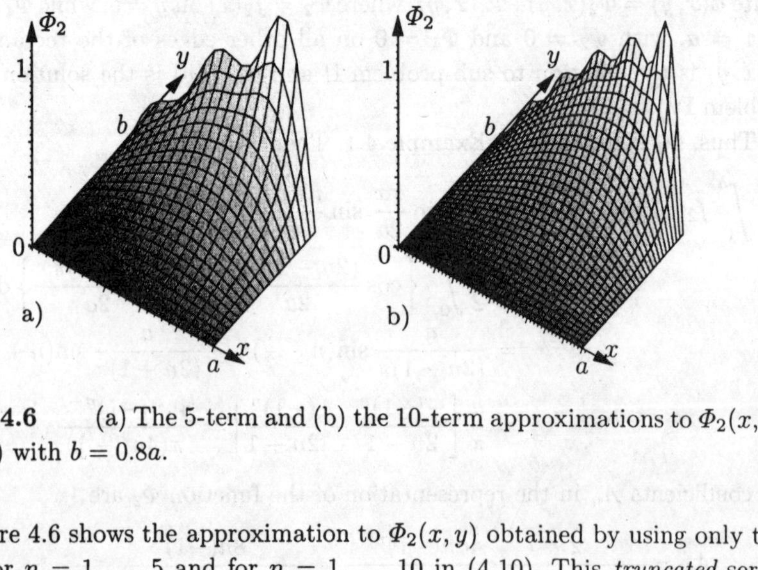

Figure 4.6 (a) The 5-term and (b) the 10-term approximations to $\Phi_2(x,y)$ of (4.10) with $b = 0.8a$.

Figure 4.6 shows the approximation to $\Phi_2(x,y)$ obtained by using only the terms for $n = 1,\ldots,5$ and for $n = 1,\ldots,10$ in (4.10). This *truncated* series gives a good approximation to ϕ except near the point $(x, y) = (a, b)$ where the boundary conditions on Φ_2 are discontinuous. While the approximation at *any chosen interior point* of the rectangle improves as the number of terms retained in the series is increased, there is always a (diminishingly small) region of oscillatory overshoot. This is a manifestation of the important feature known as the *Gibbs phenomenon* (Brown and Churchill, 1996, p.93), whereby the Fourier series for a discontinuous periodic function attempts to accommodate a finite jump at a *point of discontinuity.*

In Example 4.2, truncated approximations to the function $\Phi_4(x, y)$ (and consequently also those for $\phi(x, y)$) have a similar behaviour near (a, b). However, the boundary conditions for $\phi(x, y)$ are continuous. The Gibbs phenomenon occurs here only because the boundary conditions for Φ_2 and for Φ_4 individually are discontinuous. A slight rearrangement avoids this difficulty. A polynomial solution to $\nabla^2\phi = 0$ taking the required value $\phi(a, b) = 1$, as well as satisfying $\phi(x, 0) = 0 = \phi(0, y)$, is readily identified. One choice is $\phi = xy/(ab)$. Then, after writing $\phi = xy/(ab) + \theta(x, y)$, it is seen that $\theta(x, y)$ satisfies Laplace's equation and has boundary conditions which are zero at all four corners of the rectangle, namely:

$$\theta(x, b) = \sin\frac{\pi x}{2a} - \frac{x}{a} \equiv F_2(x)\,,\ \theta(a, y) = \sin\frac{\pi y}{2b} - \frac{y}{b} \equiv G_2(y)\,,$$

$$\theta(x, 0) = 0 = \theta(0, y)\,.$$

Computation of equivalent coefficients A_n in a representation for $\Theta_2(x, y)$ uses

$$\begin{aligned} A_n \sinh\frac{n\pi b}{a} &= \frac{2}{a}\int_0^a \left(\sin\frac{\pi x}{2a} - \frac{x}{a}\right) \sin\frac{n\pi x}{a}\,\mathrm{d}x = \frac{2}{\pi}\left\{\frac{4n(-1)^{n-1}}{4n^2-1} + \frac{(-1)^n}{n}\right\} \\ &= \frac{2(-1)^{n+1}}{\pi n(4n^2-1)}\,. \end{aligned}$$

Using similar coefficients in a function $\Theta_4(x, y)$ which accommodates the boundary data $G_2(y)$, then combining, yields

$$\begin{aligned} \phi(x, y) = \frac{xy}{ab} &+ \sum_{n=1}^{\infty} \frac{2(-1)^{n-1}}{\pi n(4n^2-1)} \cdot \frac{\sinh n\pi y/a}{\sinh n\pi b/a} \sin\frac{n\pi x}{a} \\ &+ \sum_{n=1}^{\infty} \frac{2(-1)^{n-1}}{\pi n(4n^2-1)} \cdot \frac{\sinh n\pi x/b}{\sinh n\pi a/b} \sin\frac{n\pi y}{b}\,. \end{aligned} \tag{4.11}$$

Expression (4.11) is an alternative, but preferable, representation for the solution to Example 4.2, as may be seen by the fact that its coefficients have an $O(n^{-3})$ decay as n increases. This not only means that good numerical accuracy is obtained even by truncating to a small number of terms. It also ensures that the Fourier sine series represent continuous periodic functions, so that no oscillatory behaviour typical of the Gibbs phenomenon arises.

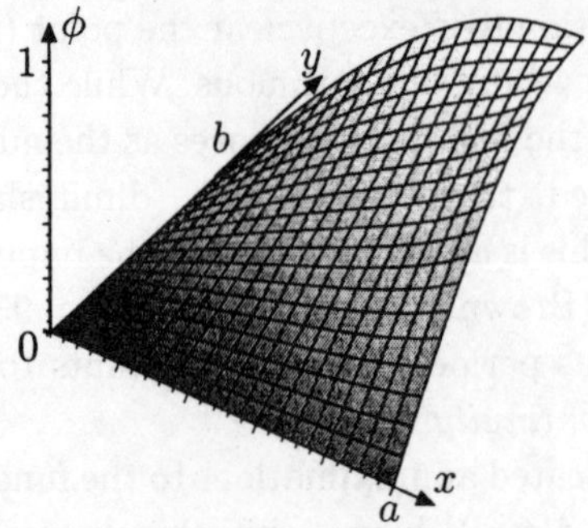

Figure 4.7 The approximation to $\phi(x, y)$ using just the terms $n = 1, 2, 3$ and 4 in each of the sums in Equation (4.11).

4.3 Poisson's Equation

The *method of separation of variables* can be extended so as to treat PDEs in which a distributed *source* (or supply rate) is specified. An important example is Poisson's equation

$$\nabla^2 \phi = F(\boldsymbol{x}) , \tag{4.12}$$

which governs, for example, the electrostatic potential in a material with specified distribution of electric charge (when $F = (4\pi\varepsilon_0)^{-1}\rho(\boldsymbol{x})$, with $\rho(\boldsymbol{x})$ the charge density), or the gravitational potential within a star having variable density $\rho(\boldsymbol{x}) = G^{-1}F(\boldsymbol{x})$.

Example 4.3

Suppose that within the cube $0 < x < a$, $0 < y < a$, $0 < z < a$, the electrostatic potential satisfies $\nabla^2\phi = C$ with boundary condition $\phi = V$ on all faces of the cube, with C and V each constant.

This problem has reflectional symmetry with respect to all three of the planes $x = \frac{1}{2}a$, $y = \frac{1}{2}a$ and $z = \frac{1}{2}a$ and under cyclic interchange of x, y and z, facts which may be exploited during solution. The first step is to identify one *particular solution* to $\nabla^2\phi = C$, preferably preserving all these symmetries. Since $\mathrm{d}^2\{x(x-a)\}/\mathrm{d}x^2 = 2 = \text{const.}$, an obvious choice is $\phi = \frac{1}{6}C(x^2 + y^2 + z^2 - ax - ay - az)$. Then, writing $\phi = \psi(x, y, z) + \frac{1}{6}C(x^2 + y^2 + z^2 - ax - ay - az) + k$ gives $\nabla^2\psi = 0$ for any choice of the constant k. The condition $\phi = V$ on all the boundary faces is replaced by

$$\psi(0,y,z) = \psi(a,y,z) = V - k - \tfrac{1}{6}C[y(y-a) + z(z-a)] \equiv f(y,z)$$
$$\text{over} \quad 0 < y < a\,,\; 0 < z < a\,,$$
$$\psi(x,0,z) = \psi(x,a,z) = V - k - \tfrac{1}{6}C[x(x-a) + z(z-a)] \equiv f(z,x)$$
$$\text{over} \quad 0 < x < a\,,\; 0 < z < a\,,$$
$$\psi(x,y,0) = \psi(x,y,a) = V - k - \tfrac{1}{6}C[x(x-a) + y(y-a)] \equiv f(x,y)$$
$$\text{over} \quad 0 < x < a\,,\; 0 < y < a\,.$$

A sub-problem I is first solved for $\Psi_1(x,y,x)$ satisfying

$$\begin{aligned} \nabla^2\Psi_1 &= 0 && \text{in } 0 < x < a\,,\; 0 < y < a\,,\; 0 < z < a\,,\\ \Psi_1(0,y,z) = \Psi_1(a,y,z) &= f(y,z) && \text{for } 0 < y < a\,,\; 0 < z < a\,;\\ \Psi_1(x,0,z) = \Psi_1(x,a,z) &= 0 && \text{for } 0 < x < a\,,\; 0 < z < a\,;\\ \Psi_1(x,y,0) = \Psi_1(x,y,a) &= 0 && \text{for } 0 < x < a\,,\; 0 < y < a\,, \end{aligned}$$

with inhomogeneous boundary conditions arising only for the function $X(x)$ when Ψ_1 is assumed to have the separable form $\Psi_1 = X(x)Y(y)Z(z)$. Then, it is seen from the symmetry of the problem defining $\psi(x,y,z)$ that the required solution is

$$\psi = \Psi_1(x,y,z) + \Psi_1(y,z,x) + \Psi_1(z,x,y)\,.$$

This means that, to complete the determination of ϕ, the main task is to construct the function $\Psi_1(x,y,z)$.

Experience of the Gibbs phenomenon in Example 4.2 suggests the choice $k = V$, so giving $f(0,0) = 0$, etc. and ensuring that the boundary data defining Ψ_1 is continuous at all corners of the cube. Then, since separable solutions $X(x)Y(y)Z(z)$ to Laplace's equation must satisfy Equation (4.6), while the conditions over $y = 0, a$ and $z = 0, a$ require that $Y(0) = 0 = Y(a)$ and that $Z(0) = 0 = Z(a)$, solutions of type (iii) to both $Y''(y) \propto Y$ and $Z''(z) \propto Z$ are required. They are

$$Y(y) = \sin\frac{m\pi y}{a}\,, \quad Z(z) = \sin\frac{n\pi z}{a} \qquad \text{for any integers } m \neq 0 \text{ and } n \neq 0\,.$$

Then, since $X''(x)/X(x) = -Y''(y)/Y(y) - Z''(z)/Z(z) = (m^2+n^2)\pi^2/a^2$, solutions to

$$\frac{\mathrm{d}^2X}{\mathrm{d}x^2} = p_{mn}^2 X(x) \qquad \text{with} \qquad p_{mn} \equiv (m^2+n^2)^{1/2}\pi/a$$

are required, having symmetry about $x = \frac{1}{2}a$. They are constant multiples of $X(x) = \cosh\{p_{mn}(x - \frac{1}{2}a)\}$. Notice how the separation constant $\alpha = p_{mn}^2$ for $X(x)$ involves two indices m and n, as do the separable solutions

$$\psi_{mn}(x, y, z) \equiv \cosh\{p_{mn}(x - \tfrac{1}{2}a)\} \sin\frac{m\pi y}{a} \sin\frac{n\pi z}{a}.$$

Notice also that there is no loss of generality in restricting to $m, n > 0$. Thus, the function $\Psi_1(x, y, z)$ may be represented as the superposition

$$\begin{aligned}\Psi_1(x, y, z) &= \sum_{m,n=1}^{\infty} A_{mn}\psi_{mn}(x, y, z) \\ &= \sum_{m,n=1}^{\infty} A_{mn}\cosh\{p_{mn}(x - \tfrac{1}{2}a)\} \sin\frac{m\pi y}{a} \sin\frac{n\pi z}{a}.\end{aligned}$$

As earlier, the coefficients are determined by matching to the inhomogeneous boundary conditions (i.e. those at $x = 0, a$) to give

$$\begin{aligned}f(y, z) \equiv \tfrac{1}{6}C(ay + az - y^2 - z^2) &= \sum_{m,n=1}^{\infty} A_{mn}\psi_{mn}(0, y, z) \\ &= \sum_{m,n=1}^{\infty} A_{mn}\cosh\frac{p_{mn}a}{2} \sin\frac{m\pi y}{a} \sin\frac{n\pi z}{a}.\end{aligned}$$

This is more complicated than (4.9), but may be recognized as a *double* (or *repeated*) *Fourier sine series*, periodic in both y and z and having period $2a$ in each. At any chosen value of z, this yields the result

$$\begin{aligned}\sum_{n=1}^{\infty} A_{mn}\cosh\frac{p_{mn}a}{2} \sin\frac{n\pi z}{a} &= \frac{2}{a}\int_0^a \tfrac{1}{6}C(ay + az - y^2 - z^2) \sin\frac{m\pi y}{a}\,\mathrm{d}y \\ &= \frac{C}{3a}\left\{(az - z^2)\frac{a}{m\pi}[1 - (-1)^m] - \frac{a^2}{m^2\pi^2}\int_0^a (-2)\sin\frac{m\pi y}{a}\,\mathrm{d}y\right\} \\ &= \frac{C}{3m\pi}[1 - (-1)^m]\left\{az - z^2 + \frac{2a^2}{m^2\pi^2}\right\}.\end{aligned}$$

This is, of course, another Fourier sine series, but for a function of z having period $2a$. The coefficients are likewise evaluated, using the standard half-range formula, as

$$\begin{aligned}A_{mn}\cosh\frac{p_{mn}a}{2} &= \frac{2C}{3am\pi}[1 - (-1)^m]\int_0^a \left\{az - z^2 + \frac{2a^2}{m^2\pi^2}\right\} \sin\frac{n\pi z}{a}\,\mathrm{d}z \\ &= \frac{4Ca^2}{3\pi^4}\left(\frac{1}{nm^3} + \frac{1}{mn^3}\right)[1 - (-1)^m][1 - (-1)^n].\end{aligned}$$

Observe that A_{mn} is non-zero only if *both of* m and n are odd – this is a consequence of the reflectional symmetries of the original problem. Hence,

$$\Psi_1(x,y,z) =$$

$$\tfrac{16}{3}Ca^2\pi^{-4} \sum_{m,n>0,\text{odd}} \frac{m^2+n^2}{(mn)^3} \frac{\cosh\{p_{mn}(x-\frac{1}{2}a)\}}{\cosh p_{mn}\frac{1}{2}a} \sin\frac{m\pi y}{a} \sin\frac{n\pi z}{a}. \quad (4.13)$$

The eventual solution for $\phi(x,y,z)$ is

$$\phi(x,y,z) = V + \tfrac{1}{6}C(x^2+y^2+z^2-ax-ay-az) + \Psi_1(x,y,z) + \Psi_1(y,z,x) + \Psi_1(z,x,y),$$

with $\Psi_1(x,y,z)$ defined by (4.13).

EXERCISES

4.1. Seek separable solutions $\phi = X(x)Y(y)$ to $\nabla^2\phi = 0$ in $0 < x < a$, $0 < y < b$ satisfying the conditions $\partial\phi/\partial y = 0$ on both $y = 0$ and on $y = b$ (if ϕ measures temperature, this corresponds to perfect insulation over both $x = 0$ and $x = a$, i.e. to zero heat flux).

4.2. Use a superposition of separable solutions to determine $\phi(x,y)$ for the region of Exercise 4.1, with additional boundary conditions $\phi(0,y) = ky^2(b-y)$ and $\phi(a,y) = 0$. [Note that the separable solutions of Exercise 4.1 include the case $Y(y) =$ constant.]

Evaluate the total flux over the face $x = a$ due to the field $\boldsymbol{\nabla}\phi$.

4.3. Replace the boundary condition (BC) $\phi(a,y) = 0$ in Exercise 4.2 by $\phi(a,y) = cy$. Construct additional separable solutions. Add these to the solution obtained in Exercise 4.2 and show that the total flux over $x = a$ vanishes if $c = \frac{1}{6}b^2k$.

4.4. Show that, for Poisson's equation $\nabla^2\phi = \sin\frac{\pi x}{a}\sin\frac{\pi y}{b}$ within the rectangle $0 < x < a$, $0 < y < b$, it is possible to seek a solution $\phi = k\sin\frac{\pi x}{a}\sin\frac{\pi y}{b}$. Determine k. Hence find $\phi(x,y)$ in $0 < x < a$, $0 < y < b$, subject to $\phi = 0$ around the boundary of the rectangle.

4.5. Seek solutions $\phi = X(x)Y(y)Z(z)$ to Laplace's equation within the rectangular box $0 < x < a$, $0 < y < b$, $0 < z < c$, satisfying also the conditions $\partial\phi/\partial y = 0$ at $y = 0, b$ and $\partial\phi/\partial z = 0$ at $z = 0, c$ and $\phi = 0$ at $x = 0$. Show that these yield for ϕ an expression

$$\phi(x,y,z) = A_0 x + \sum_{m=0}^{\infty}\sum_{n=0}^{\infty} A_{mn}\sinh\alpha_{mn}x \, \cos\frac{m\pi y}{b} \cos\frac{n\pi z}{c},$$

where $\alpha_{mn} = \pi \left\{(m/b)^2 + (n/c)^2\right\}^{1/2}$.

4.4 Dipole Solutions

Separation of variables and the use of Fourier series is particularly suited to construction of solutions in rectangular regions. In regions with spherical boundaries, *dipole solutions* are similarly useful. They are readily derived from the r^{-1} solution of Laplace's equation by use of the following result:

Whenever $\phi(x, y, z)$ is a solution of $\nabla^2\phi = 0$, then *each* of its derivatives $\partial\phi/\partial x$, $\partial\phi/\partial y$ and $\partial\phi/\partial z$ also satisfies Laplace's equation.

The proof, in regions where all third partial derivatives are continuous, is as follows:

$$\nabla^2\left(\frac{\partial\phi}{\partial z}\right) = \frac{\partial^2}{\partial x^2}\left(\frac{\partial\phi}{\partial z}\right) + \frac{\partial^2}{\partial y^2}\left(\frac{\partial\phi}{\partial z}\right) + \frac{\partial^2}{\partial z^2}\left(\frac{\partial\phi}{\partial z}\right)$$
$$= \frac{\partial}{\partial z}\left\{\frac{\partial^2\phi}{\partial x^2} + \frac{\partial^2\phi}{\partial y^2} + \frac{\partial^2\phi}{\partial z^2}\right\} = \frac{\partial}{\partial z}(\nabla^2\phi) = \frac{\partial}{\partial z}(0) = 0\,.$$

Hence, the choice $\phi = -r^{-1}$ leads to the result that the function

$$\frac{z}{r^3} = \frac{1}{r^2}\frac{\partial r}{\partial z} = \frac{\partial}{\partial z}(-r^{-1})$$

satisfies Laplace's equation (3.6) everywhere except at $r = 0$.

The potential zr^{-3} has a physical interpretation – it is the potential of a *dipole* located at $\boldsymbol{x} = \boldsymbol{0}$ and aligned along Oz. A *dipole* is the mathematical limit of two equal and opposite charges at small separation. Consider two charges $\pm q$ located at $\boldsymbol{x} = \pm a\boldsymbol{k}$, respectively. They give a potential

$$\phi = \frac{q}{4\pi\epsilon_0}\left(\frac{1}{|\boldsymbol{x} - a\boldsymbol{k}|} - \frac{1}{|\boldsymbol{x} + a\boldsymbol{k}|}\right)$$
$$= \frac{q}{4\pi\epsilon_0}\left(\frac{1}{\sqrt{r^2 - 2az + a^2}} - \frac{1}{\sqrt{r^2 + 2az + a^2}}\right)$$
$$= \frac{q}{4\pi\epsilon_0 r}\{[1 - 2az/r^2 + a^2/r^2]^{-1/2} - [1 + 2az/r^2 + a^2/r^2]^{-1/2}\},$$

where $r \equiv |\boldsymbol{x}| = (x^2 + y^2 + z^2)^{1/2}$. Since $|z/r| \leq 1$, then for $r \geq a$ this gives

$$\phi = \frac{2aq}{4\pi\epsilon_0}\frac{z}{r^3} + \frac{2aq}{4\pi\epsilon_0 r^2}\mathrm{O}\left(\frac{a^2}{r^2}\right).$$

In the limit in which $a \to 0$ with the product $2aq \equiv p$ held fixed (and finite), ϕ tends to the *dipole potential*

$$\frac{p}{4\pi\epsilon_0}\frac{z}{r^3}\,. \tag{4.14}$$

This is known as the electrostatic potential of a *dipole* located at $\boldsymbol{x} = \boldsymbol{0}$, having strength p and oriented parallel to Oz. The vector $\boldsymbol{p} \equiv p\boldsymbol{k}$ is known as the *dipole strength*; it is parallel to Oz.

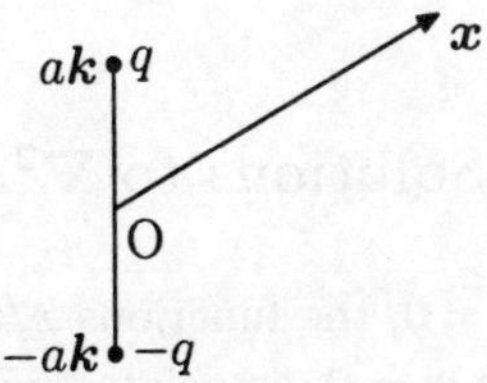

Figure 4.8 Charges $\pm q$ at locations $\pm a\boldsymbol{k}$ giving (in the limit $a \to 0$ with $2aq \equiv p$ fixed) a dipole $p\boldsymbol{k}$ at O. Its electrostatic potential is $p(4\pi\epsilon_0)^{-1}z/r^3$.

More generally, for *any* vector $\boldsymbol{p}$, the electrostatic potential

$$\frac{\boldsymbol{p}\cdot(\boldsymbol{x}-\boldsymbol{x}_0)}{4\pi\epsilon_0|\boldsymbol{x}-\boldsymbol{x}_0|^3} = \frac{-1}{4\pi\epsilon_0}\boldsymbol{p}\cdot\boldsymbol{\nabla}(|\boldsymbol{x}-\boldsymbol{x}_0|^{-1})$$

is the electrostatic potential of an electric dipole $\boldsymbol{p}$ located at the (arbitrary) position $\boldsymbol{x} = \boldsymbol{x}_0$. A mathematically similar, but physically distinct, concept is the *scalar magnetic potential*

$$\phi_{\mathrm{m}} = \frac{\boldsymbol{m}\cdot(\boldsymbol{x}-\boldsymbol{x}_0)}{4\pi|\boldsymbol{x}-\boldsymbol{x}_0|^3} = \frac{-1}{4\pi}\boldsymbol{m}\cdot\boldsymbol{\nabla}(|\boldsymbol{x}-\boldsymbol{x}_0|^{-1})$$

due to a *magnetic dipole* $\boldsymbol{m}$ located at $\boldsymbol{x} = \boldsymbol{x}_0$. It is a fundamental element in

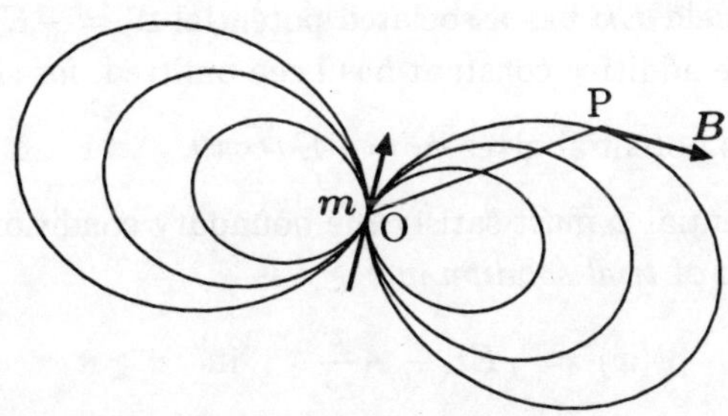

Figure 4.9 Magnetic field lines and the *magnetic induction* vector $\boldsymbol{B}$ at a typical location P, due to a magnetic dipole $\boldsymbol{m}$ at O.

the study of static magnetic fields (the resulting *magnetic induction* is given by $\boldsymbol{B} = -\mu_0\boldsymbol{\nabla}\phi_{\mathrm{m}}$, where the constant μ_0 is the permeability of free space). Note (in Figure 4.9) how the field pattern is similar to that produced by the familiar bar magnet aligned parallel to $\boldsymbol{m}$ (and often made visible as a pattern of iron filings). In books on electromagnetism, it is explained how this dipole field arises also from electric current flowing in an infinitesimal loop lying in the plane normal to $\boldsymbol{m}$ – or equivalently from electrons moving rapidly around small orbits in that plane.

4.4.1 Uses of Dipole Solutions to $\nabla^2\phi = 0$

For Laplace's equation $\nabla^2\phi = 0$, the functions x/r^3, y/r^3 and z/r^3 are known as the *unit dipole solutions* at $\boldsymbol{x} = \mathbf{0}$, associated with alignments parallel to Ox, Oy and Oz, respectively. They are useful in constructing solutions to Laplace's equation in regions bounded by spheres centred at O and with uniform fields *at infinity*, i.e. at large distances from the spheres. This arises because the *dipole potential*

$$\frac{z}{r^3} = \frac{\cos\theta}{r^2}$$

and the potential $\phi = z = r\cos\theta$ (which corresponds to the uniform field $-\boldsymbol{\nabla}\phi = -\boldsymbol{k}$) depend on the spherical polar coordinate θ through the same geometric factor $\cos\theta$. The simplest example is as follows:

Example 4.4

Find the electrostatic field $\boldsymbol{E} = -\boldsymbol{\nabla}\phi$ outside an earthed conductor located at $r = a$, when $\boldsymbol{E} \to E\boldsymbol{k}$ as $|\boldsymbol{x}| \to \infty$.

i) The uniform field $E\boldsymbol{k}$ has associated potential $\phi_1 = -Ez$, since $-\boldsymbol{\nabla}\,\phi_1 = E\boldsymbol{k}$ (a possible additive constant has been omitted, for simplicity).

ii) On $r = a$, this potential gives $\phi_1 = -Ea\cos\theta$.

iii) Since the potential ϕ must satisfy the boundary condition $\phi = 0$ on $r = a$, a natural form of *trial solution* in $r \geq a$ is

$$\phi(\boldsymbol{x}) = -Ez + A\frac{z}{r^3} \qquad \text{in} \quad r \geq a$$

(i.e. a *superposition* of the potential of the uniform field $E\boldsymbol{k}$ and of a dipole of unknown strength). This certainly satisfies Laplace's equation (except at $r = 0$). Also, as $r \to \infty$, the contribution $z/r^3 = r^{-2}\cos\theta = \mathrm{O}(r^{-2}) \to 0$, (hence, it gives a correction to the uniform field $E\boldsymbol{k}$ which vanishes at large distances, i.e. $-\boldsymbol{\nabla}\phi \to E\boldsymbol{k}$, as required).

iv) Then, setting $\phi = 0$ on $r = a$ gives

$$0 = \phi = -Ea\cos\theta + Aa^{-2}\cos\theta\,.$$

Since θ is arbitrary, this equation is satisfied only when

$$Aa^{-2} = Ea \qquad \text{so giving} \qquad A = a^3E\,.$$

Thus, the potential everywhere in $r \geq a$ is

$$\phi = -Ez + Ea^3\frac{z}{r^3}\,. \tag{4.15}$$

Example 4.4 shows that the potential ϕ *outside* a spherical conductor is the sum of the potential of the externally applied *uniform field* $E\boldsymbol{k}$ and of the potential of a suitable dipole $4\pi\epsilon_0 Ea^3\boldsymbol{k}$ imagined to be at the centre of the sphere. The corresponding electrostatic field $\boldsymbol{E}$, outside the earthed conducting sphere is

$$\boldsymbol{E} = -\nabla\phi = E\boldsymbol{k}\left(1 - \frac{a^3}{r^3}\right) + 3Ea^3\frac{z\boldsymbol{x}}{r^5}\,.$$

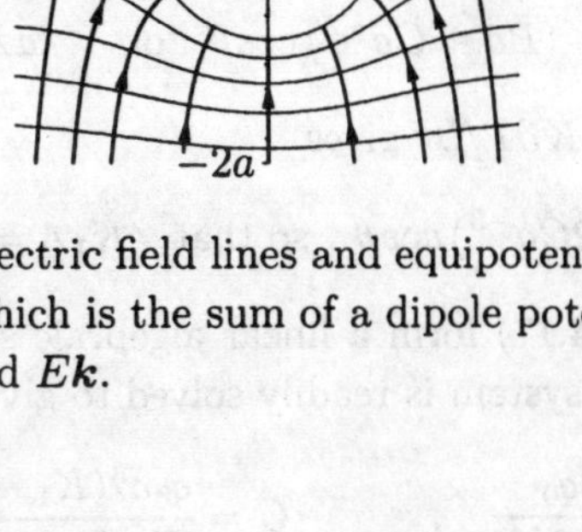

Figure 4.10 The electric field lines and equipotentials in the plane $y = 0$, for the solution (4.15) which is the sum of a dipole potential and the potential $-Ez$ for the uniform field $E\boldsymbol{k}$.

4.4.2 Spherical Inclusions

Dipole solutions may be combined with potentials linear in z to solve important problems in electrostatics, in steady electric current flow, in steady heat flow etc., where a spherical region $r < a$ (an *inclusion*) has different properties from its surroundings $r > a$. For example, from (4.5) it is seen that the heat flux in a region of *thermal conductivity* K is given by $\boldsymbol{q} = -K\boldsymbol{\nabla}T$, while if the temperature $T(\boldsymbol{x})$ is *steady* it satisfies $\nabla^2 T = 0$. Suppose that, in $r < a$ the conductivity is $K = K_1$, while for $r > a$ the conductivity is $K = K_2$. If, for $r \gg a$, the temperature gradient becomes uniform, how does the spherical inclusion affect the heat flow?

Clearly, at each point of the sphere $r = a$ the temperature must be continuous. Also, at each such point, the *outward* normal to the sphere is $\boldsymbol{n} \equiv \boldsymbol{x}/a$ and the radial component of heat flux is

$$\boldsymbol{n}\cdot\boldsymbol{q} = -K\boldsymbol{n}\cdot\boldsymbol{\nabla}T = -K\frac{\partial T}{\partial r}\,.$$

It is continuous across $r = a$ (i.e. it has the same limiting values as $r \to a$ for $r < a$ and for $r > a$). Take axes so that, as $r \to \infty$, the heat flux tends to $\boldsymbol{q}_0 = q_0\boldsymbol{k}$, with $T \to T_0 - q_0 z/K_2$ (thus, Oz is aligned with the *far-field* heat flux).

In each of $r \leq a$ and $r \geq a$, seek expressions for $T - T_0$ as a linear combination of just the special solutions z and zr^{-3} of $\nabla^2 T = 0$. But, as $r \to 0$, T must remain finite; hence, in $r \leq a$ the temperature contains no term zr^{-3} and so has the form $T = T_0 + Az$. In $r \geq a$, it may have the form $T = T_0 + Bz + Cz/r^3 = T_0 + (Br + Cr^{-2})\cos\theta$. However, as $r \to \infty$, it is easily checked (see Exercise 4.6) that $\boldsymbol{q} = -K_2\,\mathrm{grad}\,T = -K_2\boldsymbol{\nabla}T \to -K_2B\boldsymbol{k}$. Hence, B must satisfy $-K_2B = q_0$, so yielding $B = -q_0/K_2$.

As $r \to a$ in $r < a$, this gives $T \to T_0 + Aa\cos\theta$ while $\partial T/\partial r = A\cos\theta$. For $r > a$, as $r \to a$ it may be seen that $T \to T_0 + (Ba + Ca^{-2})\cos\theta$, so that $\partial T/\partial r \to (B - 2Ca^{-3})\cos\theta$. Continuity of T then gives

$$T_0 + Aa\cos\theta = T_0 + (Ba + Ca^{-2})\cos\theta \quad \text{or} \quad aA = aB + a^{-2}C\,. \qquad (4.16)$$

Similarly, continuity of $K\partial T/\partial r$ gives

$$K_1A\cos\theta = K_2(B - 2Ca^{-3})\cos\theta \quad \text{so that} \quad K_1A = K_2(B - 2a^{-3}C)\,. \qquad (4.17)$$

Equations (4.16) and (4.17) form a linear algebraic system relating A and C to B $(= -q_0/K_2)$. The system is readily solved to give

$$A = \frac{-3q_0}{K_1 + 2K_2}\,, \qquad C = \frac{q_0a^3(K_1 - K_2)}{K_2(K_1 + 2K_2)}\,.$$

Hence, within the inclusion $r < a$, the temperature is $T_0 - 3q_0z/(K_1 + 2K_2)$, so that the heat flux is uniform but altered from the *far-field* value $\boldsymbol{q}_0$ by the multiplicative factor $3K_1/(K_1 + 2K_2)$. Outside the inclusion, the temperature is altered from the distribution which is linear in z by addition of the dipole of strength C.

Similar calculations apply for steady currents in electric conductors. In these, an electric field $\boldsymbol{E} = -\boldsymbol{\nabla}\phi$ gives rise to an *electric current density* (or rate of flow of charge) $\boldsymbol{J} = \sigma_c \boldsymbol{E} = -\sigma_c \boldsymbol{\nabla}\phi$ (this is known as Ohm's law). The material parameter σ_c is known as the *electrical conductivity.* For example, if current $\boldsymbol{J} = J\boldsymbol{i}$ flows uniformly and parallel to the Ox axis in a wire of cross-sectional area A, then $\partial\phi/\partial x = -J/\sigma_c$. Thus, between $x = x_0$ and $x = x_0 + L$ there is a *potential difference* (a *voltage*) $V \equiv \phi(x_0) - \phi(x_0 + L)$, while the total electric current flowing along the wire is

$$I \equiv JA = A\sigma_c\left(\phi(x_0 + L) - \phi(x_0)\right)/L = \sigma_c AV/L\,.$$

The *resistance* (typically measured in ohms) of length L of the wire is $V/I = L(\sigma_c A)^{-1}$; the resistance per unit length is inversely proportional to both area and conductivity.

Example 4.5

Find the electric potential $\phi(\boldsymbol{x})$ corresponding to steady current $\boldsymbol{E} = -\boldsymbol{\nabla}\phi$ when the conductivity in $r < a$ is $\sigma_c = \sigma_1$ and in $r > a$ is $\sigma_c = \sigma_2$, while $\phi + E_0 z \to 0$ as $|z| \to \infty$.

Solve $\nabla^2\phi = 0$ in $r < a$ and in $r > a$, with both ϕ and $\sigma_c \partial\phi/\partial r$ continuous at $r = a$ and with $\phi \sim -E_0 z$ as $|z| \to \infty$. Take

$$\phi = \begin{cases} Az & r \le a\,, \\ -E_0 z + Bz/r^3 & r \ge a\,, \end{cases}$$

(so that $\phi \sim -E_0 z$ and $\boldsymbol{E} \to E_0 \boldsymbol{k}$ as $|z| \to \infty$). Then

$$\frac{\partial\phi}{\partial r} = \begin{cases} A\cos\theta & r < a\,, \\ -E_0\cos\theta - 2r^{-3}B\cos\theta & r > a\,. \end{cases}$$

At $r = a$, continuity of ϕ and of $\sigma_c \partial\phi/\partial r$ give

$$A = -E_0 + a^{-3}B\,, \qquad \sigma_1 A = -\sigma_2(E_0 + 2a^{-3}B)\,.$$

Solving gives $A = -3\sigma_2 E_0/(\sigma_1 + 2\sigma_2)$ and $B = (\sigma_1 - \sigma_2)a^3 E_0/(\sigma_1 + 2\sigma_2)$, so that

$$\phi(\boldsymbol{x}) = \begin{cases} \dfrac{-3\sigma_2}{\sigma_1 + 2\sigma_2} E_0 z & r \le a\,, \\ -E_0 z + \dfrac{\sigma_1 - \sigma_2}{\sigma_1 + 2\sigma_2}\dfrac{a^3}{r^3} E_0 z & r \ge a\,. \end{cases}$$

[N.B. As $\sigma_2/\sigma_1 \to 0$, the potential in $r \geq a$ approaches the potential (cf. (4.15)) surrounding a perfect spherical conductor at $r = a$, with $\phi \to 0$ throughout $r \leq a$. In the limit $\sigma_1/\sigma_2 \to 0$, the inclusion becomes a *perfect insulator*, with $-\sigma_2 \partial\phi/\partial r = \boldsymbol{J} \cdot \boldsymbol{n} \to 0$ as $r \to a+$, so that no current flows normally to the spherical boundary.]

EXERCISES

4.6. Check, by direct differentiation, that (a) $\boldsymbol{\nabla}(z/r^3) = r^{-3}\boldsymbol{k} - 3zr^{-5}\boldsymbol{x}$ and (b) $\nabla^2(z/r^3) = 0$ (for $r \neq 0$).

4.7. Confirm that the potential $\boldsymbol{p} \cdot \boldsymbol{\nabla} r^{-1}$ is a linear combination of the potentials of dipoles at $\boldsymbol{x} = \boldsymbol{0}$ aligned along Ox, Oy and Oz.

4.8. In spherical polar coordinates (r, θ, ψ), Laplace's equation takes the standard form

$$\nabla^2\phi \equiv \frac{\partial^2\phi}{\partial r^2} + \frac{2}{r}\frac{\partial\phi}{\partial r} + \frac{1}{r^2 \sin\theta}\frac{\partial}{\partial\theta}\left(\sin\theta \frac{\partial\phi}{\partial\theta}\right) + \frac{1}{r^2\sin^2\theta}\frac{\partial^2\phi}{\partial\psi^2} = 0\,.$$

Using the identities $x = r\cos\theta\,\cos\psi$, $y = r\cos\theta\,\sin\psi$ and $z = r\sin\theta$, express each of x/r^3, y/r^3 and z/r^3 in terms of the polar coordinates. *Verify* that each satisfies Laplace's equation. (They each are *separable solutions* in spherical polars. Moreover, their dependence on r is through the factor r^{-2}. Many solutions to $\nabla^2\phi = 0$ of the form $\phi = r^n\Theta(\theta)\Psi(\psi)$ exist – the function $\Theta(\theta)\Psi(\psi)$ is known as a *spherical harmonic* of degree n. It is readily shown that the function $r^{-n-1}\Theta(\theta)\Psi(\psi)$ also satisfies Laplace's equation.)

4.9. If electric current flows in the region $r > a$ having electrical conductivity σ_2, with $\phi \sim -E_0 z$ as $|z| \to \infty$ and with the region $r \leq a$ being a perfect insulator (so that $\partial\phi/\partial r = 0$ at $r = a$), find the potential ϕ and electric field $\boldsymbol{E} = \boldsymbol{\nabla}\phi$ in $r \geq a$.

4.10. If the perfect insulator within $r \leq a$ of Exercise 4.9 is coated by a layer $a < r < b$ having conductivity σ_1, while the conductivity in $r > b$ remains as $\sigma_c = \sigma_2$, find ϕ in both $a \leq r \leq b$ and in $r \geq b$ when $\phi \sim -E_0 z$ as $r \to \infty$. Confirm that, on the sphere $r = a$, the maximum value of the electric field strength $|r^{-1}\partial\phi/\partial\theta|$ is

$$\frac{9\sigma_2}{2(2\sigma_2 + \sigma_1) + (\sigma_2 - \sigma_1)a^3/b^3} E_0\,.$$

5

Motion of an Elastic String

5.1 Tension and Extension; Kinematics and Dynamics

The motion of an elastic string provides a good example of how to describe and how to model continuous media by using balance laws. It also illustrates the phenomena of *standing waves* and of *travelling waves*, which occur widely, in many different contexts.

An elastic string is a very slender solid body, which changes its length when subject to *tension*. Its lateral thickness is negligible compared with dimensions characterizing its length or curvature. Suppose that a sample of string has *natural* length l (its length when no tensile force is applied). Then, as the *tension* T (a force) is increased, this sample stretches to length λl. The quantity λ is known as the *stretch ratio*. The relationship

$$T = T(\lambda) \tag{5.1}$$

between T and λ varies from material to material (and is proportional to the cross-sectional area of the string). It is an example of a *constitutive relation*, a rule relating physical cause to physical effect. Experiment shows that T increases monotonically (and often virtually linearly) with λ for $\lambda \geq 1$, with typical behaviour as in Figure 5.1 (since a string crumples when compressed, treating a solid as a string is appropriate only for $\lambda \geq 1$).

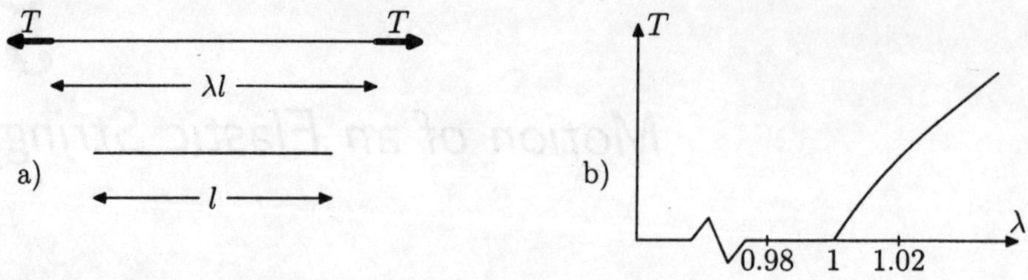

Figure 5.1 a) Stretched and *natural* lengths of a portion of string. b) A typical form for the *constitutive relation* $T = T(\lambda)$.

Postulate 1. Since a string has negligible thickness, its configuration is described by the equation for a moving curve

$$\boldsymbol{x} = \boldsymbol{r}(X, t)\,. \tag{5.2}$$

Here, X is a *material coordinate* (or *Lagrangian coordinate*), which labels the material which is at distance X along the string when the string occupies its *reference* (unstretched) *configuration* (taken, without loss of generality, as $\boldsymbol{x} = X\boldsymbol{i}$ and with $\lambda = 1$, $T = 0$).

In a general motion of the string, the stretch ratio λ (and consequently the tension T) varies with distance along the string and with time. We first analyse the geometry of the deformed string. At time t, let P and P′ denote the points on the string (5.2) which are labelled by coordinates X and $X+\delta X$, respectively. They correspond to material points which, when the string is in its reference configuration, are separated by distance δX. Since $\overrightarrow{\mathrm{OP}} = \boldsymbol{r}(X, t)$, while $\overrightarrow{\mathrm{OP}'} = \boldsymbol{r}(X + \delta X, t)$, their current relative separation is

$$\overrightarrow{\mathrm{PP}'} = \delta\boldsymbol{x} \approx \frac{\partial \boldsymbol{r}}{\partial X}(X, t)\delta X\,.$$

The distance s along the deformed string, measured from the material point $X = 0$, must satisfy

$$\delta s \approx \frac{\partial s}{\partial X}\delta X = \left|\frac{\partial \boldsymbol{r}}{\partial X}\right| \delta X$$

(the parameter X varies along the curve, while t is taken to have a constant value). Then, since the stretch ratio at P is $\lambda = \lim \delta s/\delta X$, we have $\lambda = |\partial \boldsymbol{r}/\partial X|$. If $\boldsymbol{t}$ denotes a unit vector tangential to the curve at P (a *unit tangent*), then

$$\frac{\partial \boldsymbol{r}}{\partial X} = \lambda \boldsymbol{t} \qquad \left(\text{i.e.}\quad \overrightarrow{\mathrm{PP}'} \approx \frac{\partial \boldsymbol{r}}{\partial X}\delta X\right).$$

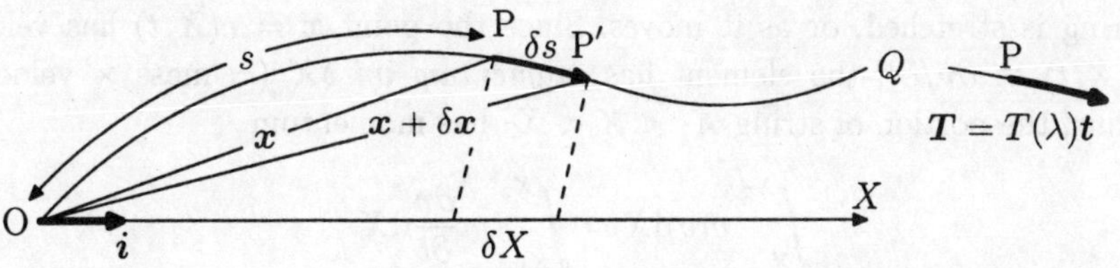

Figure 5.2 A typical string configuration, showing tension $\boldsymbol{T}$ parallel to the unit tangent $\boldsymbol{t}$.

The second important ingredient of a string model is

Postulate 2. At point P, the force on the portion OP of the string (due to the adjacent portion PQ) is *tangential to* the string, with magnitude $T(\lambda)$. It is

$$T(\lambda)\boldsymbol{t} \equiv \boldsymbol{T}(X,t) \qquad \text{(say)}. \tag{5.3}$$

The portion PQ experiences an equal but opposite force $-T(\lambda)\boldsymbol{t}$ acting at P, due to the portion OP (Newton's first law states that *action* and *reaction* are equal and opposite). In this model, the string can support only a *tension* of magnitude $T(\lambda)$ (*shear forces* and *bending moments* are negligible, because the string is so slender).

5.1.1 Dynamics

The motion of a portion $X_1 < X < X_2$ of the string is determined (from Newton's laws of motion), because the *resultant force* must equal the *rate of change of momentum.* Since the force at $X = X_2$ is $\boldsymbol{T}(X_2, t)$, while the force at $X = X_1$ is $-\boldsymbol{T}(X_1, t)$, the resultant force is (see Figure 5.3)

$$\boldsymbol{T}_2 - \boldsymbol{T}_1 = \boldsymbol{T}(X_2, t) - \boldsymbol{T}(X_1, t).$$

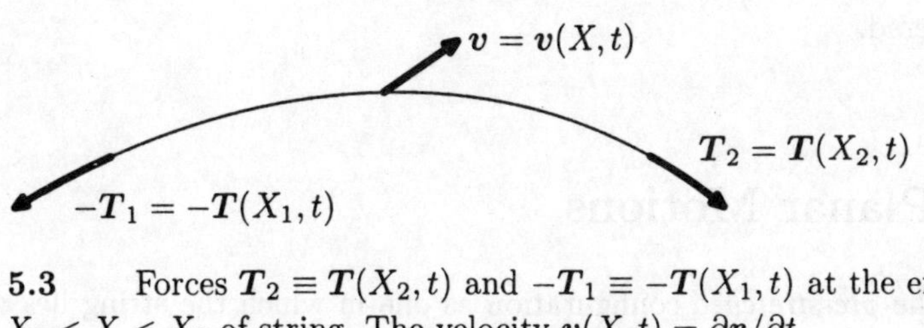

Figure 5.3 Forces $\boldsymbol{T}_2 \equiv \boldsymbol{T}(X_2,t)$ and $-\boldsymbol{T}_1 \equiv -\boldsymbol{T}(X_1,t)$ at the ends of a portion $X_1 < X < X_2$ of string. The velocity $\boldsymbol{v}(X,t) = \partial \boldsymbol{r}/\partial t$.

Let the string have mass per unit length $\hat{m}$ when unstretched, so that a small element labelled by δX has mass $\hat{m}\,\delta X$. This mass does not alter as the string is stretched, or as it moves. Since the point $\boldsymbol{x} = \boldsymbol{r}(X,t)$ has velocity $\boldsymbol{v}(X,t) \equiv \partial\boldsymbol{r}/\partial t$, the element has *momentum* $\boldsymbol{v}\hat{m}\,\delta X$ (= mass × velocity). Thus, the portion of string $X_1 < X < X_2$ has momentum

$$\int_{X_1}^{X_2} \hat{m}\boldsymbol{v}\,\mathrm{d}X = \int_{X_1}^{X_2} \hat{m}\frac{\partial \boldsymbol{r}}{\partial t}\mathrm{d}X .$$

Equating the time derivative of this to the resultant force gives

$$\frac{\mathrm{d}}{\mathrm{d}t}\int_{X_1}^{X_2} \hat{m}\frac{\partial \boldsymbol{r}}{\partial t}\mathrm{d}X = \boldsymbol{T}_2 - \boldsymbol{T}_1 = \boldsymbol{T}(X_2,t) - \boldsymbol{T}(X_1,t) = \int_{X_1}^{X_2}\frac{\partial \boldsymbol{T}}{\partial X}\mathrm{d}X . \quad (5.4)$$

(Note how the fundamental theorem of calculus has again been used. Also, other *external forces* such as gravity and air resistance have been neglected.) Since, even though the material points move (i.e. vary position with time) the range of integration $X_1 < X < X_2$ remains fixed, the term on the left-hand side may be treated as in (1.5), so giving

$$\int_{X_1}^{X_2}\left\{\frac{\partial \boldsymbol{T}}{\partial X} - \frac{\partial}{\partial t}\left(\hat{m}\frac{\partial \boldsymbol{r}}{\partial t}\right)\right\}\mathrm{d}X = 0 . \quad (5.5)$$

Of course, in this derivation, the values X_1 and X_2 are adjustable, so that at *each* point of the string the integrand must vanish. This yields (using (5.3)) the equation of motion:

$$\hat{m}\frac{\partial^2 \boldsymbol{r}}{\partial t^2} = \frac{\partial \boldsymbol{T}}{\partial X} = \frac{\partial}{\partial X}\left(T(\lambda)\boldsymbol{t}\right) = \frac{\partial}{\partial X}\left(\frac{T(\lambda)}{\lambda}\frac{\partial \boldsymbol{r}}{\partial X}\right) . \quad (5.6)$$

Equation (5.6) governs the *vector* function $\boldsymbol{r}(X,t)$, but is difficult to analyse exactly, since the *longitudinal* (or *stretching*) motions are coupled to the *transverse* motions. However, in practice, string motion frequently involves only small displacements from some straight *pre-stretched* state, in which the stretch ratio λ_0 and associated tension $T_0 = T(\lambda_0)$ are uniform. This allows considerable simplification, particularly when only motions in the plane of Oxy are considered.

5.2 Planar Motions

Take the *pre-stretched* configuration as one in which the string lies along the Ox axis with stretch ratio λ_0 (and axial tension $T_0 \equiv T(\lambda_0)$), so that $\boldsymbol{x} = \lambda_0 X\boldsymbol{i}$. This is an *equilibrium configuration* satisfying (5.6); the velocity $\boldsymbol{v}$ is zero, while λ and $\boldsymbol{t}$ are constants. When the string is further deformed, the point P undergoes a *small displacement* $u\boldsymbol{i}+v\boldsymbol{j}$ from the pre-stretched location $\boldsymbol{x} = \lambda_0 X\boldsymbol{i}$, so that

$$\boldsymbol{r}(X,t) = x\boldsymbol{i} + y\boldsymbol{j} = (\lambda_0 X + u)\boldsymbol{i} + v\boldsymbol{j}$$

(with $u = u(X,t)$ and $v = v(X,t)$ small). The unit tangent $\boldsymbol{t}$ and stretch ratio

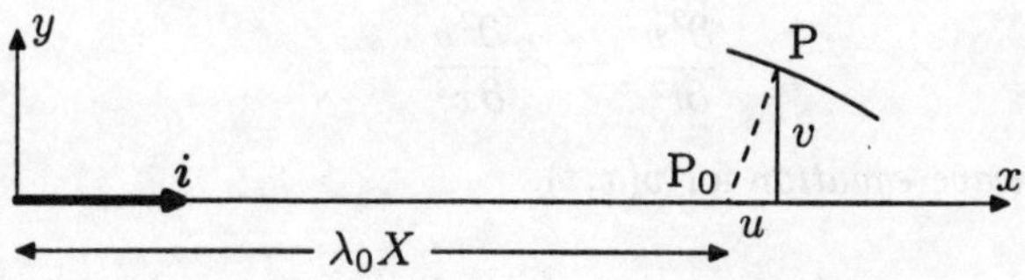

Figure 5.4 Small displacements (u, v) in a planar motion.

λ are calculated from

$$\lambda\boldsymbol{t} = \frac{\partial \boldsymbol{r}}{\partial X} = \boldsymbol{i}\left(\lambda_0 + \frac{\partial u}{\partial X}\right) + \boldsymbol{j}\frac{\partial v}{\partial X} = (\lambda_0 + u_X)\boldsymbol{i} + v_X\boldsymbol{j}$$

(where subscripts X and t denote partial differentiation), so yielding

$$\lambda = \{(\lambda_0 + u_X)^2 + v_X^2\}^{1/2} \approx \lambda_0 + u_X\,.$$

Here the fact that $|u_X| \ll 1$ and $|v_X| \ll 1$ has been exploited, i.e. the derivatives of displacement are small, since displacements are small. To this approximation, $\boldsymbol{t}$ is found as

$$\boldsymbol{t} = \lambda^{-1}\{(\lambda_0 + u_X)\boldsymbol{i} + v_X\boldsymbol{j}\} \approx \boldsymbol{i} + \lambda_0^{-1}v_X\boldsymbol{j}$$

(the term u_X affects $\boldsymbol{t}$ only at $O(u_X^2)$, i.e. at least quadratically in terms of the small quantities). Inserting the above into the equation for string motions (5.6) and working only to the accuracy of terms *linear* in u, v and their derivatives (since $T(\lambda) \approx T_0 + T'(\lambda_0)u_X$) gives

$$\boldsymbol{i}\frac{\partial^2 u}{\partial t^2} + \boldsymbol{j}\frac{\partial^2 v}{\partial t^2} = \frac{1}{\hat{m}}\frac{\partial}{\partial X}\left(T_0 + T'(\lambda_0)u_X\right)\boldsymbol{i} + \frac{1}{\hat{m}}\frac{\partial}{\partial X}\left(\frac{T(\lambda_0)}{\lambda_0}v_X\right)\boldsymbol{j}\,.$$

(Since the *extension*, usually defined as $\lambda - 1$, becomes $\lambda_0 - 1 + u_X$, this shows that the longitudinal component of force $T_0 + T'(\lambda_0)u_X$ is linear in the change in extension.) Considering separately the $\boldsymbol{i}$ and $\boldsymbol{j}$ components of this equation gives equations for the *transverse* and *longitudinal* motions:
Transverse motions (displacements parallel to $\boldsymbol{j}$) satisfy

$$\frac{\partial^2 v}{\partial t^2} = \frac{T_0}{\hat{m}\lambda_0}\frac{\partial^2 v}{\partial X^2}\,. \tag{5.7}$$

Longitudinal motions (displacements parallel to $\boldsymbol{i}$) satisfy

$$\frac{\partial^2 u}{\partial t^2} = \frac{T'(\lambda_0)}{\hat{m}} \frac{\partial^2 u}{\partial X^2} . \tag{5.8}$$

The two equations (5.7) and (5.8) have the same mathematical form – they each may be simply related to the important equation

$$\frac{\partial^2 v}{\partial t^2} = c^2 \frac{\partial^2 v}{\partial x^2} \tag{5.9}$$

known as the *wave equation* for $v(x, t)$.

5.2.1 Small Transverse Motions

Using the approximation $x = \lambda_0 X$ in (5.7) gives

$$v_{tt} = \frac{\lambda_0 T_0}{\hat{m}} v_{xx} = c^2 v_{xx} \tag{5.10}$$

which is of the form (5.9) when c is chosen as $c = \sqrt{T_0/m}$, with $m \equiv \hat{m}/\lambda_0$ being the mass per unit length in the pre-stretched state. This describes, for example, the transverse vibrations of a string on a musical instrument.

5.2.2 Longitudinal Motions

In (5.8), the choice $c = \sqrt{\lambda_0 T'(\lambda_0)/m}$ shows that the longitudinal displacement u also satisfies a *wave equation*, namely

$$u_{tt} = \frac{\lambda_0 T'(\lambda_0)}{m} u_{xx} = c^2 u_{xx} . \tag{5.11}$$

However, the value c for in this equation typically much exceeds that in (5.10), since the *elastic modulus* $T'(\lambda_0)$ (particularly for metals) much exceeds the ratio T_0/λ_0 (the slope in Figure 5.1(b) is large).

5.3 Properties of the Wave Equation

We now investigate some of the properties of (5.9), especially as they relate to motions of a string.

i) Equation (5.9) allows various special solutions:
(a) $v = \cos k(x - ct)$,
(b) $v = \sin k(x - ct)$,
(c) $v = f(x - ct)$ (whenever f possesses a second derivative f'').
This describes a waveform of *general shape* and travelling at speed c in the direction of *increasing* x.
(d) $v = g(x + ct)$ – another general waveform, but travelling at *speed* c in the direction of *decreasing* x. For general continuous functions f and g, each of the expressions (c) and (d) describes a travelling wave.

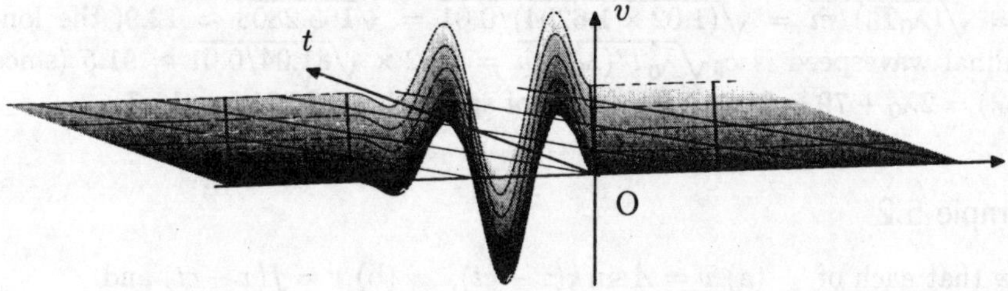

Figure 5.5 A typical waveform travelling at speed c.

ii) The (positive) parameter c in (5.9) is known as the *wavespeed*; since, when t is increased by t_0, the *only* effect on the graph of v versus x in each of (a), (b) and (c) is a translation to the right through a distance ct_0 – in (d) there is an equal translation to the left.
For transverse motions the wavespeed is $c = c_\perp = \sqrt{T_0/m}$, so that $c_\perp^2 = T_0/m =$ tension/(mass per unit of prestretched length).
For longitudinal motions, $c^2 = c_\parallel^2 = (\lambda_0 T'(\lambda_0))/m$, where the mass per pre-stretched length equals $m = \hat{m}/\lambda_0$.

iii) The *wave equation* (5.9) possesses *separable solutions* (cf. Section 4.2)

$$v(x,t) = Y(x)\, Z(t)$$

i.e. a *product* of functions of the individual variables. Insertion into (5.9) gives

$$Y(x)\,\ddot{Z}(t) = c^2\, Y''(x)\, Z(t)\,,$$

where dots and primes denote differentiation. This may be rearranged, as in (4.5), so as to *separate* terms having x-dependence from those having t-dependence, thus:

$$\frac{\ddot{Z}(t)}{Z(t)} = c^2 \frac{Y''(x)}{Y(x)} = \text{constant}\,. \tag{5.12}$$

(The left-hand side cannot depend on x, while the right-hand side cannot depend on t, so each must be constant.)

Example 5.1

Suppose that, in certain units, a string has constitutive law $T = (80+\lambda)(\lambda-1)$ (for $\lambda \geq 1$) and has $\hat{m} = 0.01$. Find the *wavespeeds* for transverse motions and for longitudinal motions, when $\lambda_0 = 1.02$. Show that their ratio is approximately $1:7$.

Since $T_0 = T(\lambda_0) = 81.02 \times 0.02 = 1.6204$, the transverse wavespeed is $c_\perp = \sqrt{(\lambda_0 T_0)/\hat{m}} = \sqrt{(1.02 \times 1.6204)/0.01} = \sqrt{165.2808} \approx 12.9$; the longitudinal wavespeed is $c_\parallel \sqrt{\lambda_0^2 T'(\lambda_0)/\hat{m}} = 1.02 \times \sqrt{81.04/0.01} \approx 91.5$ (since $T'(\lambda_0) = 2\lambda_0 + 79 = 81.04$). The ratio of speeds is $12.9 : 91.5 \approx 1:7$.

Example 5.2

Show that each of (a) $v = A\sin k(x-ct)$, (b) $v = f(x-ct)$ and (c) $v = A\exp\{-B(x+ct)^2\}$ satisfies the wave equation (5.9).

(a) Since $\partial^2 v/\partial x^2 = -k^2 A\sin k(x-ct)$ and $\partial^2 v/\partial t^2 = -(-kc)^2 A\sin k(x-ct)$, then $v_{tt} = c^2 v_{xx}$.

(b) Here, $v_{xx} = \partial(f'(x-ct))/\partial x = f''(x-ct)$ and $v_{tt} = \partial(-cf'(x-ct))/\partial t = (-c)^2 f''(x-ct) = c^2 v_{xx}$, so that v satisfies (5.9).

(c) $v_x = A(-2B(x+ct))\exp\{-B(x+ct)^2\}$,
$v_t = A(-2Bc(x+ct))\exp\{-B(x+ct)^2\}$; so that
$v_{xx} = -2AB\exp\{-B(x+ct)^2\} + A(-2B(x+ct))^2\exp\{-B(x+ct)^2\}$. Then,
$v_{tt} = -2ABc^2\exp\{-B(x+ct)^2\} + A(-2Bc(x+ct))^2\exp\{-B(x+ct)^2\} = c^2 v_{xx}$.

Example 5.3

Show that each of (a) $\sin 10x \cos 10ct$, (b) $\cos 2x \cos 2ct$ and (c) $\sin 2x \cos 2ct$ satisfies (5.9).

(a) $v_{xx} = -100\sin 10x \cos 10ct$ and $v_{tt} = -(10c)^2 \sin 10x \sin 10ct = c^2 v_{xx}$.
(b) $v_{xx} = -4\cos 2x \cos 2ct$, $v_{tt} = \cos 2x(-4c^2)\cos 2ct = c^2 v_{xx}$.
(c) $v_{tt} = \sin 2x(-4c^2)\cos 2ct = c^2(-4\sin 2x)\cos 2ct = c^2 v_{xx}$.

5.3.1 Standing Waves

Many separable solutions to (5.9) may be found satisfying also the *boundary conditions* corresponding to zero displacement at either end of the string, namely $v(0,t) = 0$ and $v(a,t) = 0$. Substituting $v = Y(x)Z(t)$ gives the requirement $Y(0)Z(t) = 0 = Y(a)Z(t)$ for all t, which reduces to $Y(0) = 0 = Y(a)$.

Since (5.12) implies that Y''/Y is constant, $Y(x)$ must satisfy the ODE

$$Y''(x) = \alpha Y(x) .$$

There are three cases, depending on whether the constant α is positive, zero or negative. Solutions in the case $\alpha > 0$ are inappropriate since they may be written as $Y = A\cosh\sqrt{\alpha}x + B\sinh\sqrt{\alpha}x$, so that the condition $Y(0) = 0$ enforces the choice $A = 0$, while $Y(a) = 0$ then requires that $B = 0$; i.e. $Y(x) \equiv 0$. Likewise, the case $\alpha = 0$ gives only the *trivial solution* ($Y(x) \equiv 0$), since the only solution $Y = A+Bx$ which equals zero at both $x = 0$ and $x = a$ is $Y \equiv 0$. However, the case $\alpha < 0$ yields *sinusoidal* behaviour for $Y(x)$. Solutions satisfying the condition $Y(0) = 0$ must have the form $Y(x) = B\sin\sqrt{-\alpha}x$, which also allows $Y(a) = 0$ whenever $\sin\sqrt{-\alpha}a = 0$. Thus, if $\sqrt{-\alpha}$ is an integer multiple of π/a (i.e. $\alpha = -(n\pi/a)^2$), there exist permissible solutions

$$Y(x) = \sin\frac{n\pi x}{a} \equiv Y_n(x)$$

satisfying $Y''(x) = -(n\pi/a)^2 Y(x)$, $Y(0) = Y(a) = 0$ (cf. solutions (4.6) to Laplace's equation). The parameter B has here been *chosen* as 1, since $Z(t)$ has yet to be analysed.

Inserting the choice $\alpha = -(n\pi/a)^2$ into Equation (5.12) gives

$$\frac{\ddot{Z}(t)}{Z(t)} = c^2\frac{Y''(x)}{Y(x)} = c^2\alpha = -\left(\frac{n\pi c}{a}\right)^2 .$$

For *each* integer n, the *general solution* for $Z(t)$ may be written in two alternative forms

$$\begin{aligned} Z &= A_n \cos\frac{n\pi c}{a}t + B_n \sin\frac{n\pi c}{a}t \\ &= a_n \cos(\omega_n t - \phi_n) , \end{aligned}$$

where $\omega_n \equiv n\pi c/a$ is a (radian) frequency, while the *arbitrary constants* a_n and ϕ_n are related to A_n and B_n through $A_n = a_n\cos\phi_n$ and $B_n = a_n\sin\phi_n$.

As special cases, we see that *each* of

$$v(x,t) = \sin\frac{n\pi x}{a}\cos\frac{n\pi ct}{a} \qquad \text{and} \qquad v(x,t) = \sin\frac{n\pi x}{a}\sin\frac{n\pi ct}{a}$$

describes a *standing wave* solution of (5.9), i.e. a *mode of vibration* of the string with fixed ends and which has (radian) frequency $n\pi c/a$.

The general *standing wave solution* to (5.9) satisfying also the *boundary conditions* $v(0,t) = 0 = v(a,t)$ is

$$v = a_n \cos(\omega_n t - \phi_n)\sin\frac{n\pi x}{a} , \qquad (5.13)$$

where $\omega_n \equiv n\pi c/a$ is the *radian frequency*, a_n is the *amplitude* and ϕ_n is a *phase* (ϕ_n/ω_n determines a time delay relative to the oscillatory function $\cos \omega_n t$). At *each* $x \in [0, a]$, the displacement v oscillates sinusoidally with *period*

$$\frac{2a}{nc} \quad \left(= \frac{2\pi}{\omega_n}\right)$$

between the extreme values $\pm a_n \sin(n\pi x/a)$. Figure 5.6 shows these displacements $v = \pm a_n \sin(n\pi x/a)$ which have maximum modulus and shows that the standing wave has *nodes* at $x = ra/n$, for $r = 1, 2, \ldots, n-1$, i.e. (internal) points where the displacement v is zero at all times. At each instant, the solution (5.13) describes a displacement (transverse or longitudinal, as appropriate) proportional to those shown in Figure 5.6. The motions at all points of the string are said to be *in phase.* For transverse vibrations, a fast snapshot photograph, or stroboscopic image, of a violin string resonating at a pure tone would illustrate these shapes.

Observe that, since c is the *wavespeed*, the period $2c^{-1} \times a/n$ is the time required for travel at speed c from one node to a neighbouring node and then back. Hence, the period is inversely proportional to n.

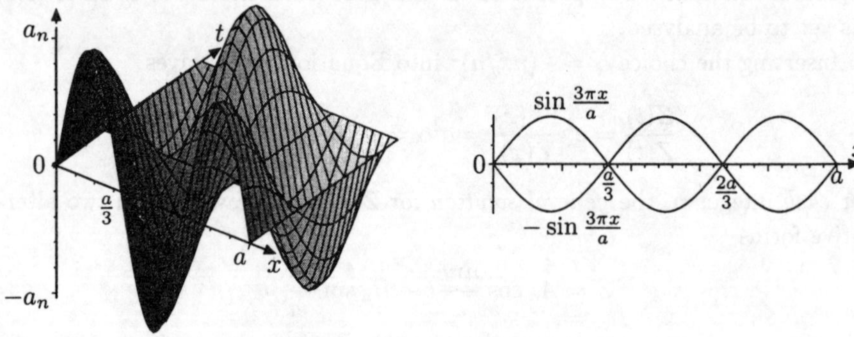

Figure 5.6 The standing wave (5.13) for $n = 3$, showing the nodes at $x = a/3$ and $x = 2a/3$.

5.3.2 Superposition of Standing Waves

Any $v(x, t)$ formed as a linear combination of standing waves (5.13) still satisfies both the wave equation (5.9) and the conditions at $x = 0, a$. Specifically, it is readily checked that for *all* choices of A_n and B_n the *superposition*

$$v(x,t) = \sum_{n=1}^{\infty} (A_n \cos \omega_n t + B_n \sin \omega_n t) \sin \frac{n\pi x}{a} \tag{5.14}$$

satisfies both $v_{tt} = c^2 v_{xx}$ and $v(0,t) = 0 = v(a,t)$, when $\omega_n = n\pi c/a$ (provided, of course, that the relevant infinite series converge).

If $v(x,t)$ describes transverse displacement, the corresponding (transverse) velocity of a point of the string is

$$\frac{\partial v}{\partial t} \equiv v_t(x,t) = \sum_{n=1}^{\infty} (\omega_n B_n \cos \omega_n t - \omega_n A_n \sin \omega_n t) \sin \frac{n\pi x}{a} \,. \tag{5.15}$$

In the representation (5.14) for $v(x,t)$, the *coefficients* A_n are related to the *initial displacement* (set $t = 0$ in (5.14)) by

$$v(x,0) = \sum_{n=1}^{\infty} A_n \sin \frac{n\pi x}{a} \,,$$

while the coefficients B_n are related to the *initial velocity* (set $t = 0$ in (5.15)) by

$$v_t(x,0) = \sum_{n=1}^{\infty} \omega_n B_n \sin \frac{n\pi x}{a} \,.$$

Each of these is a Fourier sine series. As in Section 4.2, the coefficients in a *Fourier sine series*

$$f(x) = \sum_{n=1}^{\infty} b_n \sin \frac{n\pi x}{a} \,, \qquad x \in (0,a)$$

are evaluated using the *half-range formula*

$$b_n = \frac{2}{a} \int_0^a f(x) \sin \frac{n\pi x}{a} \,.$$

Hence, in (5.14), the formulae

$$A_n = \frac{2}{a} \int_0^a v(x,0) \sin \frac{n\pi x}{a} \,\mathrm{d}x \,, \quad B_n = \frac{2}{n\pi c} \int_0^a v_t(x,0) \sin \frac{n\pi x}{a} \,\mathrm{d}x \tag{5.16}$$

relate the coefficients A_n to the *initial displacement* distribution and relate the coefficients B_n to the *initial velocity* distribution (i.e. x-dependence at $t = 0$).

For *all* functions $v(x,0)$ and $v_t(x,0)$ (and consequent coefficients (5.16)) each term in (5.14) has period $2\pi/\omega_1 = 2a/c$ (since $\omega_n 2\pi/\omega_1 = 2n\pi$). Hence, whatever the initial conditions, transverse displacements of a string repeat periodically with this period.

Example 5.4

If a string is stretched from $x = 0$ to $x = a$ at tension T_0 with mass/unit length m and its transverse displacements $v(x,t)$ satisfy $v_{tt} = T_0 m^{-1} v_{xx}$, with $v(0,t) = 0 = v(a,t)$, show that solutions of the form $v(x,t) = Z(t) \sin n\pi x/a$ may exist for each $n = 0, 1, 2, \ldots$. By writing $T_0/m = c^2$, determine possible forms for $Z(t)$.

Since $v(0,t) = Z(t) \sin 0 = 0$ and $v(a,t) = Z(t) \sin n\pi = 0$ the displacements at $x = 0, a$ both vanish, for each integer n. Also, $v_{tt} = \ddot{Z}(t) \sin n\pi x/a$ and $v_{xx} = -(n\pi/a)^2 Z(t) \sin n\pi x/a$, so that substituting into the wave equation gives $\ddot{Z}(t) + (n\pi/a)^2 Z(t) = 0$. This has general solution $Z(t) = A_n \cos(n\pi ct/a) + B_n \sin(n\pi ct/a)$.

EXERCISES

5.1. Using separable solutions $v(x,t) = \sin(n\pi x/a) Z_n(t)$ of the equation $v_{tt} = c^2 v_{xx}$ and boundary conditions $v(0,t) = 0 = v(a,t)$ governing transverse displacements $v(x,t)$ of a stretched string, determine $Z_n(t)$ and so show that motions starting from rest ($v_t(x,0) \equiv 0$ in $0 \le x \le a$) may be represented as

$$v = \sum_{n=1}^{\infty} A_n \sin \frac{n\pi x}{a} \cos \frac{n\pi ct}{a}.$$

If, at $t = 0$, the string is released from rest with shape $v(x,0) = kx(a-x)$ in $0 \le x \le a$, determine the coefficients A_n.

Confirm that the motion is periodic. What is the fundamental period?

5.2. If a string has uniform tension T_0 but non-uniform mass distribution $m(x)$ in its prestretched state, the governing equation for small transverse motions is

$$\frac{\partial^2 v}{\partial t^2} = \frac{T_0}{m(x)} \frac{\partial^2 v}{\partial x^2}.$$

Confirm that the string allows standing wave vibrations of the form

$$v = V(x) \cos(\omega t - \alpha)$$

in which the *phase* α is arbitrary, while $V(x)$ satisfies an ordinary differential equation. Derive this ODE. Determine the boundary conditions for V at $x = 0$ and $x = a$, if $v(0,t) = 0 = v(a,t)$ for all t.

5.3. Show that if the transverse displacement $v(x,t)$ of a string satisfies $v_{tt} = c^2 v_{xx}$, $v(0,t) = 0 = v(a,t)$ and is initially zero ($v(x,0) = 0$), then it has the representation

$$v = \sum_{n=1}^{\infty} B_n \sin\frac{n\pi ct}{a} \sin\frac{n\pi x}{a}.$$

If, at $t = 0$, the central portion $\frac{1}{3}a < x < \frac{2}{3}a$ is impulsively set into motion so that the *initial velocity* is

$$v_t(x,0) = \begin{cases} V & \frac{1}{3}a < x < \frac{2}{3}a, \\ 0 & 0 < x < \frac{1}{3}a \quad \text{and} \quad \frac{2}{3}a < x < a, \end{cases}$$

determine the coefficients B_n by using an appropriate half-range formula.

5.4. Show that the full equation (5.6) of string motions possesses solutions $\boldsymbol{r} = x(X,t)\boldsymbol{i}$ in which displacements are purely longitudinal. By writing $\partial x/\partial X = \lambda$ and $\partial x/\partial t = U$ (the longitudinal component of velocity), show that (5.6) is equivalent to

$$\hat{m}\frac{\partial U}{\partial t} = \frac{\partial T}{\partial X}, \qquad \frac{\partial \lambda}{\partial t} = \frac{\partial U}{\partial X} \quad \text{with} \quad T = T(\lambda).$$

5.5. For planar solutions satisfying (5.6), write $\boldsymbol{t} = \boldsymbol{i}\cos\psi + \boldsymbol{j}\sin\psi$ and $\boldsymbol{n} = -\boldsymbol{i}\sin\psi + \boldsymbol{j}\cos\psi$ (so ψ is the inclination of the string to the Ox axis and $\boldsymbol{n}$ is a unit normal to the string). Deduce that $\partial\boldsymbol{t}/\partial t = \boldsymbol{n}\,\partial\psi/\partial t$ and $\partial\boldsymbol{n}/\partial t = -\boldsymbol{t}\,\partial\psi/\partial t$.

After writing $\partial\boldsymbol{r}/\partial t = \boldsymbol{v} = U\boldsymbol{t} + V\boldsymbol{n}$ (so that U and V are velocity components parallel and normal to the deformed string), show that the acceleration $\partial\boldsymbol{v}/\partial t$ may be expressed as $(\partial U/\partial t - V\partial\psi/\partial t)\boldsymbol{t} + (\partial V/\partial t + U\partial\psi/\partial t)\boldsymbol{n}$. Likewise, show that $\partial(T\boldsymbol{t})/\partial X = \boldsymbol{t}\partial T/\partial X + \boldsymbol{n}T\partial\psi/\partial X$. By inserting these into Equation (5.6), show that

$$\hat{m}\left(\frac{\partial U}{\partial t} - V\frac{\partial \psi}{\partial t}\right) = T'(\lambda)\frac{\partial \lambda}{\partial X}, \quad \hat{m}\left(\frac{\partial V}{\partial t} + U\frac{\partial \psi}{\partial t}\right) = T(\lambda)\frac{\partial \psi}{\partial X}.$$

Find the two additional equations relating U, V, λ and ψ which follow from the identity $\partial\boldsymbol{v}/\partial X = \partial^2\boldsymbol{r}/\partial X\partial t = \partial(\lambda\boldsymbol{t})/\partial t$.

5.4 D'Alembert's Solution, Travelling Waves and Wave Reflections

Use of the trigonometric identity

$$\sin\theta\,\cos\phi = \tfrac{1}{2}\left[\sin(\theta-\phi)+\sin(\theta+\phi)\right]$$

shows that a standing wave such as (5.13) may be regarded as the sum of two *travelling waves.* Indeed, this particular solution

$$\begin{aligned} v(x,t) &= a_n \sin\frac{n\pi x}{a}\cos\left(\frac{n\pi ct}{a}-\phi_n\right) \\ &= \tfrac{1}{2}a_n\sin\left[\frac{n\pi}{a}(x-ct)+\phi_n\right] + \tfrac{1}{2}a_n\sin\left[\frac{n\pi}{a}(x+ct)-\phi_n\right] \end{aligned}$$

is seen to be the superposition of two *travelling waves*, each of which is sinusoidal. One has the form $f(x-ct)$ describing a waveform travelling *to the right* at speed c, while the second has the form $g(x+ct)$ of a waveform travelling *to the left* at speed c.

It turns out (see Exercise 5.6) that the more general solution (5.14) also is the superposition of a *right-travelling* wave $f(x-ct)$ and a *left-travelling* wave $g(x+ct)$. Earlier it was shown that $v=f(x-ct)$ and $v=g(x+ct)$ each individually satisfies the wave equation (5.9). By superposition, it follows that so also does their sum

$$v(x,t) = f(x-ct)+g(x+ct) \tag{5.17}$$

(provided only that second derivatives exist). Equation (5.17) is *D'Alembert's representation* for solutions to (5.9). Moreover, it can be shown (see Exercise 5.7) that *every* solution to (5.9) may be written in the form (5.17). Thus, every solution may be regarded as the *sum* of a *waveform* $f(x-ct)$ which is travelling to the right and a *waveform* $g(x+ct)$ which is travelling to the left, each moving with speed c.

The waveforms *need not be sinusoidal.* For all *initial shapes* described by $f(x)$ and $g(x)$, the waveforms do not alter as the waves travel (except when reflected at the ends $x=0,a$).

5.4.1 Wave Reflections

If D'Alembert's solution (5.17) is used (instead of (5.14)) to determine $v(x,t)$ when the *initial displacement* $v(x,0)=P(x)$ (say) and *initial velocity* $v_t(x,0)=$

$Q(x)$ (say) are specified in $0 \le x \le a$, then $f(x)$ and $g(x)$ are found within $0 \le x \le a$ from

$$f(x) + g(x) = v(x,0) = P(x)\,, \qquad -c\{f'(x) - g'(x)\} = v_t(x,0) = Q(x)$$

as $f(x) = \frac{1}{2}\{P(x) - c^{-1}q(x)\}$, $g(x) = \frac{1}{2}\{P(x) + c^{-1}q(x)\}$, where $q(x)$ is an indefinite integral of $Q(x)$. Notice that the *constant of integration* included within $q(x)$ (and hence within $f(x - ct)$ and $g(x - ct)$) does not affect $v(x,t)$.

To find v for later times t, we need to extend the range of definition of $f(\xi)$ into $\xi < 0$ and that of $g(\eta)$ into $\eta > a$. Figure 5.7 indicates the lines $\xi = x - ct =$ constant and $\eta = x + ct =$ constant, within $0 \le x \le a$ for $t \ge 0$. The boundary conditions $v = 0$ at both $x = 0$ and $x = a$ provide the necessary information. Putting $x = 0$ in (5.17), gives

$$0 = v(0,t) = f(-ct) + g(ct) \qquad \text{for all } ct \ge 0\,.$$

Thus, $f(-\eta) = -g(\eta)$ for all $\eta \ge 0$, so that

$$f(x - ct) = -g(ct - x) \qquad \text{whenever } ct \ge x\,. \tag{5.18}$$

Similarly, setting $x = a$ in (5.17) gives

$$0 = v(a,t) = f(a - ct) + g(a + ct) \quad \text{for all } ct \ge 0\,,$$

so that $g(\eta) = -f(2a - \eta)$ for all values $\eta \ge a$.

Combining this with $g(\eta) = -f(-\eta)$ for $\eta \ge 0$ gives

$$f(-\eta) = f(2a - \eta) \qquad \text{for all } -\eta \le -a\,,$$

so showing that the required extension of $f(\xi)$ for negative ξ is periodic of period $2a$. Consequently, $g(\eta)$ is also periodic of period $2a$, for $\eta \ge 0$.

The function $g(\eta)$ is first defined for $a < \eta \le 2a$ (the first reflection from $x = a$) by $g(\eta) = -f(2a - \eta)$ (since $a > 2a - \eta > 0$), then once $g(\eta)$ is known over an interval $0 < \eta \le 2a$ of length equal to its period it is extended to all $\eta \ge 0$ using the periodicity condition $g(\eta) = g(\eta - 2a)$. The function $f(x - ct)$ is then found for all $x - ct \le a$ from (5.18).

Example 5.5

When a harp string of stretched length a is plucked at position $x = \frac{1}{3}a$ and then released, transverse displacements are set up with initial conditions $v(x,0) = 3bx/a$ in $0 \le x \le \frac{1}{3}a$, $v(x,0) = 3b(a - x)/2a$ in $\frac{1}{3}a \le x \le a$ and with $v_t(x,0) = 0$ for all $0 \le x \le a$. Determine the functions $f(x - ct)$ and $g(x + ct)$ in D'Alembert's representation for $0 \le x - ct \le a$, $0 \le x + ct \le a$, respectively. Determine their odd periodic extensions $\tilde{f}(x - ct)$ and $\tilde{g}(x + ct)$. Explain how

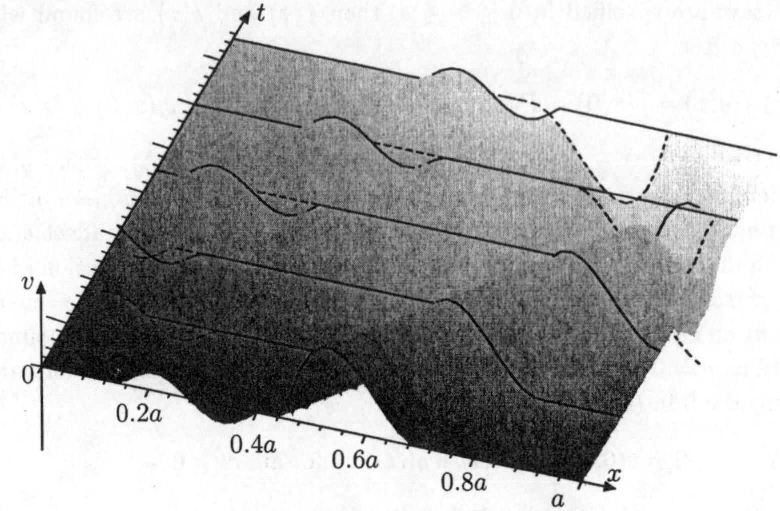

Figure 5.7 Reflection of right-travelling $f(x-ct)$ —·—·— and left-travelling $g(x+ct)$ – – – – –– waveforms, with boundary conditions $f+g=0$ at both $x=0$ and $x=a$.

these extended functions yield the D'Alembert solution, drawing attention to the loci $x-ct=(\pm\frac{1}{3}a+2na)$, for all integers n. Sketch the shape of the string at times $t=a/6c$, $a/3c$, $a/2c$, $2a/3c$ and a/c.

Write $x-ct=\xi$ and $x+ct=\eta$, so that $v=f(\xi)+g(\eta)$, $v_t=-cf'(\xi)+cg'(\eta)$. At $t=0$, it follows that $\xi=x=\eta$, so that $0=c\{g'(x)-f'(x)\}$. This allows the choice $g(x)=f(x)=\frac{1}{2}v(x,0)$ for $0\le x\le a$. Thus $f(\xi)$ and $g(\eta)$ are found in $0\le\xi\le a$ and $0\le\eta\le a$, respectively.

Let $\tilde{v}(x)$ be the odd periodic extension of $v(x,0)$; then $\tilde{f}(\xi)=\frac{1}{2}\tilde{v}(\xi)$ and $\tilde{g}(\eta)=\frac{1}{2}\tilde{v}(\eta)$ (as shown). Writing $v(x,t)=\frac{1}{2}\{\tilde{v}(x-ct)+\tilde{v}(x+ct)\}$, gives $v(x,0)=\tilde{v}(x)$ (as required), $v(0,t)=\frac{1}{2}\{\tilde{v}(-ct)+\tilde{v}(ct)\}=0$ (since $\tilde{v}(x)$ is odd) and $v(a,t)=\frac{1}{2}\{\tilde{v}(a-ct)+\tilde{v}(a+ct)\}=\frac{1}{2}\{\tilde{v}(-a-ct)+\tilde{v}(a+ct)\}=0$ (since $\tilde{v}$ has period $2a$ and is odd). Since, for $-a\le x\le a$, $\tilde{v}(x)$ has constant derivative ($3b/a$ or $-3b/2a$) except at $x=\pm\frac{1}{3}a$, the function $\tilde{v}(\xi)$ has constant derivative except at locations $\xi=\pm\frac{1}{3}a+2na$. Similarly, the lines $\eta=\pm\frac{1}{3}a+2na$ are the loci across which the derivative of $\tilde{v}(\eta)$ has a jump. The string shapes at the requested times are shown in Figure 5.8, together with a view of successive string shapes during the first period $0\le t\le 2a/c$ of vibration.

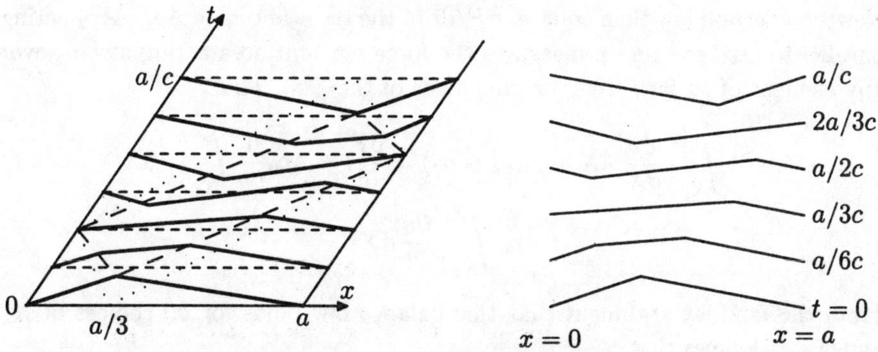

Figure 5.8 Vibrations of a harp string plucked at $x = \frac{1}{3}a$.

5.5 Other One-dimensional Waves

The wave equation (5.9) describes many phenomena other than displacements and vibrations of a string. Important examples are vibrations of gas or liquid in tubes (musical instruments, oil, water and pneumatic supply lines, etc.) and electric currents and voltages in telegraphy and for high-voltage power transmission. To analyse such phenomena, separable solutions, standing waves and D'Alembert's solution all prove useful.

5.5.1 Acoustic Vibrations in a Tube

Suppose that a tube of cross-sectional area A is filled with gas which, when at rest at pressure p_0, has uniform density ρ_0. If the gas undergoes small longitudinal motions, gas which has reference position X moves to $x = x(X,t)$. It undergoes a *displacement* $u \equiv x(X,t) - X$ (see Figure 5.9) and its density changes to $\rho = \rho(X,t)$. However, the gas originally within the interval $(X, X+\delta X)$ has mass $A\rho_0\delta X = A\rho\delta x$ and momentum $vA\rho\delta x$. Note that $\partial x/\partial X = \rho_0/\rho$ (cf. $\lambda_0 = \hat{m}/m$ in Section 5.2.1).

At any position $x = x(X,t)$, the gas has axial velocity $v = \partial x/\partial t$, so that

$$\frac{\partial v}{\partial X} = \frac{\partial^2 x}{\partial X \partial t} = \frac{\partial}{\partial t}\left(\frac{\rho_0}{\rho}\right) = \frac{-\rho_0}{\rho^2}\frac{\partial \rho}{\partial t}\,. \tag{5.19}$$

The axial momentum within the interval $x_1 \equiv x(X_1,t) < x < x(X_2,t) \equiv x_2$ is

$$P \equiv \int_{x_1}^{x_2} v\,A\rho\,\mathrm{d}x = \int_{X_1}^{X_2} A\,v\rho_0\,\mathrm{d}X = A\rho_0\int_{X_1}^{X_2} v\,\mathrm{d}X\,.$$

Newton's second law then equates $\mathrm{d}P/\mathrm{d}t$ to the resultant force $Ap_1 - Ap_2$ acting parallel to Ox (pressure p measures the force per unit area acting inward over any element of surface (real, or imagined) of the gas). Thus,

$$-A\int_{X_1}^{X_2}\frac{\partial p}{\partial X}\,\mathrm{d}X = A(p_1 - p_2) = \frac{\mathrm{d}P}{\mathrm{d}t} = A\rho_0\frac{\mathrm{d}}{\mathrm{d}t}\int_{X_1}^{X_2} v\,\mathrm{d}X$$
$$= A\rho_0\int_{X_1}^{X_2}\frac{\partial v}{\partial t}\,\mathrm{d}X\,.$$

From the familiar argument that this balance law holds for *all* choices of X_1 and X_2, it follows that

$$\rho_0\frac{\partial v}{\partial t} = -\frac{\partial p}{\partial X}\,. \tag{5.20}$$

The system (5.19), (5.20) is completed by invoking a constitutive relation between pressure and density, typically taken as $p = p(\rho)$.[1]

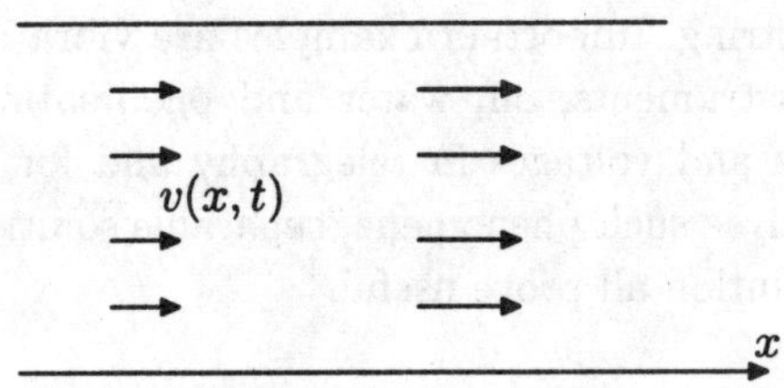

Figure 5.9 Longitudinal gas oscillations within a tube.

In acoustics, the fractional change in pressure is very small (the human ear experiences pain if $\Delta p/p_0 > 0.2\%$). Writing

$$\rho = \rho_0 + \tilde{\rho}\quad,\qquad p = p_0 + \tilde{p} \approx p(\rho_0) + p'(\rho_0)\tilde{\rho}\quad,\qquad p_0 \equiv p(\rho_0)$$

gives $\partial x/\partial X \approx 1$, so that, by retaining only terms linear in $\tilde{\rho}$ and v and by using x and t as independent variables, Equations (5.19) and (5.20) are linearized as

$$\rho_0\frac{\partial v}{\partial x} = -\frac{\partial \tilde{\rho}}{\partial t}\,,\qquad \rho_0\frac{\partial v}{\partial t} = -p'(\rho_0)\frac{\partial \tilde{\rho}}{\partial x}\,. \tag{5.21}$$

[1] Historically, there was some confusion in choosing this. Newton used Boyle's law $p = R\rho T$ between pressure p, density ρ and temperature T, with R the *universal gas constant*, but he took T to be constant. However, acoustic vibrations are rapid. Compressing a gas causes warming – try inflating a cycle tyre. It is necessary to consider *adiabatic* changes in ρ, i.e. changes in which heat cannot escape. This relates T to ρ and so yields a different relation $p = p(\rho)$. For so-called *perfect* gases this is taken as $p \propto \rho^\gamma$, where $\gamma \approx 1.4$, for familiar gas mixtures, such as air.

Eliminating v from these gives

$$p'(\rho_0)\frac{\partial^2\tilde{\rho}}{\partial x^2} = -\rho_0\frac{\partial^2 v}{\partial x\partial t} = \frac{\partial^2\tilde{\rho}}{\partial t^2}\,, \tag{5.22}$$

while eliminating $\tilde{\rho}$ gives

$$\frac{\partial^2 v}{\partial t^2} = -\frac{p'(\rho_0)}{\rho_0}\frac{\partial^2\tilde{\rho}}{\partial t\partial x} = p'(\rho_0)\frac{\partial^2 v}{\partial x^2}\,. \tag{5.23}$$

Hence, density increment $\tilde{\rho}(x,t)$, pressure increment $\tilde{p}(x,t)$ and associated gas velocity $v(x,t)$ *each* satisfy the *wave equation* (5.9). The relevant wavespeed is the *acoustic wavespeed* $c = \sqrt{p'(\rho_0)}$, calculated using the derivative of the constitutive law $p = p(\rho)$ and evaluated at the reference density ρ_0 (cf. the *extensional modulus* $T'(\lambda_0)$ arising in the speed of longitudinal waves in strings).

Example 5.6

In a long tube extending over $0 \le x \le L$ and having reference density ρ_0, the velocity v and density increment $\tilde{\rho}$ satisfy

$$\rho_0\frac{\partial v}{\partial x} = -\frac{\partial\tilde{\rho}}{\partial t}\,,\quad \rho_0\frac{\partial v}{\partial t} = -c^2\frac{\partial\tilde{\rho}}{\partial x}\,,\quad \tilde{p} = c^2\tilde{\rho}\,,$$

where $c = \sqrt{p'(\rho_0)}$ is the acoustic speed. Show that $v_{tt} = c^2 v_{xx}$. Find separable solutions $v = X(x)T(t)$ which satisfy boundary conditions corresponding to a closed end at $x = 0$ and an open end at $x = L$, namely $v(0,t) = 0$ and $\tilde{p}(L,t) = 0$.

Since $\rho_0 v_{tt} = -c^2\tilde{\rho}_{tx} = -c^2\tilde{\rho}_{xt} = \rho_0 c^2 v_{xx}$, then $v_{tt} = c^2 v_{xx}$. Separable solutions $v = X(x)T(t)$ must satisfy $T''(t)/T = c^2 X''(x)/X =$ constant. The boundary condition $v(0,t) = 0$ gives $X(0) = 0$, while the condition $\tilde{p}(L,t) = 0$ gives $v_x(L,t) = -\rho_0^{-1}\tilde{\rho}_t(L,t) = -\tilde{p}_t(L,t)/(c^2\rho_0) = 0$, which requires that $X'(L) = 0$. Solving $X''(x) = q^2X(x)$ and $X(0) = 0$ gives $X(x) = A\sinh qx$ which is compatible with $X'(L) = 0$ only if $A = 0$, so that $X \equiv 0$. This gives only the *trivial solution* $v(x,t) = 0$. The case $X''(x) = 0$ gives, with $X(0) = 0$, only the case $X = Ax$. Again, the only choice satisfying the condition $X'(L) = 0$ is $A = 0$, which yields just the *trivial solution* $v = 0$. The third possibility $X''(x) + q^2X = 0$ gives, with $X(0) = 0$, the form $X = a\sin qx$. The condition $X'(L) = 0$ then gives $qA\cos qL = 0$. For $q \ne 0$, $A \ne 0$, this allows $qL = \frac{1}{2}\pi + n\pi = (n + \frac{1}{2})\pi$. The corresponding separable solutions are

$$v_n(x,t) = \sin\left\{(n+\tfrac{1}{2})\tfrac{\pi x}{L}\right\}\left[A_n\cos\left\{(n+\tfrac{1}{2})\tfrac{\pi ct}{L}\right\} + B_n\sin\left\{(n+\tfrac{1}{2})\tfrac{\pi ct}{L}\right\}\right]\,,\qquad n = 0,1,\ldots,$$

where A_n, B_n are arbitrary constants.

The solution to Example 5.6 shows that the *natural frequencies* of oscillation of a column of air in a tube with an open end (modelled as a location where pressure is constant, i.e. $\tilde{p} = 0$) are *odd* multiples of $\pi c/(2L)$; hence they are in the ratio $1 : 3 : 5 : \ldots$ (rather than the ratio $1 : 2 : 3 : \ldots$ for vibrations of a string with fixed ends or, indeed, for a tube which is closed at both ends). The general solution, obtained by superposition, is

$$v(x,t) = \sum_{n=1}^{\infty} \sin\left\{(n+\tfrac{1}{2})\frac{\pi x}{L}\right\} \left[A_n \cos\left\{(n+\tfrac{1}{2})\frac{\pi ct}{L}\right\} + B_n \sin\left\{(n+\tfrac{1}{2})\frac{\pi ct}{L}\right\}\right] \qquad (5.24)$$

in which the coefficients A_n and B_n are related to the initial velocity and initial pressure gradient through expressions

$$v(x,0) = \sum_{n=1}^{\infty} A_n \sin\left\{(n+\tfrac{1}{2})\tfrac{\pi x}{L}\right\}$$

$$\rho_0^{-1}\tilde{p}_x(x,0) = v_t(x,0) = \sum_{n=1}^{\infty} \left\{(n+\tfrac{1}{2})\tfrac{\pi c}{L}\right\} B_n \sin\left\{(n+\tfrac{1}{2})\tfrac{\pi x}{L}\right\}.$$

While each right-hand side is a Fourier sine series in x, each has period $4L$ even though the tube extends only over $0 \le x \le L$. However, the right-hand side of (5.24) is a periodic function in x which is symmetric with respect to $x = L$ (since it involves only $\sin m\pi x/2L$ for m *odd*). Hence, by defining initial conditions throughout $0 \le x \le 2L$ to be compatible with this reflectional symmetry (i.e. in $L \le x \le 2L$, take $v(x,t) = v(2L - x, t)$) and using the standard formula gives (for example)

$$A_n = \frac{1}{2L}\int_0^{2L} v(x,0)\sin\left\{(n+\tfrac{1}{2})\tfrac{\pi x}{L}\right\}\,\mathrm{d}x = \frac{1}{L}\int_0^{L} v(x,0)\sin\left\{(n+\tfrac{1}{2})\tfrac{\pi x}{L}\right\}\,\mathrm{d}x\,,$$

with a similar expression for the coefficients $(n+\frac{1}{2})\pi c B_n/L$. (Moreover, the imagined distribution of velocity in $L < x \le 2L$ ensures that no sine terms involving even multiples of $\pi x/(2L)$ are introduced.)

5.5.2 Telegraphy and High-voltage Transmission

Copper wires and cables are widely used to carry telephone messages and A.C. electric currents. Electric *current* is a flux of *electric charge* (see also §3.4). If the current $I(x,t)$ in a wire is unsteady, two effects are important; there may be a time-varying *line density* $Q(x,t)$ of electric charge (i.e. charge per unit length) and the magnetic field which always surrounds any flowing current (i.e. a flux of charge) will itself become unsteady.

Since (cf. Example 3.9) a long thin charged cylinder establishes a surrounding electric field, for which the potential (or *voltage*) $V(x,t)$ is related to the *line density* of charge by $Q = CV$, where the constant C is the *electrical capacitance* per unit length, the law of charge balance is

$$\frac{\partial I}{\partial x} = -\frac{\partial Q}{\partial t} = -C\frac{\partial V}{\partial t} \tag{5.25}$$

(when all leakage of charge to surroundings is neglected). The time-varying magnetic field (with strength $\propto I$) exerts on the charge within the conducting wire an electric field which is proportional to $\partial I/\partial t$ (this is a consequence of the famed *Faraday law of induction* (1831)). When the wire is treated as a perfect conductor (i.e. *electrical resistance* is neglected, so that no potential gradient is required in order to maintain a steady current), the resulting equation is

$$\frac{\partial V}{\partial x} = -L\frac{\partial I}{\partial t}\,. \tag{5.26}$$

Here, the constant L is the *self-inductance* of the wire per unit length.

Eliminating V between Equations (5.25) and (5.26) shows that

$$\frac{\partial^2 I}{\partial x^2} = \frac{\partial}{\partial x}\left(-C\frac{\partial V}{\partial t}\right) = -C\frac{\partial}{\partial t}\left(-L\frac{\partial I}{\partial t}\right) = LC\frac{\partial^2 I}{\partial t^2}\,.$$

Comparison with (5.9) then leads to the conclusion that the current in the wire (or transmission line) satisfies the *wave equation* (5.9), with wavespeed equal to $(LC)^{-1/2}$. Likewise, the voltage $V(x,t)$ satisfies the same wave equation, so showing that both telegraphic messages and electric power may be transmitted as given by (5.17). In particular, it is the design engineer's rôle to ensure that a transmitted signal $f(x - t/\sqrt{LC})$ does not cause unwanted reflections $g(x + t/\sqrt{LC})$ from a receiver located e.g. at $x = a$ (see Exercise 5.13).

EXERCISES

5.6. By using the trigonometric identities

$$\begin{aligned}\sin\theta\,\cos\phi &= \tfrac{1}{2}\left[\sin(\theta-\phi)+\sin(\theta+\phi)\right]\,,\\ \sin\theta\,\sin\phi &= \tfrac{1}{2}\left[\cos(\theta-\phi)-\cos(\theta+\phi)\right]\,,\end{aligned}$$

show that expression (5.14) has the form $f(x - ct) + g(x + ct)$.

5.7. For the *wave equation* (5.9) $v_{tt} = c^2 v_{xx}$ the coordinates $\xi = x - ct$ and $\eta = x + ct$ are especially significant (they are known as *characteristic coordinates*). Sketch on the x, t plane the lines $\xi =$ constant and $\eta =$ constant. Express v_t, v_x, v_{tt} and v_{xx} in terms of derivatives with respect to ξ and η. Hence show that the wave equation is equivalent to $\partial^2 v/\partial\xi\partial\eta = 0$. From this, deduce that the *most general* expression for v_η may be written as

$$v_\eta = g'(\eta) \qquad \text{for some function } g(\eta)\,.$$

Hence, demonstrate that *each* solution of the wave equation (5.9) must be expressible in D'Alembert's representation (5.17).

5.8. Show that an alternative version of D'Alembert's representation is

$$v = F(t - x/c) + G(t + x/c)\,.$$

In this representation, obtain an expression for the velocity $v_t(x,t)$. Show also that the boundary conditions $v(0,t) = 0$ and $v(a,t) = 0$ imply that $-F(t) = G(t) = G(t + 2a/c)$ for all $t \geq 0$.

5.9. If the harp string of Example 5.5 is plucked at its midpoint, so that $v(x,0) = 2bx/a$ over $0 \leq x \leq a/2$ and $v(x,0) = 2b(a-x)/a$ over $a/2 \leq x \leq a$, with $v_t(x,0) = 0$ throughout $0 \leq x \leq a$, find the appropriate functions $f(\xi)$ and $g(\eta)$ in D'Alembert's representation $v(x,t) = f(x - ct) + g(x + ct)$. Plot graphs of v versus x at times $t = 0$, $a/4c$, $a/2c$, $3a/4c$, a/c, $3a/2c$ and $2a/c$.

5.10. If transverse displacements $v(x,t)$ of a string satisfy $v_{tt} = c^2 v_{xx}$ in $0 \leq x \leq a$, with $v(0,t) = 0 = v(a,t)$ and also with $v(x,0) = P(x)$, $v_t(x,0) = Q(x)$ in $0 \leq x \leq a$, relate $f(x)$ and $g(x)$ to $P(x)$ and $Q(x)$ for $0 \leq x \leq a$.

Using the boundary conditions at $x = 0, a$, determine $f(\xi)$ over the interval $-2a \leq \xi \leq a$ and $g(\eta)$ in $0 \leq \eta \leq 3a$. Hence, confirm that $v(x,t)$ is periodic in t for $t \geq 0$.

Find $f(x)$ and $g(x)$ in $0 < x < a$ when the string is set impulsively into motion (with $P(x) \equiv 0$, $Q(x) \equiv V =$ constant for $0 < x < a$). Sketch the resulting displacements $v(x, a/(2c))$, $v(x, 3a/(4c))$ and $v(x, a/c)$.

5.11. Gas vibrations within a tube extending over $0 \leq x \leq a$ are governed by $\rho_0 v_x = -\tilde{\rho}_t$, $\rho_0 v_t = -c^2 \tilde{\rho}_x$ in $0 < x < a$. If the end $x = a$ is *open* (so that $\tilde{\rho}(a,t) = 0$) but motion at the end $x = 0$ is specified as $v(0,t) = A\sin\omega_0 t$, find the corresponding *forced oscillation* described by $v = X(x)\sin\omega_0 t$. Explain the phenomenon exhibited by this solution when $2a\omega_0/(\pi c)$ is close to an odd integer.

5.12. If the gas within the tube of Exercise 5.11 is initially at rest with $\tilde{\rho}(x,0) = 0$, write $v(x,t) = X(x)\sin\omega_0 t + \tilde{v}(x,t)$. Show that $\tilde{v}_{tt} = c^2\tilde{v}_{xx}$, $\tilde{v}(0,t) = 0$ and $\tilde{v}_x(a,t) = 0$. Find the appropriate initial conditions for $\tilde{v}$ and $\tilde{v}_t$ over $0 < x < a$ at $t = 0$.

5.13. The condition at one end ($x = a$) of an electrical transmission line is often stated in the form $V(a,t) = ZI(a,t)$, where the constant Z is known as the *impedance*. Show, using (5.25) and (5.26), that this yields the condition $ZV_x(a,t) = -LV_t(a,t)$.

Suppose that a waveform $V = f(x - t/\sqrt{LC})$ is *incident* upon the end $x = a$ from the region $x < a$. Find the corresponding *reflected wave* $g(x + t/\sqrt{LC})$. What is special about the choice $Z = \sqrt{L/C}$?

6 Fluid Flow

6.1 Kinematics and Streamlines

In a flowing fluid each cubic centimetre typically contains $O(10^{23})$ molecules. Large-scale features of its flow are well described by the study of a *velocity field* $\boldsymbol{u}(\boldsymbol{x}, t)$ which is a smooth (differentiable) function of position $\boldsymbol{x}$ and time t. Although individual molecules have wildly erratic motions and undergo frequent collisions, the velocity $\boldsymbol{u}$ describes the *average motion* of the fluid.

In any flow, all the curves obtained as solutions to the system of ODEs (in vector form)

$$\frac{\mathrm{d}\boldsymbol{x}}{\mathrm{d}t} = \boldsymbol{u}(\boldsymbol{x}, t) \tag{6.1}$$

are known as *particle paths*. They are the curves described by points which move so as to follow the velocity field $\boldsymbol{u}(\boldsymbol{x}, t)$ as time t increases. For example, suppose that P_0 (i.e. $\boldsymbol{x} = \boldsymbol{x}_0$) denotes the position occupied at time t_0, then the three-dimensional curve obtained by integrating (6.1) with initial condition $\boldsymbol{x}(t_0) = \boldsymbol{x}_0$ gives the position $\boldsymbol{x}(t)$ (denoted by P) at later times t. Consequently, in a typical flow, there is a 3-parameter family of particle paths – one corresponding to positions emanating from each position at the reference time t_0. The particle paths may be imagined as the trajectories of small tracer particles (dust specks or plankton) carried along by the surrounding fluid.

Streamlines form another important set of curves associated with the velocity field $\boldsymbol{u}(\boldsymbol{x}, t)$. However, they correspond to the field $\boldsymbol{u}$ at a chosen instant t.

Imagine a photograph of a flowing fluid taken at time t with exposure insufficiently fast to prevent slight blurring of moving specks. At each point, a flow direction is then apparent. The curves which are everywhere tangential to the flow direction $\boldsymbol{u}(\boldsymbol{x}, t)$ are defined by

$$\frac{\mathrm{d}\boldsymbol{x}}{\mathrm{d}\alpha} = \boldsymbol{u}(\boldsymbol{x}, t),$$

in which α is a parameter which varies along each curve, while t is a parameter held fixed during the integration (it is, after all, the label of the instant at which the photograph is taken).

Of course, in an *unsteady flow* (one in which $\partial \boldsymbol{u}/\partial t \neq 0$), the blurred photographs taken at different instants will reveal that flow direction depends on time. Consequently the streamlines vary with time. They also differ from the particle paths. In the important special case of *steady flows* (i.e. $\boldsymbol{u} = \boldsymbol{u}(\boldsymbol{x})$, so that $\partial \boldsymbol{u}/\partial t = 0$), particle paths coincide with streamlines.

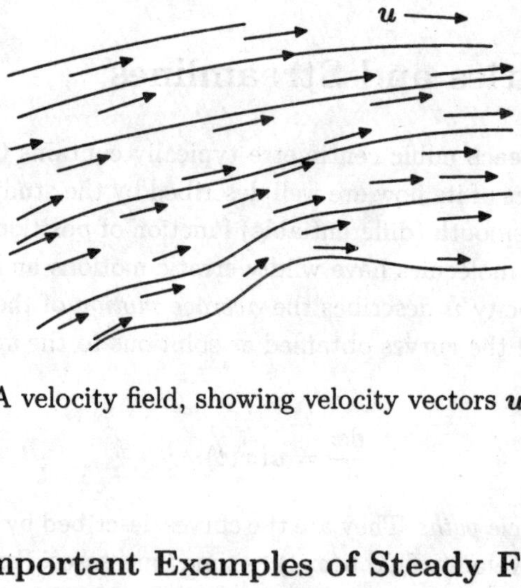

Figure 6.1 A velocity field, showing velocity vectors $\boldsymbol{u}$ and streamlines.

6.1.1 Some Important Examples of Steady Flow

Whenever the velocity field is independent of time (so that $\partial \boldsymbol{u}/\partial t = 0$), the velocity may be written as $\boldsymbol{u} = \boldsymbol{u}(\boldsymbol{x})$ and the flow is a *steady flow.* In steady flows, the particle paths obtained by integrating equations (6.1) with $\boldsymbol{x}(t_j) = \boldsymbol{x}_0$ at various times $t = t_j$, $j = 1, 2, \ldots$ are identical. They also are *streamlines.*

(Notice that the same family of curves is defined also as solutions to $\mathrm{d}\boldsymbol{x}/\mathrm{d}t = f(\boldsymbol{x}, t)\boldsymbol{u}(\boldsymbol{x})$, where $f(\boldsymbol{x}, t)$ is *any* function which is everywhere positive. It is the *direction* of $\boldsymbol{u}$ at each point which defines the tangent to the streamline – the magnitude merely affects the time required to traverse a chosen portion of the curve.)

Example 6.1

Find the streamlines for each of the steady flows

(a) $\boldsymbol{u} = U\boldsymbol{i}$, (b) $\boldsymbol{u} = \dfrac{y}{a}U\boldsymbol{i}$ and (c) $\boldsymbol{u} = U\left(\dfrac{a^2 - r^2}{a^2}\right)\boldsymbol{k}$, with $r^2 = x^2 + y^2$

and with $r \leq a$. Describe the flow in each case.

(a) The streamlines satisfy $\mathrm{d}x/\mathrm{d}\alpha = U$, $\mathrm{d}y/\mathrm{d}\alpha = 0$ and $\mathrm{d}z/\mathrm{d}\alpha = 0$. Hence, $x = x_0 + U\alpha$, $y = y_0$, $z = z_0$. Streamlines are the straight lines parallel to the Ox axis.

Since the velocity is independent of position, the flow is said to be *uniform*. It has speed U.

(b) The streamlines are $x = x_0 + (Uy/a)\alpha$, $y = y_0$, $z = z_0$. Again, each line parallel to Ox is a streamline. While the flow direction is everywhere parallel to Ox, the flow speed is $(U/a)y$ and so is proportional to distance from the plane $y = 0$.

This flow is an example of a *Couette flow* (or *simple shear flow*). Layers of fluid at different 'heights' y are sliding over each other. Fluid in the plane $y = a$ has speed U, while fluid at $y = 0$ is at rest.

(c) Here $x = x_0$, $y = y_0$ and $z = z_0 + Ua^{-2}(a^2 - x_0^2 - y_0^2)\alpha$, so that streamlines are straight lines parallel to the Oz axis.

The velocity vector is everywhere parallel to $\boldsymbol{k}$; the speed takes its maximum value U on the Oz axis and decreases quadratically with radial distance r from it. The speed vanishes at radius $r = a$.

The flow in (c) arises when viscous fluid is driven through a pipe or tube of radius a by a pressure which varies linearly with z. It is known as *Poiseuille flow* (after the 19th-century French physiologist famous for investigating blood flow). The *velocity profiles* associated with these unidirectional flows are illustrated in Figure 6.2.

6.2 Volume Flux and Mass Flux

The velocity field $\boldsymbol{u}(\boldsymbol{x}, t)$ gives the direction and rate at which the fluid flows. It also allows calculation of the rate at which fluid leaves any specified region.

Consider a surface element having area δS and unit normal $\boldsymbol{n}$. If $\boldsymbol{x}$ denotes an interior point, the velocity there is $\boldsymbol{u}(\boldsymbol{x}, t)$. Hence, during the time interval $(t, t + \delta t)$ fluid particles initially located in the surface element each move through a *displacement* $\approx \boldsymbol{u}\,\delta t$. They sweep out a volume (see Figure 6.3) given approximately by $(\boldsymbol{u}\,\delta t) \cdot \boldsymbol{n}\,\delta S$. Thus, the rate at which fluid volume crosses the surface element is approximately $\boldsymbol{u} \cdot \boldsymbol{n}\,\delta S$. Only, the *normal component*

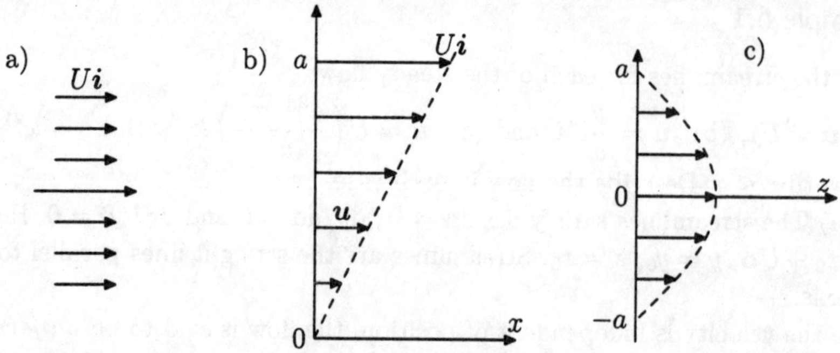

Figure 6.2 Velocity profiles as in Example 6.1 for a) uniform flow, b) Couette (shear) flow and c) Poiseuille flow, shown in the plane $x = 0$.

$\boldsymbol{u} \cdot \boldsymbol{n} \equiv u_\perp$ of the velocity $\boldsymbol{u}$ is responsible for the *flux* of fluid volume across any unit area of surface located at $\boldsymbol{x}$ and having unit normal $\boldsymbol{n}$ (the tangential components do not contribute). Notice the similarity with the heat flux vector $\boldsymbol{q}$ of Chapter 2. There, $\boldsymbol{q} \cdot \boldsymbol{n}\, \delta S$ measures the rate at which heat flows across the surface element; here, $\boldsymbol{u} \cdot \boldsymbol{n}\, \delta S$ measures the rate at which fluid volume crosses the same surface element. Thus, the fluid velocity $\boldsymbol{u}$ may be regarded also as the *volume flux vector.*

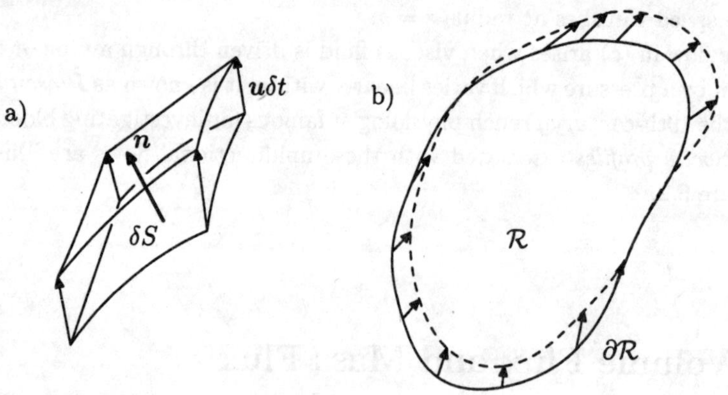

Figure 6.3 a) Volume element swept out during time interval δt. b) Mass flux through a portion of the bounding surface $\partial\mathcal{R}$ of $\mathcal{R}$.

Similarly, since $\rho\boldsymbol{u} \cdot \boldsymbol{n}\, \delta S$ measures the *rate of mass flow* across the *surface element* $\boldsymbol{n}\, \delta S$ (i.e. an element of surface having area δS and unit normal $\boldsymbol{n}$). The vector field $\rho\boldsymbol{u}$ is known as the *mass flux* vector; its importance is that, for any chosen surface $\mathcal{S}$, the rate at which mass crosses $\mathcal{S}$ is

$$\iint_{\mathcal{S}} \rho \boldsymbol{u} \cdot \boldsymbol{n} \, \mathrm{d}S$$

(where $\rho \boldsymbol{u} \cdot \boldsymbol{n} > 0$ when fluid flows *into* the region towards which $\boldsymbol{n}$ points).

6.2.1 Incompressible Fluids

In many situations (e.g. for water, oil, etc.), the *density* ρ of a fluid is virtually unaffected by any change in pressure within the flow. These may be treated as flows of an *incompressible fluid* – i.e. one in which density at a fluid element remains constant as the point P flows along a particle path. Moreover, if the density ρ has the same value at all positions in a flow region, then the flow is said to have *uniform density*, with $\rho =$ constant.

Example 6.2

Suppose that a fluid flows incompressibly and has velocity $\boldsymbol{u} = Q\boldsymbol{x}/(4\pi r^3)$, where $r \equiv |\boldsymbol{x}|$. Show that the rate of mass outflow through *each* sphere centred at O is the same.

Choose $\mathcal{S}$ as the sphere $|\boldsymbol{x}| = r_0$ centred at O. It has outward unit normal $\boldsymbol{n} = \boldsymbol{x}/r_0$. On $\mathcal{S}$, the normal component of velocity is

$$\boldsymbol{u} \cdot \boldsymbol{n} = \frac{Q}{4\pi r_0^3}\boldsymbol{x} \cdot \frac{\boldsymbol{x}}{r_0} = \frac{Q\, r_0^2}{4\pi r_0^4} = \frac{Q}{4\pi r_0^2} \, .$$

Since this has the same value $Q/(\text{area of } \mathcal{S})$ everywhere on $\mathcal{S}$, the integration reduces to multiplication, so giving

$$\text{Outward rate of mass flow across } \mathcal{S} = \rho Q \, .$$

The flow in Example 6.2 is three-dimensional radial flow. Q is the *volume flow rate* and ρQ the *mass flow rate* due to a *source* imagined to be at $\boldsymbol{x} = \boldsymbol{0}$. Moreover, ρQ is the rate of mass outflow through a fixed sphere at *any* radius r_0.

Note the similarity with the flux of the electric field $\boldsymbol{E} = q\boldsymbol{x}/(4\pi\epsilon_0 r^3)$ due to a point charge in electrostatics (Gauss's law) and also the r^{-2} decay in Newtonian gravitation. In incompressible fluid flow, an unsteady version of Example 6.2 describes motion in the vicinity of a pulsating bubble (see Exercise 6.5).

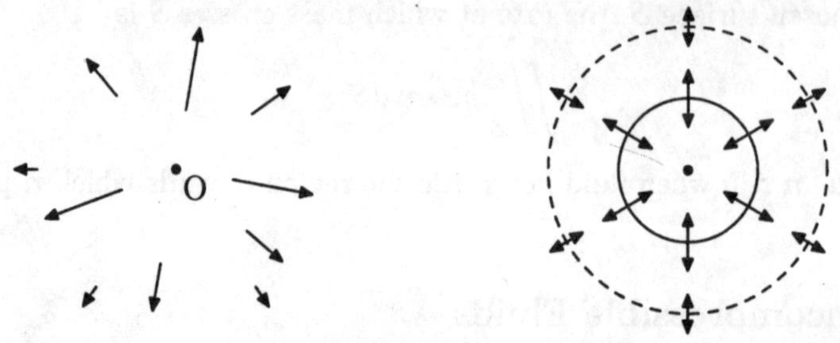

Figure 6.4 a) Radially symmetric flow. b) Flow due to a pulsating bubble.

6.2.2 Mass Conservation

Let $\mathcal{R}$ be any fixed region of space occupied by fluid, with bounding surface $\partial\mathcal{R}$ and outward unit normal $\boldsymbol{n}$. Then, the rate of mass outflow from $\mathcal{R}$ is

$$\iint_{\partial\mathcal{R}} \rho\boldsymbol{u}\cdot\boldsymbol{n}\,\mathrm{d}S\,.$$

Since mass can be neither created nor destroyed, this rate of mass outflow must vanish, so giving the *law of mass conservation*

$$\iint_{\partial\mathcal{R}} \rho\boldsymbol{u}\cdot\boldsymbol{n}\,\mathrm{d}S = 0 \tag{6.2}$$

for any region $\mathcal{R}$ of fluid flow. For incompressible fluids, in which $\rho =$ constant, this simplifies to

$$\iint_{\partial\mathcal{R}} \boldsymbol{u}\cdot\boldsymbol{n}\,\mathrm{d}S = 0\,, \tag{6.3}$$

often known as the *law of volume conservation.*

Since Equation (6.3) holds for all regions $\mathcal{R}$ within an incompressible fluid, the reader should not be surprised that there is an equivalent point form expressed, like (1.8), in terms of partial derivatives. Those who know the divergence theorem of vector calculus (see Section 6.6 and Matthews, 1998, Ch. 5) will realize that (6.3) may be rewritten as

$$\iiint_{\mathcal{R}} \operatorname{div}\boldsymbol{u}\,\mathrm{d}V = 0\,. \tag{6.4}$$

Then, since the boundary of $\mathcal{R}$ may be varied at will, the integrand must vanish identically, so yielding

$$\operatorname{div}\boldsymbol{u} = \boldsymbol{\nabla}\cdot\boldsymbol{u} = \frac{\partial u}{\partial x} + \frac{\partial v}{\partial y} + \frac{\partial w}{\partial z} = 0\,, \tag{6.5}$$

where u, v and w denote the components of the velocity $\boldsymbol{u}$, which therefore has the form $\boldsymbol{u} = u\boldsymbol{i}+v\boldsymbol{j}+w\boldsymbol{k}$. Equation (6.5) is often called the *continuity equation* (the reason for this name is largely historical; the name *volume conservation equation* better reflects its rôle as the point form of (6.3)).

6.3 Two-dimensional Flows of Incompressible Fluids

The two-dimensional version of (6.5) may be derived directly from Green's theorem in the plane (i.e. without knowledge of the divergence theorem):

$$\oint_C P(x,y)\mathrm{d}x + Q(x,y)\mathrm{d}y = \iint_{\mathcal{D}} \left(\frac{\partial Q}{\partial x} - \frac{\partial P}{\partial y}\right) \mathrm{d}x\,\mathrm{d}y\,,$$

which applies for any functions P, Q which are differentiable over the connected region $\mathcal{D}$ of the Oxy plane, having boundary $\mathcal{C} = \partial\mathcal{D}$.

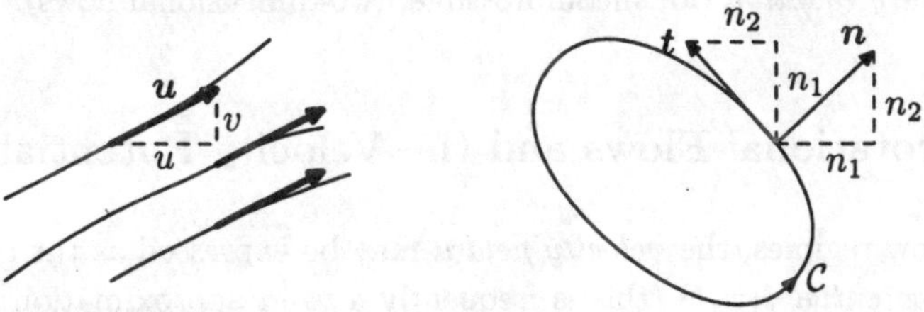

Figure 6.5 a) Two-dimensional velocity field. b) Unit tangent $\boldsymbol{t}$ and unit normal $\boldsymbol{n}$ to a closed loop $\mathcal{C}$.

6.3.1 The Continuity Equation

In a two-dimensional flow, the velocity may be written as $\boldsymbol{u} = u\boldsymbol{i} + v\boldsymbol{j}$ (i.e. Oz is aligned perpendicularly to the flow field). In the Oxy plane, let domain $\mathcal{D}$ be bounded by the closed curve $\mathcal{C} = \partial\mathcal{D}$, with outward unit normal $\boldsymbol{n} = n_1\boldsymbol{i}+n_2\boldsymbol{j}$ and *unit tangent* (taken *anti-clockwise* around $\mathcal{C}$)

$$\boldsymbol{t} = n_1\boldsymbol{j} - n_2\boldsymbol{i}\,.$$

Then, if s measures distance around $\mathcal{C}$ from some chosen reference point, the definition

$$\boldsymbol{t} = \frac{\mathrm{d}\boldsymbol{x}}{\mathrm{d}s}$$

implies that $d\boldsymbol{x} \equiv \boldsymbol{i}dx + \boldsymbol{j}dy = (-n_2\boldsymbol{i} + n_1\boldsymbol{j})ds$, so that

$$dx = -n_2 ds\,, \qquad dy = n_1 ds\,.$$

Hence, evaluating the rate of volume flow Q_C out of a region $\mathcal{R}$ having unit length along Oz and having cross-section $\mathcal{D}$ gives

$$\begin{aligned} Q_C \equiv \oint_C \boldsymbol{u}\cdot\boldsymbol{n}\,ds &= \oint_C (u\boldsymbol{i} + v\boldsymbol{j})\cdot(n_1\boldsymbol{i} + n_2\boldsymbol{j})ds \\ &= \oint_C (u\,n_1 ds + v\,n_2 ds) = \oint_C u dy - v dx \\ &= \iint_{\mathcal{D}} \left(\frac{\partial u}{\partial x} + \frac{\partial v}{\partial y}\right) dx\,dy\,. \end{aligned} \tag{6.6}$$

Then, since the cross-section $\mathcal{D}$ is arbitrary, the integrand must vanish everywhere within the flow, so yielding the two-dimensional form of (6.5)

$$\frac{\partial u}{\partial x} + \frac{\partial v}{\partial y} = 0\,, \tag{6.7}$$

the *continuity equation* (for incompressible, two-dimensional flows).

6.3.2 Irrotational Flows and the Velocity Potential

In many flow regimes, the *velocity field* $\boldsymbol{u}$ may be expressed as the gradient of a *velocity potential* $\phi(\boldsymbol{x}, t)$ (this is frequently a valid approximation in regions where *viscosity* is unimportant, which is widely true except in the vicinity of solid boundaries). In three dimensions this gives

$$\boldsymbol{u} = \text{grad}\,\phi = \boldsymbol{\nabla}\phi,$$

while, in two dimensions, the velocity components u, v may be written as

$$u = \frac{\partial \phi}{\partial x}\,, \qquad v = \frac{\partial \phi}{\partial y}\,. \tag{6.8}$$

If expressions (6.8) are substituted into (6.7), it is found that the *velocity potential* ϕ satisfies *Laplace's equation*

$$\nabla^2\phi \equiv \frac{\partial^2\phi}{\partial x^2} + \frac{\partial^2\phi}{\partial y^2} = 0\,. \tag{6.9}$$

Also, by eliminating ϕ from the two equations in (6.8) by cross-differentiation $(\partial\phi_y/\partial x - \partial\phi_x/\partial y = 0)$, it follows that the equations governing the velocity components for two-dimensional flow are[1]

$$\frac{\partial v}{\partial x} - \frac{\partial u}{\partial y} = 0 \qquad \text{(irrotationality)} \qquad (6.10)$$

$$\frac{\partial u}{\partial x} + \frac{\partial v}{\partial y} = 0 \qquad \text{(incompressibility).} \qquad (6.11)$$

Notice that (6.10) and (6.11) imply that both u and v satisfy the two-dimensional Laplace equation, just as the *velocity potential* ϕ does. Indeed, differentiating (6.11) with respect to x and (6.10) with respect to y, then equating mixed derivatives of v yields

$$\frac{\partial^2 u}{\partial x^2} = -\frac{\partial}{\partial x}\left(\frac{\partial v}{\partial y}\right) = -\frac{\partial}{\partial y}\left(\frac{\partial v}{\partial x}\right) = -\frac{\partial^2 u}{\partial y^2}.$$

Similarly, by eliminating the mixed derivative of u after taking the y derivative of (6.10) and the x derivative of (6.11) gives $v_{yy} = -v_{xx}$, so that

$$\nabla^2 u \equiv \frac{\partial^2 u}{\partial x^2} + \frac{\partial^2 u}{\partial y^2} = 0\,, \qquad \nabla^2 v \equiv \frac{\partial^2 v}{\partial x^2} + \frac{\partial^2 v}{\partial y^2} = 0\,.$$

Example 6.3

Show that the velocity potential $\phi(x,y) = \dfrac{q}{4\pi}\log\{(x-x_1)^2+(y-y_1)^2\}$ describes flow radially away from the line at $x = x_1, y = y_1$. Find the volume outflow rate per unit length of the line.

Since $r_1 = \sqrt{(x-x_1)^2 + (y-y_1)^2}$ is the distance from the line $x = x_1$, $y = y_1$, the velocity potential is

$$\phi = \frac{q}{2\pi}\log r_1\,.$$

It is readily checked using $\partial\phi/\partial x = (q/2\pi)(x-x_1)/r_1^2$, etc. that ϕ satisfies Laplace's equation (6.9). Also, ϕ depends only on the perpendicular distance r_1 from the axis $x = x_1$, $y = y_1$. The corresponding velocity components are

$$u = \frac{q}{2\pi}\frac{x-x_1}{r_1^2}\,, \qquad v = \frac{q}{2\pi}\frac{y-y_1}{r_1^2}\,.$$

If $\boldsymbol{x}$ and $\boldsymbol{x}_1$ are used to denote two-dimensional vectors $\boldsymbol{x} = x\boldsymbol{i} + y\boldsymbol{j}$ and $\boldsymbol{x}_1 = x_1\boldsymbol{i} + y_1\boldsymbol{j}$, the velocity is

$$\boldsymbol{u} = \frac{q}{2\pi}\frac{\boldsymbol{x}-\boldsymbol{x}_1}{|\boldsymbol{x}-\boldsymbol{x}_1|^2}\,.$$

[1] Those readers familiar with Stokes's theorem in *vector calculus* will know that if $\oint_C \boldsymbol{u}\cdot \mathrm{ds} = 0$ for *all* closed curves C in some region $\mathcal{R}$, then $\operatorname{curl}\boldsymbol{u} = \boldsymbol{0}$ within $\mathcal{R}$, so that a velocity potential ϕ exists, with $\boldsymbol{u} = \operatorname{grad}\phi$. Observe that, by convention, there is no negative sign in the definition of velocity potential.

To calculate the volume outflow rate, consider unit length of a cylinder of radius r_1 centred on $x = x_1$, $y = y_1$. At all points of the curved surface, the velocity is normal to the surface and the speed is $q/(2\pi r_1)$. Hence the outward rate of volume flow is q. (There is no flow through the flat ends of the cylinder.)

The flow in Example 6.3 is said to be due to a *line source* of strength q. A practical example would be the flow away from a porous pipe of outer radius a with axis along the line $x = x_1$, $y = y_1$ and with q being the rate of volume outflow per unit length (the rate of flow through unit length of any cylinder centred on this axis and having radius greater than a).

Example 6.4

Find the velocity field due to two parallel *line sources* of equal strength q located at $x = 0$, $y = \pm a$. Show that this field may be used to describe the flow due to a single line source parallel to a rigid impermeable wall lying in the plane $y = 0$.

The velocity potential is

$$\phi(x, y) = \frac{q}{4\pi}\{\log[x^2 + (y-a)^2] + \log[x^2 + (y+a)^2]\}$$

(this is a *superposition* of two potentials from Example 6.3).
The corresponding velocity field is

$$\boldsymbol{u} = \frac{q}{2\pi}\left\{\frac{x\boldsymbol{i} + (y-a)\boldsymbol{j}}{x^2 + (y-a)^2} + \frac{x\boldsymbol{i} + (y+a)\boldsymbol{j}}{x^2 + (y+a)^2}\right\}.$$

The configuration is symmetric about $y = 0$ and has $v = 0$ everywhere on $y = 0$.
Thus, if a rigid, impermeable wall is imagined to be at $y = 0$, the flow velocity has zero component normal to it. The flow in the region $y > 0$ is thus consistent with that due to a line source at $x = 0$, $y = a$ and parallel to a rigid, impermeable wall.

(The line source at $(x, y) = (0, -a)$ is outside the relevant flow region. It is located at the mirror image of the physical line source. The *method of images* is a widely used technique of applied mathematics for constructing solutions in certain regions by imagining that there exist certain sources at *image points* outside the physically relevant region.)

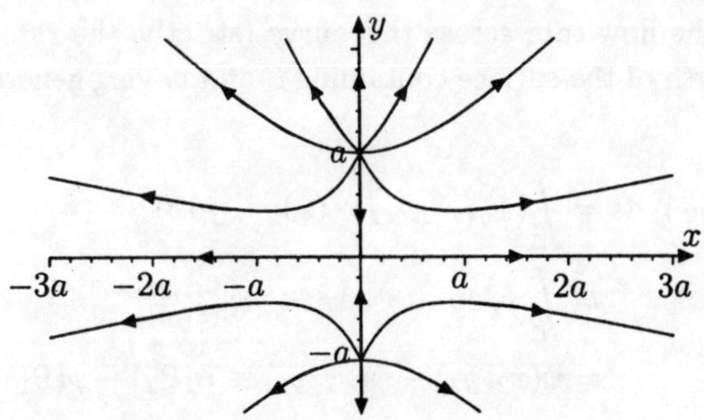

Figure 6.6 Flow due to parallel line sources at $x = 0, y = \pm a$.

6.3.3 The Stream Function

Equation (6.7) is automatically satisfied if u and v are given by

$$u = \frac{\partial \psi}{\partial y}, \qquad v = -\frac{\partial \psi}{\partial x}, \tag{6.12}$$

for some function $\psi(x, y)$. The function $\psi(x, y)$ is known as the *stream function*, since ψ is constant along each streamline of the flow with velocity $u\boldsymbol{i} + v\boldsymbol{j}$. To verify this, use the chain rule to evaluate the total derivative along a streamline defined by $\mathrm{d}y/\mathrm{d}x = v/u$.

This shows that

$$\frac{\mathrm{d}\psi}{\mathrm{d}x} = \frac{\partial \psi}{\partial x} + \frac{\partial \psi}{\partial y}\frac{\mathrm{d}y}{\mathrm{d}x} = -v + u\frac{\mathrm{d}y}{\mathrm{d}x} = 0,$$

so that ψ is *constant along each streamline.*

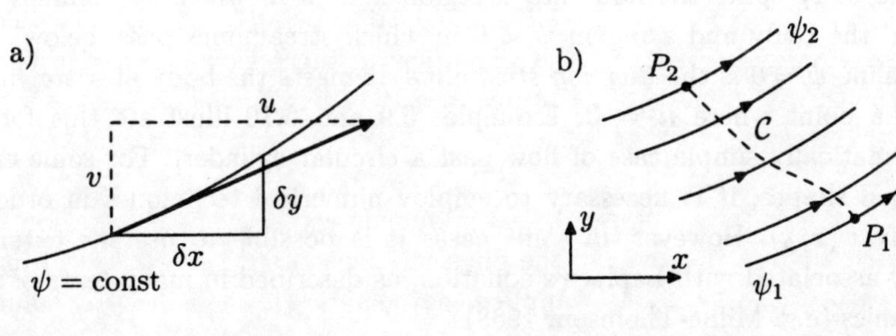

Figure 6.7 a) Streamline and velocity vector. b) Stream function and volume flow rate across $\mathcal{C}$.

Moreover, the value of $\psi(x, y)$ is related to volume flow rate: Let $\mathcal{C}$ denote a curve joining two points P_1 and P_2 in the Oxy plane with coordinates (x_1, y_1)

and (x_2, y_2). The flow rate across this curve (strictly, the rate of volume flow across unit length of the surface containing C and having generators parallel to Oz) is

$$\begin{aligned}\int_C (u\boldsymbol{i} + v\boldsymbol{j}) \cdot \boldsymbol{n}\,\mathrm{d}s &= \int_C (\psi_y \boldsymbol{i} - \psi_x \boldsymbol{j}) \cdot (\boldsymbol{i}\mathrm{d}y - \boldsymbol{j}\mathrm{d}x) \\ &= \int_C \psi_y \mathrm{d}y + \psi_x \mathrm{d}x = [\psi]_{(x_1,y_1)}^{(x_2,y_2)} \\ &= \psi(x_2, y_2) - \psi(x_1, y_1) \equiv \psi(\mathrm{P}_2) - \psi(\mathrm{P}_1) = \psi_2 - \psi_1 .\end{aligned}$$

Here, if C has unit tangent $\boldsymbol{t}$ and unit normal $\boldsymbol{n}$, then, since $\boldsymbol{t}\,\mathrm{d}s = \boldsymbol{i}\mathrm{d}x + \boldsymbol{j}\mathrm{d}y$, it follows that $\boldsymbol{n}\,\mathrm{d}s = \boldsymbol{i}\mathrm{d}y - \boldsymbol{j}\mathrm{d}x$. Hence, the difference between the values of $\psi(x, y)$ at any two points equals the flow rate across *any* curve joining them (observe that this difference is zero whenever P_1 and P_2 lie on the same streamline).

The representation (6.12), when combined with (6.10), implies that the stream function ψ satisfies Laplace's equation

$$\frac{\partial^2 \psi}{\partial x^2} + \frac{\partial^2 \psi}{\partial y^2} = 0 . \tag{6.13}$$

A typical problem concerning two-dimensional fluid flow is to find the flow past a cylindrical surface represented by a closed curve C and having a specified behaviour at large distances from the body. This is equivalent to identifying a suitable solution $\psi(x, y)$ of (6.13). For example, flow past a streamlined body, or aerofoil shape such as illustrated in Figure 6.8, will have stream function ψ which is constant on C. Moreover, ψ must satisfy Laplace's equation outside the body and, if the flow at large distances from the body is $\boldsymbol{u} \to U\boldsymbol{i}$, then $\psi \to Uy+$ constant as $x^2 + y^2 \to \infty$.

The body splits the flow into a region $\psi > 0$ in which streamlines pass 'above' the body and a region $\psi < 0$ in which streamlines pass 'below'. The streamline $\psi = 0$ is the *dividing streamline.* It meets the body at a *stagnation point*, a point where $\boldsymbol{u} = \boldsymbol{0}$. Examples 6.9 and 6.10 illustrate this for the mathematically simple case of flow past a circular cylinder. For some cross-sectional shapes, it is necessary to employ numerical techniques in order to identify $\psi(x, y)$. However, in some cases it is possible to use the extensive theory associated with Laplace's equation, as described in many texts of fluid mechanics (e.g. Milne-Thomson, 1968).[2]

[2] *Complex variable theory* proves very useful. Since the *velocity potential* $\phi(x, y)$ and *stream function* $\psi(x, y)$ are related through

$$\frac{\partial \phi}{\partial x} = \frac{\partial \psi}{\partial y}, \qquad \frac{\partial \phi}{\partial y} = -\frac{\partial \psi}{\partial x},$$

which have the same form as the *Cauchy – Riemann equations* (Osborne, 1999), the

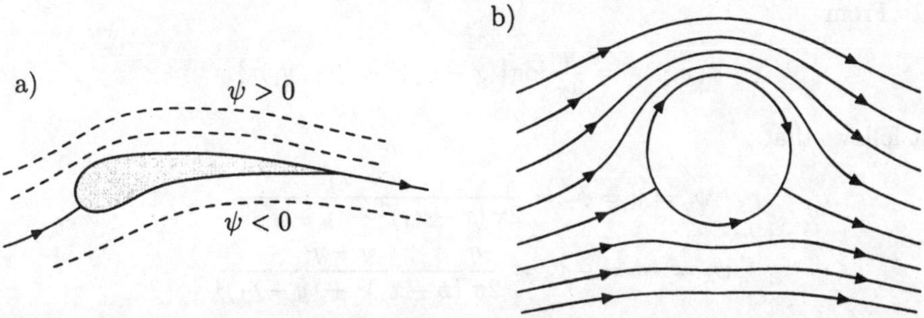

Figure 6.8 Stagnation points and dividing streamlines for flow a) past an aerofoil section, b) past a circular cylinder.

Example 6.5

Show that $\psi = kxy$ satisfies Laplace's equation. Find the corresponding velocity field. Show that in the region $y > 0$ it describes a flow impinging on a wall at $y = 0$. Identify any stagnation points and dividing streamlines. Determine the expression for the velocity potential $\phi(x, y)$.

Since $\partial^2\psi/\partial x^2 = 0$ and $\partial^2\psi/\partial y^2 = 0$, then $\nabla^2\psi = 0$.
The velocity $\boldsymbol{u}$ has components $u = kx$, $v = -ky$. Also, both $y = 0$ and $x = 0$ correspond to $\psi = 0$. Thus, $\psi(x, y)$ is the stream function for a flow which gives no flux through any portion of $y = 0$, with $v < 0$ in $y > 0$.
Any stagnation point must be identified by the property $\boldsymbol{u} = \boldsymbol{0}$, which arises only at $(x, y) = (0, 0)$. Hence, the origin is the only stagnation point.
In $y > 0$, the dividing streamline is defined by the axis $x = 0$.
The corresponding velocity potential follows immediately from $\phi_x = \psi_y = kx$ and $\phi_y = -\psi_x = -ky$, which yield $\phi = \frac{1}{2}k(x^2 - y^2) + \phi_0$.

The flow in Example 6.5 is an example of a *stagnation point flow.* Taylor expansion of $\psi(x, y)$ near any stagnation point (at which $\psi_y = u = 0$, $\psi_x = -v = 0$) will have dominant terms giving u and v linear in the coordinates, as here.

Example 6.6

Find the stream function $\psi(x, y)$ corresponding to the velocity potential and velocity components of Example 6.3.

---

function $\phi(x, y) + \imath\psi(x, y)$ is an *analytic function* of the complex variable $z \equiv x + \imath y$. The function $f(z) \equiv \phi + \imath\psi$ is known as the *complex potential.* Its real part $\phi(x, y)$ and imaginary part $\psi(x, y)$ are *conjugate harmonic functions* and its derivative $f'(z) = \phi_x + \imath\psi_x = \psi_y - \imath\phi_y$ is the *complex velocity* $u - \imath v$.

From

$$\phi(x,y) = \frac{q}{4\pi}\log\{(x-x_1)^2+(y-y_1)^2\}$$

it follows that

$$\psi_y = u = \phi_x = \frac{q}{2\pi}\frac{x-x_1}{(x-x_1)^2+(y-y_1)^2}$$

$$\psi_x = -v = -\phi_y = \frac{-q}{2\pi}\frac{y-y_1}{(x-x_1)^2+(y-y_1)^2}\,.$$

Integrating the first (holding x constant) gives

$$\psi = \frac{q}{2\pi}\tan^{-1}\left(\frac{y-y_1}{x-x_1}\right)+\Psi(x)\,,$$

which agrees with the second only if $\Psi'(x)=0$. Hence, the stream function must have the form

$$\psi = \frac{q}{2\pi}\tan^{-1}\left(\frac{y-y_1}{x-x_1}\right)+\text{constant}\,.$$

The streamlines (as to be expected for radial two-dimensional flow) are the family of straight half-lines through $(x,y)=(x_1,y_1)$. Moreover, ψ is proportional to the angle at which such a half-line is inclined to the axis Ox.

Note that angle is a multi-valued function of (x,y), so that care is required. For example, the stream function corresponding to the flow of Example 6.4 is

$$\psi = \frac{q}{2\pi}\left\{\tan^{-1}\left(\frac{y-a}{x}\right)+\tan^{-1}\left(\frac{y+a}{x}\right)\right\}.$$

Whereas the second term in the braces may be taken to lie within $(0,\pi)$ for all $y\geq 0$, the first must necessarily increase by 2π for every anti-clockwise circuit around $(x,y)=(0,a)$. Nevertheless, the formula $\psi=$ constant unambiguously defines the streamlines.

EXERCISES

6.1. In the steady velocity field $\boldsymbol{u}=2U\boldsymbol{j}+U\boldsymbol{k}$, show that each streamline lies in a plane of constant x. Determine the equation of a typical streamline. Evaluate the flow speed $|\boldsymbol{u}|$ at a typical point $\boldsymbol{x}$ and so describe the flow field.

6.2. For each of the following two-dimensional velocity fields $\boldsymbol{u}(\boldsymbol{x}) = u\boldsymbol{i} + v\boldsymbol{j}$, indicate the flow direction at typical points (x, y) of the Oxy plane by short bars, determine the equations of the streamlines and add some streamlines to your sketches:

(a) $u = \dfrac{qx}{x^2+y^2}, \quad v = \dfrac{qy}{x^2+y^2}$;

(b) $u = \dfrac{-\gamma y}{x^2+y^2}, \quad v = \dfrac{\gamma x}{x^2+y^2}$;

(c) $\boldsymbol{u} = A(x\boldsymbol{i} - y\boldsymbol{j})$, where q, γ and A are constants.

6.3. For the velocity field $\boldsymbol{u} = q\dfrac{x\boldsymbol{i}+y\boldsymbol{j}}{x^2+y^2} + W\boldsymbol{k}$, show that each streamline lies in a plane containing the Oz axis (i.e. $y/x =$ constant). Writing $x^2 + y^2 = r^2$, determine a relationship between r and z along each streamline. Sketch some typical streamlines which lie in the Oxz plane. Describe how streamlines in other planes $y/x =$ constant are related to these.

Can you interpret the flow as a superposition of a uniform flow and one of those in Exercise 6.2?

6.4. For the velocity field of Exercise 6.2(c), determine the rate at which fluid volume is transported across each of the surfaces: (a) the portion $|x| < a$, $0 < z < 1$ of the plane $y = b$, (b) the portion $0 < y < b$, $0 < z < 1$ of $x = a$.

Extend this procedure to evaluate the total rate of volume flow out of all faces of the rectangular box $|x| < a$, $0 < y < b$, $0 < z < 1$. (Could the answer have been anticipated by evaluating $\operatorname{div} \boldsymbol{u}$ for Exercise 6.2(c)?)

6.5. A pulsating bubble centred at O has radius $R(t)$. It is surrounded by incompressible fluid which moves radially.

Show that at radius $r = R$ the outward speed is $\dot{R}(t) \equiv \mathrm{d}R/\mathrm{d}t$. Hence, deduce that at radius r the radial component of velocity is $R^2\dot{R}(t)/r^2$. Confirm that the velocity field is the gradient of the (unsteady) velocity potential $\phi = -R^2\dot{R}(t)/r$.

6.6. Confirm that each of the functions Ux, Vy, $A(x^2 - y^2)$, $\log r$ and θ satisfies the two-dimensional Laplace equation $\nabla^2\phi = \phi_{xx} + \phi_{yy} = 0$, where r and θ are polar coordinates such that $x = r\cos\theta$, $y = r\sin\theta$.

Using each of these functions in turn as a velocity potential $\phi(x, y)$, determine the corresponding velocity fields $\boldsymbol{u} = \boldsymbol{\nabla}\phi$. Find also the stream function $\psi(x, y)$ in each case.

6.7. Confirm that if $\phi(x, y)$ is the velocity and $\psi(x, y)$ is the stream function for a two-dimensional, incompressible flow then

(a) the streamlines are orthogonal to the equipotentials (ϕ = constant),

(b) the flow speed $|\boldsymbol{u}|$ at any point may be evaluated *either* as $|\boldsymbol{\nabla}\phi|$ *or* as $|\boldsymbol{\nabla}\psi|$,

(c) if a function $\phi(x, y)$ is the velocity potential for one flow, it may alternatively be used as the stream function for a different flow.

6.4 Pressure in a Fluid

In a fluid at rest, there is a *pressure* $p(\boldsymbol{x}, t)$ at each point. Pressure is the force per unit area acting on a surface element at $\boldsymbol{x}$ having *any* orientation. Imagine that through the point $\boldsymbol{x}$ there is a surface which has unit normal $\boldsymbol{n}$, pointing from the region labelled by 1 to the region labelled by 2. Then, over a portion of the surface containing $\boldsymbol{x}$ and having area δS, the fluid on side 2 exerts on the fluid on side 1 a force approximately equal to $-p\boldsymbol{n}\,\delta S$. (Of course, the fluid on side 1 exerts an equal and opposite force $p\boldsymbol{n}\,\delta S$ on the fluid on side 2. Newton's laws state that *action* and *reaction* are equal and opposite.)

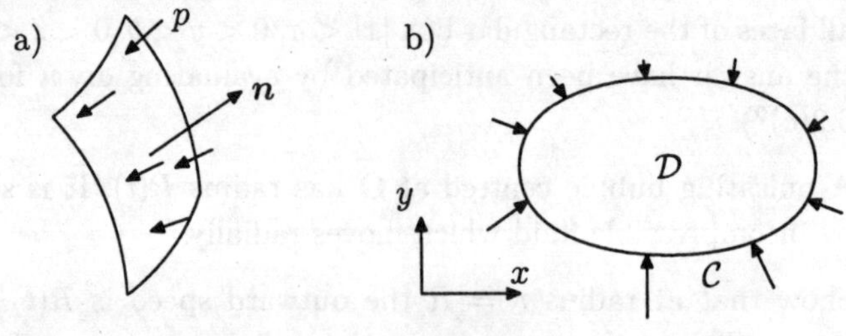

Figure 6.9 Pressure loading a) over a surface element and b) around the boundary $\mathcal{C}$ of the cross-section $\mathcal{D}$ of a cylinder.

6.4.1 Resultant Force

The integral of $-p\boldsymbol{n}\,\mathrm{d}S$ over the boundary of a closed region measures the total force (the *resultant force*) due to pressure. A useful alternative expression for the resultant force involves the volume integral of the pressure gradient. In the two-dimensional case, this is verified as follows:

Consider unit length of a cylinder lying within the fluid, having axis parallel to Oz and having cross-section $\mathcal{D}$ which is the interior of a closed curve C in the Oxy plane. Then, over this cylindrical surface, the pressure resultant is

$$\begin{aligned}\boldsymbol{F} &= \oint_C -p\boldsymbol{n}\mathrm{d}s = -\oint_C p(n_1\boldsymbol{i} + n_2\boldsymbol{j})\mathrm{d}s \\ &= -\boldsymbol{i}\oint_C p\,\mathrm{d}y + \boldsymbol{j}\oint_C p\,\mathrm{d}x \\ &= -\boldsymbol{i}\iint_{\mathcal{D}} \frac{\partial p}{\partial x}\mathrm{d}x\,\mathrm{d}y + \boldsymbol{j}\iint_{\mathcal{D}}\left(-\frac{\partial p}{\partial y}\right)\mathrm{d}x\,\mathrm{d}y\,.\end{aligned}$$

Here, both the identities $\mathrm{d}x = -n_2\mathrm{d}s$, $\mathrm{d}y = n_1\mathrm{d}s$ and *Green's theorem in the plane* have been used. Since $\operatorname{grad} p = \boldsymbol{\nabla} p = \boldsymbol{i}\partial p/\partial x + \boldsymbol{j}\partial p/\partial y$, this resultant force due to pressure may be expressed as

$$\boldsymbol{F} = -\iint_{\mathcal{D}} \boldsymbol{\nabla} p\,\mathrm{d}x\,\mathrm{d}y\,. \tag{6.14}$$

6.4.2 Hydrostatics and Archimedes' Principle

Hydrostatics is the study of forces acting within a fluid at rest. It is important in determining the buoyancy and stability of ships and marine structures and the forces and moments acting upon sluices and lock gates. To determine the pressure $p(\boldsymbol{x})$, it is necessary to take account of the gravitational force acting on a typical element of mass $\rho\,\delta V$.

Again consider the two-dimensional case. Take Oy to be vertically upward, so that this force is $-\rho g\boldsymbol{j}\delta V$, where g is the *gravitational acceleration* (gravity acts downwards and is assumed uniform over the length scales being considered). Then, for the fluid within unit length of the cylinder defined by a cross-section $\mathcal{D}$ and filled with fluid, the resultant force due to gravity is

$$\boldsymbol{F}_G = \iint_{\mathcal{D}} -\rho g\boldsymbol{j}\,\mathrm{d}S = -g\boldsymbol{j}\iint_{\mathcal{D}} \rho\,\mathrm{d}x\mathrm{d}y\,. \tag{6.15}$$

Since, in equilibrium, the resultant force $\boldsymbol{F} + \boldsymbol{F}_G$ must vanish, the resultant due to pressure must exactly counterbalance the resultant force due to gravity. After use of (6.14) for $\boldsymbol{F}$ this gives

$$\iint_{\mathcal{D}} (\rho g\boldsymbol{j} + \boldsymbol{\nabla} p)\mathrm{d}x\,\mathrm{d}y = \boldsymbol{0}$$

which, after use of the familiar argument that the boundary $\mathcal{C} = \partial\mathcal{D}$ of $\mathcal{D}$ may be altered, yields

$$\nabla p + \rho g \boldsymbol{j} = \boldsymbol{0}, \tag{6.16}$$

everywhere within the fluid.

This has two important consequences. Firstly, the pressure field on and outside $\mathcal{C}$ is unaffected if the fluid within the cylinder is replaced by a solid body, so that the *buoyancy force* $\boldsymbol{F}$ on unit length of such a body due to the surrounding fluid is

$$\boldsymbol{F} = -\boldsymbol{F}_G = \boldsymbol{j} \iint_{\mathcal{D}} \rho g \, \mathrm{d}x\mathrm{d}y \,. \tag{6.17}$$

Since the double integral exactly equals the *weight* of the fluid within unit length of the cylinder having cross-section $\mathcal{D}$, Equation (6.17) expresses *Archimedes' principle* – the buoyancy (upwards) has exactly the same magnitude as the weight (downwards) of fluid displaced.

Secondly, Equation (6.17) is readily integrated to show that p is linear in y. Indeed, by taking components as

$$\frac{\partial p}{\partial x} = 0\,, \quad \frac{\partial p}{\partial y} = -\rho g\,,$$

it is found that $p = p_0 - \rho g y$. As is familiar, pressure is independent of the horizontal coordinate(s), but increases linearly with depth.[3]

Example 6.7

Suppose that the sluice gate of a barrage is described by the arc $x = -1+2\cos\theta$, $y = 2\sin\theta$ for $-\frac{1}{3}\pi \le \theta \le 0$ and hydrostatic pressure $p = -\rho g y$ acts on the outward face. Find the resultant force acting on unit length of the sluice gate.

Express the boundary curve as $\boldsymbol{x} = \boldsymbol{i}(-1 + 2\cos\theta) + \boldsymbol{j}2\sin\theta$, so that $\mathrm{d}\boldsymbol{x} = 2(-\boldsymbol{i}\sin\theta + \boldsymbol{j}\cos\theta)$. Thus, $\boldsymbol{t} = -\boldsymbol{i}\sin\theta + \boldsymbol{j}\cos\theta$, $\mathrm{d}s = 2\mathrm{d}\theta$ and $n_1 = \cos\theta$, $n_2 = \sin\theta$ (i.e. $\boldsymbol{n} = \boldsymbol{i}\cos\theta + \boldsymbol{j}\sin\theta$). Also, the surface pressure is $p = -2\rho g\sin\theta$.

The force on unit length of the gate is

$$\begin{aligned}\boldsymbol{F} &= \int_{-\pi/3}^{0} -p\boldsymbol{n}2\mathrm{d}\theta = 2\int_{-\pi/3}^{0} \rho g 2\sin\theta\{\boldsymbol{i}\cos\theta + \boldsymbol{j}\sin\theta\}\mathrm{d}\theta \\ &= \rho g\left[-\boldsymbol{i}\cos 2\theta + \boldsymbol{j}(2\theta - \sin 2\theta)\right]_{-\pi/3}^{0} = \rho g\left\{-\tfrac{3}{2}\boldsymbol{i} + (\tfrac{2}{3}\pi - \tfrac{\sqrt{3}}{2})\boldsymbol{j}\right\}.\end{aligned}$$

---

[3] $1\mathrm{m}^3$ of water has mass 10^3kg, so, since $g = 9.81\mathrm{m/sec}^2$, then $\rho g = 9.81 \times 10^3$ kg/(m sec^2). Since 1 atmosphere $= 1.01325 \times 10^5$kg/(m sec)^2, the pressure *doubles* at depth ≈ 10 m below the water surface!

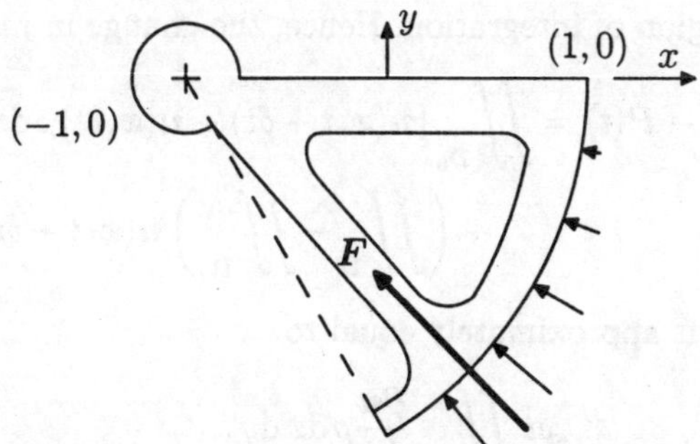

Figure 6.10 Pressure distribution over the sluice gate of Example 6.7, showing resultant force $\boldsymbol{F}$ and its line of action.

[N.B. Since the pressure loading at all points of the gate is directed towards the centre of the circle (at $x = -1$, $y = 0$), the resultant force $\boldsymbol{F}$ must also act along a line through that point. If the gate is hinged at that location, the fluid loading exerts no *moment* about the hinge. This is good design practice, since to maintain the position of the gate no torque is then needed.]

6.4.3 Momentum Density and Momentum Flux

In a moving fluid the pressure distribution is more complicated, since an additional contribution to pressure gradient is required to account for the change of velocity as a fluid element moves along a particle path. This is demonstrated here for the two-dimensional velocity field $\boldsymbol{u} = u\boldsymbol{i} + v\boldsymbol{j}$, by considering the collection of fluid particles within $0 < z < 1$ which at $t = 0$ have x and y coordinates lying in the region $\mathcal{D}_0$. Note that, at a later instant $t + \delta t$, the x and y coordinates will lie in a slightly different region $\mathcal{D}$.

In a volume element δV the mass is $\rho\,\delta V$ and velocity is $\boldsymbol{u}$, so that the momentum is $\boldsymbol{u}\rho\,\delta V$ (i.e. the *momentum density* is $\rho\boldsymbol{u}$). Hence the momentum of the relevant collection of fluid particles is evaluated at time t as

$$\boldsymbol{P}(t) = \iint_{\mathcal{D}_0} \boldsymbol{u}(\boldsymbol{x}, t)\rho\, \mathrm{d}x\, \mathrm{d}y\,,$$

while at the slightly later time $t + \delta t$ it is

$$\boldsymbol{P}(t + \delta t) = \iint_{\mathcal{D}} \boldsymbol{u}(\boldsymbol{x}, t + \delta t)\rho\, \mathrm{d}x\, \mathrm{d}y\,.$$

The second expression differs not only in the argument t within the integrand but also in the region of integration. Hence, the change in momentum is

$$\begin{aligned}\boldsymbol{P}(t+\delta t)-\boldsymbol{P}(t) &= \iint_{\mathcal{D}_0}\{\boldsymbol{u}(\boldsymbol{x},t+\delta t)-\boldsymbol{u}(\boldsymbol{x},t)\}\rho\,\mathrm{d}x\,\mathrm{d}y \\ &\quad + \left(\iint_{\mathcal{D}} - \iint_{\mathcal{D}_0}\right)\boldsymbol{u}(\boldsymbol{x},t+\delta t)\rho\,\mathrm{d}x\,\mathrm{d}y\,.\end{aligned}$$

The first integral is approximately equal to

$$\delta t\iint_{\mathcal{D}_0}\frac{\partial \boldsymbol{u}}{\partial t}\rho\,\mathrm{d}x\,\mathrm{d}y\,,$$

while the second is seen to be an integral around a narrow strip adjacent to the boundary $\partial\mathcal{D}_0$. Since, during this short time interval, the distance moved normal to $\partial\mathcal{D}_0$ by the fluid is $\boldsymbol{u}\cdot\boldsymbol{n}\,\delta t = u_\perp\delta t$, this second contribution is approximately given by

$$\oint_{\partial\mathcal{D}_0}\rho\boldsymbol{u}\,(\boldsymbol{u}\cdot\boldsymbol{n})\mathrm{d}s = \oint_{\partial\mathcal{D}_0}\rho\boldsymbol{u}\,u_\perp\mathrm{d}s\,.$$

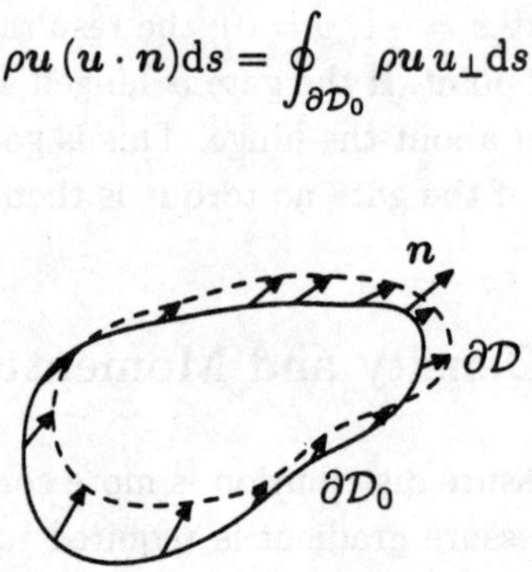

Figure 6.11 Initial and displaced boundaries to the cross-section.

Combining these statements and then taking the limit $\delta t\to 0$ gives

$$\begin{aligned}\dot{\boldsymbol{P}}(t) = \lim_{\delta t\to 0}\left(\frac{\boldsymbol{P}(t+\delta t)-\boldsymbol{P}(t)}{\delta t}\right) &= \iint_{\mathcal{D}_0}\rho\frac{\partial\boldsymbol{u}}{\partial t}\mathrm{d}x\,\mathrm{d}y \\ &\quad + \oint_{\partial\mathcal{D}_0}\rho\boldsymbol{u}(un_1+vn_2)\mathrm{d}s\,. \qquad (6.18)\end{aligned}$$

Since the component of velocity normal to the boundary is $\boldsymbol{u}\cdot\boldsymbol{n} = u_\perp = un_1+vn_2$, it is seen that the boundary integral in (6.18) is no more than the rate at which momentum is swept across the boundary $\partial\mathcal{D}_0$. This integral may be rewritten, using the two identities $n_1\mathrm{d}s = \mathrm{d}y$, $n_2\mathrm{d}s = -\mathrm{d}x$, since, for example,

$$\oint_{\partial\mathcal{D}_0}\rho u\boldsymbol{u}\,n_1\,\mathrm{d}s = \oint_{\partial\mathcal{D}_0}\rho u\boldsymbol{u}\,\mathrm{d}y = \iint_{\mathcal{D}_0}\frac{\partial}{\partial x}(\rho u\boldsymbol{u})\mathrm{d}x\,\mathrm{d}y\,.$$

Here Green's theorem has been applied to each component of the vector integrand $\rho u u$. A similar manipulation applied to the integrand $\rho u v$ gives

$$\oint_{\partial\mathcal{D}_0} \rho u v\, n_2\, \mathrm{d}s = -\oint_{\partial\mathcal{D}_0} \rho u v\, \mathrm{d}x = \iint_{\mathcal{D}_0} \frac{\partial}{\partial y}(\rho u v)\mathrm{d}x\, \mathrm{d}y\,.$$

Hence, the time-derivative of momentum in (6.18) may be expressed as

$$\dot{\boldsymbol{P}}(t) = \iint_{\mathcal{D}_0} \left(\rho\frac{\partial \boldsymbol{u}}{\partial t} + \frac{\partial}{\partial x}(\rho u \boldsymbol{u}) + \frac{\partial}{\partial y}(\rho v \boldsymbol{u})\right) \mathrm{d}x\, \mathrm{d}y\,. \tag{6.19}$$

From Newton's second law, this quantity $\dot{\boldsymbol{P}}(t)$ must equal the total force acting at time t on the fluid within the cross section $\mathcal{D}_0$. The force is of two types: The resultant of pressure acting over $\partial\mathcal{D}_0$ and the resultant of the *body force* acting, like gravitation or electrostatic attraction, on each element of mass. Thus, if $\boldsymbol{b}$ measures the body force per unit mass (cf. the gravitational acceleration $-g\boldsymbol{j}$ in (6.15)), the total force is

$$\dot{\boldsymbol{P}}(t) = \oint_{\partial\mathcal{D}_0} -p\boldsymbol{n}\, \mathrm{d}s + \iint_{\mathcal{D}_0} \rho\boldsymbol{b}\, \mathrm{d}x\, \mathrm{d}y = \iint_{\mathcal{D}_0} (-\boldsymbol{\nabla} p + \rho\boldsymbol{b})\, \mathrm{d}x\, \mathrm{d}y\,.$$

Thus, combining these two expressions for $\dot{\boldsymbol{P}}(t)$ gives

$$\iint_{\mathcal{D}_0} \left(\rho\frac{\partial \boldsymbol{u}}{\partial t} + \frac{\partial}{\partial x}(\rho u \boldsymbol{u}) + \frac{\partial}{\partial y}(\rho v \boldsymbol{u}) + \boldsymbol{\nabla} p - \rho\boldsymbol{b}\right) \mathrm{d}x\, \mathrm{d}y = \boldsymbol{0}\,,$$

which applies for general shapes $\mathcal{D}_0$. Familiar reasoning tells us that the integrand must be identically zero, which gives

$$\rho\left(\frac{\partial \boldsymbol{u}}{\partial t} + \frac{\partial}{\partial x}(u\boldsymbol{u}) + \frac{\partial}{\partial y}(v\boldsymbol{u})\right) + \boldsymbol{\nabla} p - \rho\boldsymbol{b} = \boldsymbol{0}\,, \tag{6.20}$$

since ρ = constant. Moreover, after use of the continuity equation in two dimensions (6.7), namely $\partial u/\partial x + \partial v/\partial y = 0$, in the second and third terms in (6.20) this equation is simplified to

$$\rho\left(\frac{\partial \boldsymbol{u}}{\partial t} + u\frac{\partial \boldsymbol{u}}{\partial x} + v\frac{\partial \boldsymbol{u}}{\partial y}\right) = -\boldsymbol{\nabla} p + \rho\boldsymbol{b}\,. \tag{6.21}$$

On the left-hand side, the expression in brackets is just the fluid acceleration, as fluid flows at velocity $\boldsymbol{u}$ along a trajectory. Hence, Equation (6.21) equates the rate of change of fluid momentum to the sum of the negative of the *pressure gradient* and the body force per unit volume $\rho\boldsymbol{b}$. It is known as the *Euler equation* for two-dimensional incompressible, inviscid flow. Equivalent equations in component form are obtained by taking the $\boldsymbol{i}$ and $\boldsymbol{j}$ components and writing $\boldsymbol{b} = b_1\boldsymbol{i} + b_2\boldsymbol{j}$; namely the Euler equations

$$\frac{\partial u}{\partial t} + u\frac{\partial u}{\partial x} + v\frac{\partial u}{\partial y} = \frac{-1}{\rho}\frac{\partial p}{\partial x} + b_1\,, \tag{6.22}$$

$$\frac{\partial v}{\partial t} + u\frac{\partial v}{\partial x} + v\frac{\partial v}{\partial y} = \frac{-1}{\rho}\frac{\partial p}{\partial y} + b_2\,. \tag{6.23}$$

It is readily seen that when the fluid is at rest ($\boldsymbol{u} = \boldsymbol{0}$), these simply relate the pressure gradient to the body force per unit volume.

6.5 Bernoulli's Equation

6.5.1 The Material (Advected) Derivative

The combination of derivatives appearing on the left-hand sides of Equations (6.21)–(6.23) appears frequently, and naturally, in fluid mechanics. It is the time rate of change as measured at a fluid particle moving with velocity $\boldsymbol{u}$. It is known as the *advected derivative* (sometimes as the convected derivative, or material derivative) and is usually written as

$$\frac{\mathrm{D}}{\mathrm{D}t} \equiv \frac{\partial}{\partial t} + u\frac{\partial}{\partial x} + v\frac{\partial}{\partial y} \qquad \left(= \frac{\partial}{\partial t} + \boldsymbol{u}\cdot\boldsymbol{\nabla} \right). \tag{6.24}$$

Use of this compact notation allows (6.22) and (6.23) to be written as

$$\frac{\mathrm{D}u}{\mathrm{D}t} = \frac{-1}{\rho}\frac{\partial p}{\partial x} + b_1\,, \qquad \frac{\mathrm{D}v}{\mathrm{D}t} = \frac{-1}{\rho}\frac{\partial p}{\partial y} + b_2\,.$$

In *steady flow*, although the derivative $\partial\boldsymbol{u}/\partial t$ is zero, the advected derivative is usually non-zero. For example, the *acceleration* reduces to

$$\frac{\mathrm{D}\boldsymbol{u}}{\mathrm{D}t} = u\frac{\partial \boldsymbol{u}}{\partial x} + v\frac{\partial \boldsymbol{u}}{\partial y} = (\boldsymbol{u}\cdot\boldsymbol{\nabla})\boldsymbol{u}\,,$$

showing that, as fluid moves, the fact that velocity varies with position gives rise to acceleration. Thus, in Euler's equations (6.22) and (6.23), pressure gradient and body force are balanced in steady flow by changes in *momentum flux*.

Example 6.8

Find the pressure distribution within the two-dimensional flow having uniform density ρ and stream function $\psi = -\gamma \log \sqrt{x^2 + y^2}$, when no body forces act.

Let $r \equiv \sqrt{x^2+y^2}$, so that the velocity components are $u = -\gamma y/r^2$ and $v = \gamma x/r^2$. Then $u_x = 2\gamma xyr^{-2} = -v_y$ and $u_y = -\gamma r^{-2} + 2\gamma y^2 r^{-4} = \gamma(y^2 - x^2)r^{-4} = v_x$, so that

$$\frac{\partial p}{\partial x} = -\rho(uu_x + vu_y) = -\rho\gamma^2(2xy^2 - xy^2 + x^3)r^{-3} = -\rho\gamma^2\frac{x}{r^4}$$

and, similarly, $\partial p/\partial y = -\rho\gamma^2 y/r^4 = -\frac{1}{2}\rho\gamma^2 \partial r^{-2}/\partial y$. Thus, in this flow with circular streamlines, the pressure is

$$p = -\frac{\rho\gamma^2}{2(x^2+y^2)} + \text{constant}.$$

6.5.2 Bernoulli's Equation and Dynamic Pressure

A deduction of great practical and historical importance is Bernoulli's equation, which relates pressure to flow speed in steady flows.

Consider two-dimensional, irrotational flow, so that in Equations (6.22) and (6.23) the velocity components may be written as $u = \phi_x$ and $v = \phi_y$. This leads to the pair of equations

$$\begin{aligned} \phi_x\phi_{xx} + \phi_y\phi_{yx} &= -\rho^{-1}p_x + b_1, \\ \phi_x\phi_{xy} + \phi_y\phi_{yy} &= -\rho^{-1}p_y + b_2. \end{aligned}$$

Since both left-hand sides may be recognized as partial derivatives of the quantity $\frac{1}{2}\left(\phi_x^2 + \phi_y^2\right)$, these equations may be rewritten as

$$b_1 = \frac{\partial}{\partial x}\left\{\tfrac{1}{2}\left(\phi_x^2+\phi_y^2\right) + \frac{p}{\rho}\right\}, \qquad b_2 = \frac{\partial}{\partial x}\left\{\tfrac{1}{2}\left(\phi_x^2+\phi_y^2\right) + \frac{p}{\rho}\right\}.$$

These two equations are self-consistent if, and only if, the body force components may be expressed in the form $b_1 = -\partial\Omega/\partial x$, $b_2 = -\partial\Omega/\partial y$ for some suitable *potential* $\Omega(x, y)$. (In other words, the flow can remain irrotational only if the total body force per unit mass is the gradient of a scalar. Gravitational body forces are an important example of this type.) In such cases, the Euler equations (6.22) and (6.23) are equivalent to

$$\boldsymbol{\nabla}\left\{\tfrac{1}{2}\left(\phi_x^2+\phi_y^2\right) + \rho^{-1}p + \Omega\right\} = \mathbf{0}, \tag{6.25}$$

which imply the result

$$\rho^{-1}p + \tfrac{1}{2}\boldsymbol{u}\cdot\boldsymbol{u} + \Omega(\boldsymbol{x}) = \text{const.} \tag{6.26}$$

This is *Bernoulli's equation*, which shows a simple relationship between pressure p, flow speed $q \equiv |\boldsymbol{u}|$ and the potential $\Omega(\boldsymbol{x})$ associated with the body force

$\boldsymbol{b} = -\nabla\Omega$. When body forces may be neglected, this relationship simplifies further as

$$p + \tfrac{1}{2}\rho q^2 = \text{constant}\,, \tag{6.27}$$

so showing how the pressure is directly related to the square of the flow speed. (Note how, in Example 6.8, the flow speed is γr^{-1}, so that the expression for the pressure agrees with (6.27).)

An important use of (6.27) (which holds also for three-dimensional, steady flows with $\boldsymbol{u} = \nabla\phi$) is for flows which originate from a *uniform region*, where the speed is q_∞ and the pressure is p_∞. Equivalently, this applies to flow past a steadily moving aircraft or projectile, since, relative to axes fixed in the craft, the surrounding air appears to be approaching at a uniform velocity.

At any stagnation point on the craft (i.e. where $q = 0$), Bernoulli's equation gives

$$\frac{p_0}{\rho} + \tfrac{1}{2}0^2 = \frac{p_\infty}{\rho} + \tfrac{1}{2}q_\infty^2\,, \qquad \text{so that} \quad p_0 = p_\infty + \tfrac{1}{2}\rho q_\infty^2\,.$$

Thus, at a *stagnation point*, the *dynamic pressure* p equals p_0 and exceeds the *static pressure* p_∞ by $\frac{1}{2}\rho q_\infty^2$. (N.B. In aerodynamic applications, p_∞ refers to the pressure in the air where it is undisturbed by the aircraft and hence is regarded as at rest.) An important application is the *Pitot tube*, which is a simple device for measuring the speed of an aeroplane relative to the air. A small round-nosed tube is mounted so that it points upstream. In the tube are drilled two narrow holes. One is at the nose, the other is on the side. Both are connected to pressure gauges, but do not allow significant internal flow. The hole at the nose measures the *stagnation pressure* p_0, while at the side where the local flow speed is essentially the same as q_∞ the second tube measures the pressure p_∞. Thus the vehicle speed (relative to the air) is obtained as

$$q_\infty = \sqrt{\frac{2(p_0 - p_\infty)}{\rho}}\,. \tag{6.28}$$

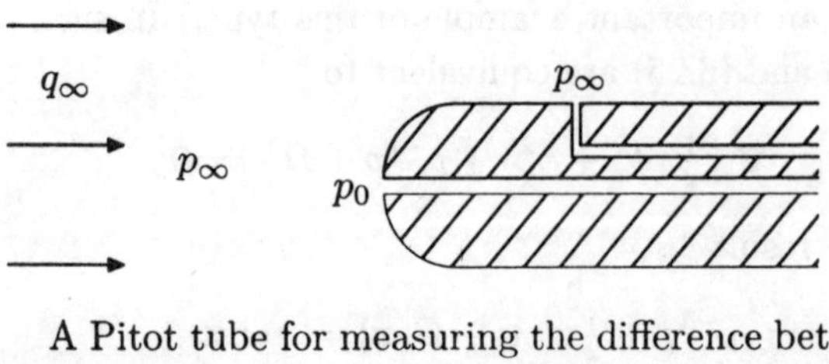

Figure 6.12 A Pitot tube for measuring the difference between static and stagnation pressures.

6.5.3 The Principle of Aerodynamic Lift

Bernoulli's equation reveals the basic principles of *lift* on an aerofoil or hydrofoil, so allowing mechanized flight. A wing is essentially a solid body designed to induce a (closely two-dimensional) flow in which the speeds *above* the body are faster than those *below* the body, so causing the pressure *below* the foil to be greater than that *above* the foil and thus giving a positive upward component of resultant force. (There are, of course, many additional complications – all fluids are viscous, while air changes its density due to changes in either pressure or temperature. In fact, the *state of stress* within a fluid cannot be described solely in terms of a pressure, since the force on a surface element within the flowing fluid is not necessarily normal to that surface. Frequently the *shear stresses* within the fluid may be written in terms of the *velocity gradient* and a material parameter known as the *viscosity.* This replaces the *Euler equations* by the famed *Navier–Stokes equations.* However, although viscous forces are very significant close to a solid boundary and substantially contribute to *aerodynamic drag*, it is often appropriate to analyse the broad features of the external flow by treating the flow as irrotational, so allowing the representation $\boldsymbol{u} = \nabla\phi$.)

Example 6.9 (Flow past a circular cylinder)

Show that both the stream functions (a) $\psi = \log r$ and (b) $\psi = -Uy(1-a^2/r^2)$, where $r^2 = x^2+y^2$, describe irrotational flows for which $\psi =$ constant on $r = a$. Deduce that, for any value γ, the stream function $\psi = -\gamma \log r - Uy(1-a^2/r^2)$ describes a flow past the cylinder having surface $r = a$. Describe the flow as $r \to \infty$.

Each of $\log r$, y and $y/r^2 = \partial(\log r)/\partial y$ satisfies Laplace's equation $\psi_{xx} + \psi_{yy} = 0$. For (a), $\psi = \log a$ on $r = a$, while for (b) $\psi = 0$. Hence, in each case, ψ is the stream function of an irrotational flow having $r = a$ as a streamline. Consequently, the superposition $\psi = -\gamma \log r - Uy(1 - a^2/r^2)$, which also has $\psi = -\gamma \log a =$ constant on $r = a$, describes a flow past the circular cylinder.

Since $\psi_x = -\gamma x/r^2 - 2Uya^2x/r^4 \to 0$ and $\psi_y = -\gamma y/r^2 - U(1 - a^2/r^2) - 2Uya^2y/r^4 \to U$ as $r \to \infty$, for *each* value of γ, the flow at large distances from the cylinder is a uniform flow $\boldsymbol{u} = -U\boldsymbol{i}$ at speed U parallel to Ox but from right to left.

Example 6.10 (Pressure and lift on a cylinder)

Use Bernoulli's law to evaluate the pressure on the cylindrical surface $r = a$ in Example 6.9. Show that the resultant force on unit length of the cylinder has magnitude $2\pi\rho U\gamma$ and is perpendicular to the incident flow direction.

Use the polar angle θ, so that $x = a\cos\theta$ and $y = a\sin\theta$ on the cylinder $r = a$. Now, $u = \psi_y = -\gamma a^{-1}\sin\theta - 2U\sin^2\theta$, while $v = -\psi_x = \gamma a^{-1}\cos\theta + 2U\sin\theta\cos\theta$. Thus $q^2 = (\sin^2\theta + \cos^2\theta)(\gamma a^{-1} + 2U\sin\theta)^2$, while Bernoulli's law gives $p - p_\infty = \frac{1}{2}(U^2 - q^2)$.

On the cylinder $\boldsymbol{n} = \boldsymbol{i}\cos\theta + \boldsymbol{j}\sin\theta$, while $\mathrm{d}s = a\mathrm{d}\theta$, so the resultant force on unit length of the cylinder is

$$\begin{aligned}\boldsymbol{F} &= \int_0^{2\pi} -p\boldsymbol{n}a\,\mathrm{d}\theta = \tfrac{1}{2}a\int_0^{2\pi}(\rho q^2 - \rho U^2 - 2p_\infty)(\boldsymbol{i}\cos\theta + \boldsymbol{j}\sin\theta)\mathrm{d}\theta \\ &= \frac{\rho a}{2}\int_0^{2\pi}\left[4U^2\sin^2\theta + \frac{4U\gamma}{a}\sin\theta + \frac{\gamma^2}{a^2} - U^2 - \frac{2p_\infty}{\rho}\right](\boldsymbol{i}\cos\theta + \boldsymbol{j}\sin\theta)\mathrm{d}\theta\,.\end{aligned}$$

Since the integrals over $0 \le \theta \le 2\pi$ of each of $\cos\theta$, $\sin\theta$, $\sin\theta\cos\theta$, $\sin^2\theta\cos\theta$ and $\sin^3\theta$ are zero, the only non-vanishing contribution to $\boldsymbol{F}$ is

$$\boldsymbol{F} = \boldsymbol{j}2\rho U\gamma\int_0^{2\pi}\sin^2\theta\,\mathrm{d}\theta = 2\pi\gamma\rho U\boldsymbol{j}\,.$$

The force per unit length has magnitude $2\pi\gamma\rho U$, it is perpendicular to the incident flow velocity $-U\boldsymbol{i}$ and directed towards the side where the flow speed is increased (i.e. positive y, when U and γ have the same sign).

Example 6.10 is just a very special case of a general formula for *lift* in two-dimensional irrotational flow past a body with cross section $\mathcal{D}$. The quantity $2\pi\gamma$ has a very special significance. It is the *circulation* $\oint_C \boldsymbol{u}\cdot\mathrm{d}\boldsymbol{x} \equiv \Gamma$ around *any* two-dimensional curve C enclosing $\mathcal{D}$ and taken counterclockwise. Blasius' law states that the resultant lift is $\rho U\Gamma$ and is perpendicular to the incident flow direction, whatever the shape of the solid cross section.

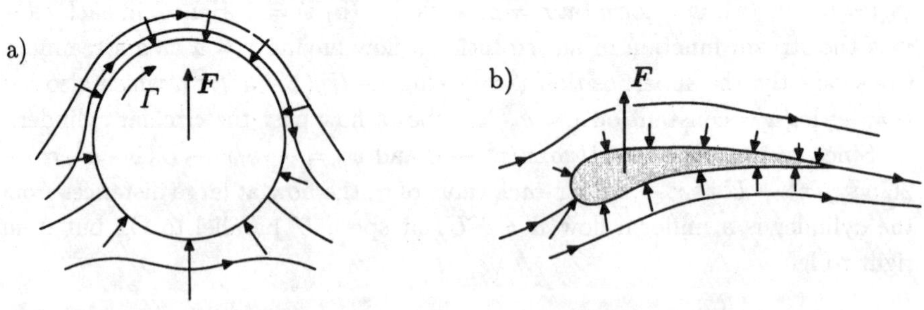

Figure 6.13 a) Flow past a circular cylinder with circulation Γ, showing pressure loading and its resultant $\boldsymbol{F}$. b) Sketch of pressure distribution around the aerofoil cross-section of Figure 6.8.

Aerodynamic lift is proportional to the circulation. For flows at moderate speeds the circulation is induced by the *camber* of the aerofoil – the concavity of the undersurface and enhanced convexity of the upper surface. In practice, the amount of circulation is determined by a balance between viscous effects in a thin region known as the *boundary layer*, which adjusts the flow so that the rearward stagnation point migrates to the sharp *trailing edge*. There are many textbooks dealing with such topics in fluid mechanics, e.g. Paterson (1983). For finding flows around cross-sections more complicated than a circular cylinder there are many techniques, both analytic and numerical. Use of complex variable theory is amply developed by Milne-Thomson (1968).

6.6 Three-dimensional, Incompressible Flows

Many of the above results for two-dimensional flows are readily generalized to three dimensions, if results from vector calculus can be quoted.

The *divergence* $\operatorname{div} \boldsymbol{v} = \boldsymbol{\nabla} \cdot \boldsymbol{v}$ of *any* differentiable vector field $\boldsymbol{v}(\boldsymbol{x})$ satisfies the *divergence theorem*:

Theorem 6.11

If $\boldsymbol{v}(\boldsymbol{x})$ is continuously differentiable within a region $\mathcal{R}$ and $\boldsymbol{n}$ is the outward unit normal to the bounding surface $\partial\mathcal{R}$, then

$$\iint_{\partial\mathcal{R}} \boldsymbol{v} \cdot \boldsymbol{n} \, \mathrm{d}S = \iiint_{\mathcal{R}} \operatorname{div} \boldsymbol{v} \, \mathrm{d}V \,. \tag{6.29}$$

Thus, $\operatorname{div} \boldsymbol{v}$ may be regarded as the *source density* associated with the vector field $\boldsymbol{v}$. The theorem may be proved by considering separately the components of $\boldsymbol{v} = u\boldsymbol{i} + v\boldsymbol{j} + w\boldsymbol{k}$.

Suppose that the surface $\partial\mathcal{R}$ is cut by any line $x =$ constant, $y =$ constant in no more than two places. Then (see Figure 6.14) it has an *upper* portion $\partial\mathcal{R}^+$, a lower portion $\partial\mathcal{R}^-$ and a projection $\mathcal{D}_z$ onto the Oxy plane. On the left-hand side of (6.29), in $\iint_{\partial\mathcal{R}} w\boldsymbol{k} \cdot \boldsymbol{n} \, \mathrm{d}S$ the quantity $\boldsymbol{k} \cdot \boldsymbol{n} \, \mathrm{d}S$ may be recognized as $\pm \mathrm{d}x \, \mathrm{d}y$ – the positive sign being taken whenever the boundary point $\boldsymbol{x}$ lies on $\partial\mathcal{R}^+$, the negative sign applying if $\boldsymbol{x}$ lies on $\partial\mathcal{R}^-$. Then,

$$\begin{aligned} \iint_{\partial\mathcal{R}} w\boldsymbol{k} \cdot \boldsymbol{n} \, \mathrm{d}S &= \iint_{\mathcal{D}_z} \left(w(x, y, z^+) - w(x, y, z^-)\right) \mathrm{d}x \, \mathrm{d}y \\ &= \iiint_{\mathcal{R}} \frac{\partial w}{\partial z} \mathrm{d}V \,, \end{aligned} \tag{6.30}$$

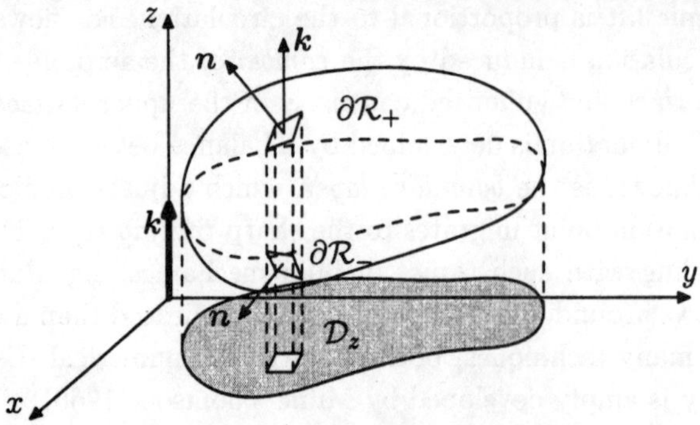

Figure 6.14 A bounding surface, showing its upper and lower portions and its projection $\mathcal{D}_z$ onto the Oxy plane.

where the fundamental theorem of calculus has been used. Result (6.30) may be extended to regions with more contorted boundary surface by suitable subdivision into simpler regions, then combining the results. Two analogous calculations concerning $\boldsymbol{ui}$ and $\boldsymbol{vj}$ are derived by projection onto the planes Oyz and Oxz. When added, these results give the divergence theorem (6.29) (see also Matthews, 1998).

6.6.1 The Continuity Equation

If ρ is the fluid density and $\boldsymbol{u}$ the fluid velocity within a region $\mathcal{R}$, the law of mass conservation (a balance between rate of mass outflow and rate of change of mass contained within $\mathcal{R}$) is

$$\frac{\mathrm{d}}{\mathrm{d}t}\iiint_{\mathcal{R}} \rho \,\mathrm{d}V + \iint_{\partial\mathcal{R}} (\rho\boldsymbol{u})\cdot\boldsymbol{n}\,\mathrm{d}S = 0\,.$$

Use of (6.29) and a familiar manipulation for time derivatives then yields

$$\iiint_{\mathcal{R}} \left(\frac{\partial\rho}{\partial t} + \boldsymbol{\nabla}\cdot(\rho\boldsymbol{u})\right)\mathrm{d}V = 0\,.$$

Varying the shape and extent of the region $\mathcal{R}$ then shows that the integrand must vanish, so yielding the *law of mass conservation*

$$\frac{\partial\rho}{\partial t} + \boldsymbol{\nabla}\cdot(\rho\boldsymbol{u}) = 0\,, \tag{6.31}$$

in a form appropriate to compressible fluids. In the special case of incompressible fluids with uniform density, all derivatives of ρ vanish, so giving

$$\boldsymbol{\nabla}\cdot\boldsymbol{u} = 0 \quad (\text{i.e.}\,\mathrm{div}\,\boldsymbol{u} = 0)\,, \tag{6.32}$$

which is just the *continuity equation* (6.5) for incompressible fluids.

6.6.2 Irrotational Flows, the Velocity Potential and Laplace's Equation

The concept of rotation within fluid flow is a generalization of the idea of *circulation* $\Gamma \equiv \oint_C \boldsymbol{u} \cdot \mathrm{d}\boldsymbol{x}$ around a closed loop C located anywhere within the fluid. If the loop is small and lies in a plane through $\boldsymbol{x}$ having unit normal $\boldsymbol{n}$ and if the loop encloses a small area A in a counterclockwise sense, denote the circulation by Γ_C. Then, as $A \to 0$, the limit of Γ_C/A can depend only on the orientation of the plane, i.e. upon $\boldsymbol{n}$. As $\boldsymbol{n}$ is varied, the limit varies, but it turns out that this limit has many of the properties of a vector. It is denoted by $\operatorname{curl} \boldsymbol{u} = \nabla \times \boldsymbol{u}$ and may be evaluated in component form as

$$\operatorname{curl} \boldsymbol{u} = \nabla \times \boldsymbol{u} = \begin{vmatrix} \boldsymbol{i} & \boldsymbol{j} & \boldsymbol{k} \\ \frac{\partial}{\partial x} & \frac{\partial}{\partial y} & \frac{\partial}{\partial z} \\ u & v & w \end{vmatrix} .$$

It is known as the *curl* of the vector field $\boldsymbol{u}$. This definition applies to *all* vector fields $\boldsymbol{u}(\boldsymbol{x}, t)$, but if $\boldsymbol{u}$ is the fluid velocity the quantity $\operatorname{curl} \boldsymbol{u}$ is known as the *vorticity*, namely $\boldsymbol{\omega} \equiv \nabla \times \boldsymbol{u} = \operatorname{curl} \boldsymbol{u}$.

Example 6.12

Suppose that a mass of fluid rotates about an axis through O, as if it were a rigid body with angular velocity $\boldsymbol{\Omega} = \omega_1 \boldsymbol{i} + \omega_2 \boldsymbol{j} + \omega_3 \boldsymbol{k}$. Given that the corresponding fluid velocity is $\boldsymbol{u} = \boldsymbol{\Omega} \times \boldsymbol{x}$, evaluate the vorticity $\boldsymbol{\omega}$ at a typical point.

The fluid velocity has components u, v and w given by

$$u\boldsymbol{i} + v\boldsymbol{j} + w\boldsymbol{k} = (\omega_2 z - \omega_3 y)\boldsymbol{i} + (\omega_3 x - \omega_1 z)\boldsymbol{j} + (\omega_1 y - \omega_2 x)\boldsymbol{k} .$$

Hence, $\boldsymbol{\omega} = \operatorname{curl} \boldsymbol{u} = (w_y - v_z)\boldsymbol{i} + (u_z - w_x)\boldsymbol{j} + (v_x - u_y)\boldsymbol{k} = 2\omega_1 \boldsymbol{i} + 2\omega_2 \boldsymbol{j} + 2\omega_3 \boldsymbol{k}$. Hence, in *rigid body flow* the vorticity is *uniform*; it equals $2\boldsymbol{\Omega}$, i.e. twice the *angular velocity.*

In a general flow, the vorticity $\boldsymbol{\omega}$ may depend upon $\boldsymbol{x}$ and t. A particularly important special case is that of *irrotational flows*, for which

$$\boldsymbol{\omega} = \nabla \times \boldsymbol{u} = \operatorname{curl} \boldsymbol{u} = \boldsymbol{0} .$$

An important result due to Kelvin is that, if a mass of fluid is initially irrotational (e.g. at rest, or in uniform flow), then it remains irrotational if viscosity may be neglected and if any body force $\boldsymbol{b}$ is the gradient of a scalar. In an irrotational flow, the *circulation* around *any* loop C is zero. Hence, between any two points P_1 and P_2 the line integral $\int \boldsymbol{u} \cdot d\boldsymbol{x}$ has a value independent of the path joining P_1 to P_2. This implies that $\boldsymbol{u}$ may be written as the gradient of a *velocity potential* $\phi(\boldsymbol{x}, t)$, i.e.

$$\boldsymbol{u} = \boldsymbol{\nabla}\phi \tag{6.33}$$

as in Section 6.3.2 (cf. also the electrostatic potential in Section 3.4). Conversely, the choice $\boldsymbol{u} = \boldsymbol{\nabla}\phi$ automatically yields $\operatorname{curl} \boldsymbol{u} = \boldsymbol{\nabla} \times \boldsymbol{u} = \mathbf{0}$, from the vector identity $\boldsymbol{\nabla} \times \boldsymbol{\nabla}\phi = \operatorname{curl}(\operatorname{grad}\phi) = 0$ which applies for all scalars ϕ.

Inserting (6.33) into the continuity equation (6.32) (or (6.5)) for incompressible flows gives

$$\operatorname{div} \boldsymbol{u} = \operatorname{div}(\operatorname{grad}\phi) = \frac{\partial^2 \phi}{\partial x^2} + \frac{\partial^2 \phi}{\partial y^2} + \frac{\partial^2 \phi}{\partial z^2} = 0 ,$$

which is a natural generalization to three dimensions of (6.9) for irrotational, incompressible flows. Thus, many types of three-dimensional flow, both steady and unsteady, may be analysed using methods applicable to Laplace's equation.

EXERCISES

6.8. The hydrostatic pressure $p(\boldsymbol{x})$ in a fluid at rest and of constant density ρ satisfies $\boldsymbol{\nabla} p + \rho g \boldsymbol{k} = \mathbf{0}$, where g is the acceleration due to gravity and $\boldsymbol{k}$ is the vertically upward unit vector. If $p = p_0$ at $\boldsymbol{x} = \mathbf{0}$, find $p(\boldsymbol{x})$ everywhere within the fluid.

Water of uniform density ρ fills a bath occupying

$$-(z + 2a) < x < z + 2a , \quad 0 < y < L , \quad -a < z < 0 .$$

Determine the mass of water in the bath. Find the resultant force on the flat bottom at $z = -a$. Find also the pressure at a typical point of each of the sloping surfaces and so determine the resultant force acting on each of these.

Confirm that the resultant of all forces acting on the surfaces of the bath is vertical and equals the weight of water in the bath plus the force exerted over $z = 0$ by the atmospheric pressure p_0.

6.9. In the flow with stream function $\psi = -\gamma \log r - U \sin\theta\,(r - a^2/r)$ (see Example 6.9), show that at points of the cylinder $r = a$ the flow speed is $q = |\boldsymbol{u}| = |\partial\psi/\partial r|$. Hence deduce that, if $|\gamma| > 2a|U|$, stagnation points arise at each of the locations given by $\sin\theta = -\gamma/(2aU)$.

If $|\gamma| > 2a|U|$, show that the only stagnation points in the flow occupying $r \geq a$ are at $x = 0$, $y = -\{\gamma + \sqrt{\gamma^2 - 4a^2U^2}\}/(2U)$.

6.10. If a two-dimensional fluid flow is described by the Euler equations without body forces

$$\frac{\partial u}{\partial t} + u\frac{\partial u}{\partial x} + v\frac{\partial u}{\partial y} = \frac{-1}{\rho}\frac{\partial p}{\partial x}\,, \quad \frac{\partial v}{\partial t} + u\frac{\partial v}{\partial x} + v\frac{\partial v}{\partial y} = \frac{-1}{\rho}\frac{\partial p}{\partial y}$$

and the continuity equation $\partial u/\partial x + \partial v/\partial y = 0$, show that if it is assumed that the velocity may be written in terms of an unsteady velocity potential $\phi = \phi(x, y, t)$ as $\mathbf{u} = \nabla\phi$, then, at each instant t, the potential $\phi(\boldsymbol{x}, t)$ must satisfy Laplace's equation $\phi_{xx} + \phi_{yy} = 0$. Show also that the Euler equations state that both the x and y partial derivatives of

$$\frac{\partial\phi}{\partial t} + \frac{1}{2}\left[\left(\frac{\partial\phi}{\partial x}\right)^2 + \left(\frac{\partial\phi}{\partial y}\right)^2\right] + \rho^{-1}p$$

are zero. Deduce that in *unsteady, irrotational* flows

(a) this quantity can depend only on t,

(b) and that there is no loss of generality in choosing this quantity to be constant.

[Note: It is ϕ_x and ϕ_y which have physical relevance. The choice (b) gives a version of Bernoulli's equation for *unsteady* flows. It proves useful, for example, in analysing water waves.]

7

Elastic Deformations

In many materials (steel, rubber, wood, etc) a *deformation* (i.e. a change of shape) can be sustained only if a system of forces (loads) is applied, so setting up internally a distribution of *stress* (i.e. force per unit area). If the body returns to its original shape when these loads are removed, the material is said to be *elastic.* The state in which no loads are applied is often taken as the *natural state*, or *reference configuration.* (The elastic string in §5.1 is a one-dimensional example which is rather degenerate, since resistance to bending is neglected. With $T(\lambda)$ chosen such that $T = 0$ whenever $\lambda = 1$, any unstretched state, even if curved, is a reference configuration if $|\partial \boldsymbol{r}/\partial X| \equiv \lambda = 1$.)

7.1 The Kinematics of Deformation

We find it convenient to introduce a *reference configuration* and then to analyse the geometry and kinematics of the subsequent *deformation*, in which a typical *material point* moves from P to P′. The position of the material point P in the reference configuration is written as $\overrightarrow{\mathrm{OP}} = X_1\boldsymbol{i} + X_2\boldsymbol{j} + X_3\boldsymbol{k} \equiv \boldsymbol{X}$ (so that $\{X_i\}$, with $i = 1, 2, 3$, are cartesian coordinates). Then, as the material is deformed, the labels (X_1, X_2, X_3) remain attached to the material point; they are *Lagrangian coordinates*, or material coordinates.

At time t, the material originally at P will be at some point P′, where

$$\overrightarrow{\mathrm{OP}}' = x_1\boldsymbol{i} + x_2\boldsymbol{j} + x_3\boldsymbol{k}\,.$$

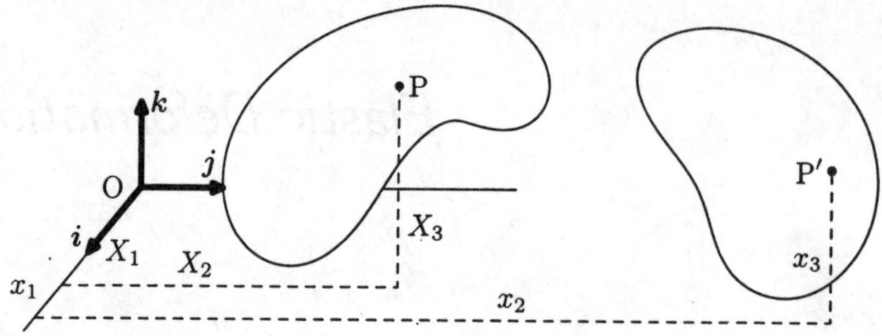

Figure 7.1 Reference and deformed configurations of a material region.

Here (x_1, x_2, x_3) are the *current*, or *Eulerian* coordinates. They will depend on X_1, X_2 and X_3 (and in unsteady deformations also on t), so that

$$\overrightarrow{\mathrm{OP}'} = \boldsymbol{r} = \boldsymbol{x}(\boldsymbol{X}, t) = x_1(\boldsymbol{X}, t)\boldsymbol{i} + x_2(\boldsymbol{X}, t)\boldsymbol{j} + x_3(\boldsymbol{X}, t)\boldsymbol{k} \tag{7.1}$$

describes the motion of each material point, or particle, labelled by $\boldsymbol{X}$.

The definitions of *displacement, particle velocity* and *acceleration* are as follows:

Displacement. This is the change in position of any particle, e.g.

$$\overrightarrow{\mathrm{PP}'} \equiv \boldsymbol{x}(\boldsymbol{X}, t) - \boldsymbol{X}.$$

The *displacement components* along $\boldsymbol{i}$, $\boldsymbol{j}$ and $\boldsymbol{k}$ are written as $u_i \equiv x_i - X_i$, with the index i taking the values $i = 1, 2, 3$ respectively.

(Particle) velocity. This is obtained simply as the derivative of $\boldsymbol{x}$ with respect to t, holding $\boldsymbol{X}$ fixed, namely the partial derivative

$$\boldsymbol{v}(\boldsymbol{X}, t) \equiv \frac{\partial \boldsymbol{x}}{\partial t}. \tag{7.2}$$

Acceleration. The acceleration of any material particle is

$$\boldsymbol{a}(\boldsymbol{X}, t) = \frac{\partial^2 \boldsymbol{x}}{\partial t^2} = \frac{\partial \boldsymbol{v}}{\partial t} \equiv a_1\boldsymbol{i} + a_2\boldsymbol{j} + a_3\boldsymbol{k} \qquad \text{(say)}.$$

Since partial differentiation with respect to t corresponds to holding each of X_1, X_2 and X_3 fixed, these formulae are simpler to use than those using the advected derivative (which is necessary when velocity and acceleration are expressed in terms of Eulerian coordinates $(x, y, z) = (x_1, x_2, x_3)$). Note also that the components a_1, a_2 and a_3 of $\boldsymbol{a}$ are

$$a_i(\boldsymbol{X}, t) = \frac{\partial^2 x_i}{\partial t^2} \qquad \text{for} \quad i = 1, 2, 3.$$

7.1.1 Deformation Gradient

In the vicinity of any chosen particle, the essential measure of deformation is the *deformation gradient.* To motivate this, consider two material points, which in the reference configuration are at neighbouring locations P and Q and which at time t occupy positions P′ and Q′. Then, if $\overrightarrow{\mathrm{PQ}} = \delta\boldsymbol{X}$, the relative displacement $\delta\boldsymbol{x}$ in the *deformed configuration* is obtained, by using the chain rule, as

$$\begin{aligned}\delta\boldsymbol{x} = \overrightarrow{\mathrm{P'Q'}} &= \boldsymbol{x}(\boldsymbol{X}+\delta\boldsymbol{X},t) - \boldsymbol{x}(\boldsymbol{X},t) \\ &\approx \frac{\partial\boldsymbol{x}}{\partial X_1}\delta X_1 + \frac{\partial\boldsymbol{x}}{\partial X_2}\delta X_2 + \frac{\partial\boldsymbol{x}}{\partial X_3}\delta X_3 .\end{aligned}$$

Resolving this vector equation into components along $\boldsymbol{i}$, $\boldsymbol{j}$ and $\boldsymbol{k}$ gives

$$\delta x_i \approx \sum_{j=1}^{3} \frac{\partial x_i}{\partial X_j}\,\delta X_j \tag{7.3}$$

for each value $i = 1, 2, 3$. At $(\boldsymbol{X}, t)$, the set of nine partial derivatives $\partial x_i/\partial X_j$ encodes all details of the linear approximation to the mapping $\overrightarrow{\mathrm{PQ}} \to \overrightarrow{\mathrm{P'Q'}}$. They describe the relative positions of *neighbouring* material particles. Collectively, they define the *deformation gradient* at the material point labelled by (X_1, X_2, X_3). The components of deformation gradient are denoted by F_{ij}, where

$$\frac{\partial x_i}{\partial X_j} \equiv F_{ij}(\boldsymbol{X}, t) .$$

If F_{ij} is treated as the ij element of a 3×3 matrix $\mathbf{F}$ and column vectors dx and $\mathsf{d}\mathbf{X}$ corresponding to the differentials $\mathrm{d}\boldsymbol{x}$ and $\mathrm{d}\boldsymbol{X}$ are introduced through

$$\mathsf{dx} \equiv \begin{pmatrix} \mathrm{d}x_1 \\ \mathrm{d}x_2 \\ \mathrm{d}x_3 \end{pmatrix} ; \quad \mathsf{d}\mathbf{X} \equiv \begin{pmatrix} \mathrm{d}X_1 \\ \mathrm{d}X_2 \\ \mathrm{d}X_3 \end{pmatrix} ; \quad \mathbf{F} \equiv \begin{pmatrix} F_{11} & F_{12} & F_{13} \\ F_{21} & F_{22} & F_{23} \\ F_{31} & F_{32} & F_{33} \end{pmatrix} , \tag{7.4}$$

the limiting form of (7.3) becomes[1]

$$\mathsf{dx} = \mathbf{F}\,\mathsf{d}\mathbf{X} . \tag{7.5}$$

Volume ratio. For a material element surrounding P, the ratio of the current volume to the reference volume is given by the Jacobian (a determinant)

$$\det\mathbf{F} = \frac{\partial(x_1, x_2, x_3)}{\partial(X_1, X_2, X_3)} \equiv J .$$

Note that, for physical reality, J can never be negative.

[1] Strictly, the deformation gradient should be treated as a *second-order cartesian tensor,* but when it is referred to a *fixed* orthonormal set of basis vectors its components may be treated as the elements of a matrix.

Example 7.1

Find $\mathbf{F}$ for the deformation in which the Eulerian coordinates are

$$x_1 = aX_1 + b\sin X_3\,, \quad x_2 = cX_1 + gX_2\,, \quad x_3 = h\cos kX_2 + qX_3\,.$$

Straightforward partial differentiation (e.g. $\partial x_1/\partial X_3 = b\cos X_3$) gives

$$\mathbf{F} = \begin{pmatrix} a & 0 & b\cos X_3 \\ c & g & 0 \\ 0 & -hk\sin kX_2 & q \end{pmatrix}.$$

Note that $J = \det\mathbf{F} = agq - bchk\cos X_3 \sin kX_2$, which satisfies the condition $J > 0$ for all X_2 and X_3 only if $agq > |bchk|$.

Example 7.2

Find $\mathbf{F}$ and $J \equiv \det\mathbf{F}$ for

$$x_1 = aX_1\cos\alpha - aX_2\sin\alpha + kX_3\,,\; x_2 = aX_1\sin\alpha + aX_2\cos\alpha\,,\; x_3 = bX_3\,.$$

Differentiating as before gives

$$\mathbf{F} = \begin{pmatrix} a\cos\alpha & -a\sin\alpha & k \\ a\sin\alpha & a\cos\alpha & 0 \\ 0 & 0 & b \end{pmatrix}; \qquad \text{while} \quad J = \det\mathbf{F} = a^2 b\,.$$

Observe that, in Example 7.2, $\mathbf{F}$ is a constant matrix; such a deformation gradient is said to be *uniform* and the material is said to be in a state of *homogeneous strain* (i.e. the deformation gradient is the same at all points).

7.1.2 Stretch and Rotation

The change of shape of a material region immediately surrounding any point is known as the *strain* at that point. It is analysed by considering Equation (7.5), which is a rule relating deformed infinitesimal *line elements* $\mathrm{d}\mathbf{X}$ at a material point to the corresponding undeformed line elements $\mathrm{d}\boldsymbol{x}$. To describe how the material initially in the vicinity of the material point labelled by (X_1, X_2, X_3), is stretched and rotated, it is convenient to consider the effect of the linear mapping associated with the matrix $\mathbf{F}$. In particular, the set of column vectors $\boldsymbol{l} = \mathbf{FL}$ obtained by considering all column vectors $\mathbf{L}$ of fixed magnitude (i.e. satisfying $\mathbf{L}^T\mathbf{L} = a^2$) describes an ellipsoid. Thus, if $\mathbf{L} = \mathrm{d}\mathbf{X}$ is the column vector corresponding to $\mathrm{d}\boldsymbol{X}$ and a is small, the corresponding material points are those which, in the reference configuration, lie on the sphere of radius a centred at $\boldsymbol{X}$. In the deformed configuration, the displacements $d\boldsymbol{x}$ relative to the point $\boldsymbol{x}(\boldsymbol{X},t)$ lie approximately on the ellipsoid represented by $\boldsymbol{l} = \mathbf{FL}$. (All small balls centred at $\boldsymbol{X}$ are deformed into solid ellipsoids each having similar shape.)

The ellipsoid is characterized by its three *principal axes* which, like the major and minor axes of an ellipse, are determined by seeking stationary values of the length $|\boldsymbol{l}|$ amongst all vectors $\mathbf{L}$ satisfying $|\mathbf{L}| = a$. As always, it is easier to consider squared lengths, using

$$\boldsymbol{l}^T\boldsymbol{l} = \mathbf{L}^T\mathbf{F}^T\mathbf{F}\mathbf{L}\,.$$

Definition: The *Cauchy–Green strain* matrix is defined by $\mathbf{G} \equiv \mathbf{F}^T\mathbf{F}$; its elements are

$$G_{ij} = \sum_{p=1}^{3} F_{pi}F_{pj} = G_{ji} \qquad (\mathbf{G}\text{ is symmetric}). \tag{7.6}$$

It turns out that the stationary values are provided by the eigenvalues and eigenvectors of $\mathbf{G}$. In deriving this result, it helps if we can use a shorthand notation:

Notation: Summation convention (Einstein). When an *index* is to be summed over the range 1–3 and occurs *exactly twice* within a product (e.g. p in (7.6)), the $\sum$ sign may be omitted (but summation is nevertheless presumed).

Thus, Equation (7.6) may be written as $G_{ij} = F_{pi}F_{pj} = F_{pj}F_{pi}$. Note also that, although this statement corresponds to a matrix multiplication, it is permissible to transpose the factors (since the positions of the various indices determine the choices of elements to be multiplied). For example, the squared distance $\boldsymbol{l}^T\boldsymbol{l} = \mathbf{L}^T\mathbf{G}\mathbf{L}$ may be written in a number of equivalent forms as

$$\boldsymbol{l}^T\boldsymbol{l} = l_il_i = l_1^2 + l_2^2 + l_3^2 = l_pl_p = L_iG_{ij}L_j.$$

Note that any repeated index should be summed over all possible values 1,2 and 3 and so cannot feature in the final expression. Thus it may be replaced (as here $i \to p$) by any other letter (so long as that letter does not appear elsewhere in the expression). Such an index is called a *dummy index*. For later convenience, the index i is removed from both the objective function $\boldsymbol{l}^T\boldsymbol{l}$ and the constraint equation $\mathbf{L}^T\mathbf{L} - a^2 = 0$. Hence, stationary values are sought for

$$f(L_1, L_2, L_3) \equiv G_{pj}L_pL_j - \mu(L_pL_p - a^2)\,,$$

in which μ is a Lagrange multiplier. Stationary values are identified by setting equal to zero the partial derivatives with respect to each of L_1, L_2, L_3 and μ. Thus, setting $\partial f/\partial L_1 = 0$ gives

$$G_{1j}L_j + G_{p1}L_p - 2\mu L_1 = 0$$

(since, for example $\partial L_p/\partial L_1 = 1$ for $p = 1$, and $= 0$ otherwise). The indicial rules then allow rewriting of the second term as $G_{p1}L_p = G_{j1}L_j = G_{1j}L_j$ (using the symmetry of $\mathbf{G}$). After division by 2, this yields

$$G_{1j}L_j = \mu L_1 .$$

Similarly, the conditions $\partial f/\partial L_2 = 0$ and $\partial f/\partial L_3 = 0$ give $G_{2j}L_j = \mu L_2$ and $G_{3j}L_j = \mu L_3$. Together these results may be summarized as $G_{ij}L_j = \mu L_i$ for each $i = 1, 2, 3$, which is exactly equivalent to the matrix statement

$$\mathbf{GL} = \mu \mathbf{L} . \tag{7.7}$$

Since (7.7) defines the three *eigenvectors* $\mathbf{L} \equiv \mathbf{L}^{(k)}$ and *eigenvalues* $\mu = \mu_k$ of the matrix $\mathbf{G}$ (for $k = 1, 2, 3$), it follows that:
The *principal axes of stretch* at the material point $\boldsymbol{X}$ are aligned along the material directions $\mathrm{d}\boldsymbol{X}$ corresponding to the *eigenvectors* $\mathbf{L}^{(1)}$, $\mathbf{L}^{(2)}$ and $\mathbf{L}^{(3)}$ of $\mathbf{G}$. Since $\mathbf{G}$ is symmetric, these eigenvectors are orthogonal (even in the case of repeated eigenvalues, an orthogonal set may be *chosen*). The corresponding eigenvalues μ_1, μ_2 and μ_3 have the property

$$|\boldsymbol{l}|^2 = \mathbf{L}^T\mathbf{GL} = \mu\mathbf{L}^T\mathbf{L} = \mu|\mathbf{L}|^2 \tag{7.8}$$

when $\boldsymbol{l} = \boldsymbol{l}^{(k)}$, $\mathbf{L} = \mathbf{L}^{(k)}$ and $\mu = \mu_k$ for $k = 1, 2, 3$.

Thus, when $|\mathbf{L}^{(k)}| = a$, corresponding to each eigenvector there is a stretched length $|\boldsymbol{l}^{(k)}| = \sqrt{\mu_k}a$, so that $\lambda_k = \sqrt{\mu_k}$ ($k = 1, 2, 3$) are the three *principal stretches* at $\boldsymbol{X}$. The *state of strain* at any point of a deformed material involves six parameters; the choice, within the reference configuration, of the orthogonal triad of principal directions $\{\mathbf{L}^{(k)}\}$ involves three parameters, while the three *principal stretches* $\{\lambda_k\}$ are another three. The remaining three parameters encoded within the *deformation gradient* matrix $\mathbf{F}$ describe a superimposed rigid body rotation of each infinitesimal ellipsoid surrounding $\boldsymbol{X}$, as represented by an appropriate orthogonal matrix $\mathbf{R}$. This will be described in the next section.

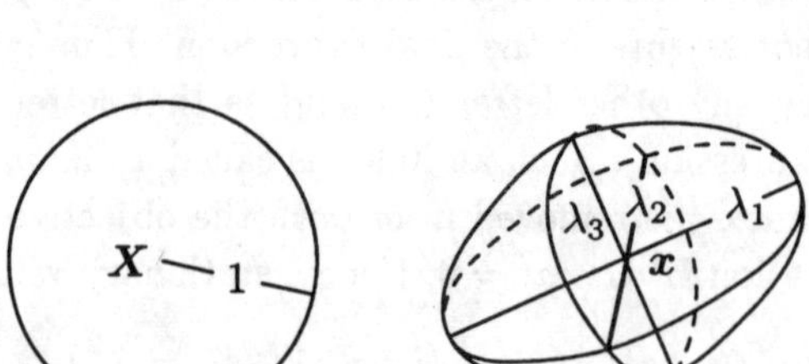

Figure 7.2 A sphere at $\boldsymbol{X}$ deformed into an ellipsoid at $\boldsymbol{x}$ with principal stretches λ_1, λ_2 and λ_3.

Example 7.3

For the deformation gradient matrix

$$\mathbf{F} = \begin{pmatrix} \cos\alpha & -\sin\alpha & -\frac{1}{3}\sin\alpha \\ \sin\alpha & \cos\alpha & \frac{1}{3}\cos\alpha \\ 0 & \frac{1}{3} & 1 \end{pmatrix},$$

find (i) the associated Cauchy–Green strain $\mathbf{G}$, (ii) its eigenvalues (and the corresponding principal stretches) and (iii) the unit eigenvectors of $\mathbf{G}$.

(i) Multiplication readily yields the Cauchy–Green strain matrix (which is symmetric)

$$\mathbf{G} = \mathbf{F}^T\mathbf{F} = \begin{pmatrix} 1 & 0 & 0 \\ 0 & \frac{10}{9} & \frac{2}{3} \\ 0 & \frac{2}{3} & \frac{10}{9} \end{pmatrix}.$$

(ii) The *eigenvalues* μ of $\mathbf{G}$ satisfy $\det(\mathbf{G} - \mu\mathbf{I}) = 0$, where $\mathbf{I}$ is the 3×3 unit matrix. Hence,

$$\begin{vmatrix} 1-\mu & 0 & 0 \\ 0 & \frac{10}{9}-\mu & \frac{2}{3} \\ 0 & \frac{2}{3} & \frac{10}{9}-\mu \end{vmatrix} = (1-\mu)\left[(\mu - \tfrac{10}{9})^2 - \tfrac{4}{9}\right] = 0,$$

so that $\mu = 1$ or $\mu = \frac{10}{9} \pm \frac{2}{3}$. If the roots are labelled as $\mu_1 = 1$, $\mu_2 = \frac{16}{9}$ and $\mu_3 = \frac{4}{9}$, the corresponding *principal stretches* are $\lambda_1 = 1$, $\lambda_2 = \frac{4}{3}$ and $\lambda_3 = \frac{2}{3}$. (N.B. The volume ratio is $J = \lambda_1\lambda_2\lambda_3 = 8/9$.)

(iii) Each *eigenvector* $\mathbf{L}^{(k)}$ of $\mathbf{G}$ is found in turn, using $(\mathbf{G} - \mu\mathbf{I})\mathbf{L}^{(k)} = \mathbf{0}$:

$$\text{For } \mu_1 = 1, \begin{pmatrix} 0 & 0 & 0 \\ 0 & \frac{1}{9} & \frac{2}{3} \\ 0 & \frac{2}{3} & \frac{1}{9} \end{pmatrix} \mathbf{L}^{(1)} = \mathbf{0} \text{ gives, as unit vectors, } \mathbf{L}^{(1)} = \pm \begin{pmatrix} 1 \\ 0 \\ 0 \end{pmatrix}.$$

$$\text{For } \mu_2 = \tfrac{16}{9}, \quad \begin{pmatrix} \frac{-2}{3} & 0 & 0 \\ 0 & \frac{-2}{3} & \frac{2}{3} \\ 0 & \frac{2}{3} & \frac{-2}{3} \end{pmatrix} \mathbf{L}^{(2)} = \mathbf{0} \quad \text{gives} \quad \mathbf{L}^{(2)} = \pm \begin{pmatrix} 0 \\ \frac{1}{\sqrt{2}} \\ \frac{1}{\sqrt{2}} \end{pmatrix}.$$

$$\text{For } \mu_3 = \tfrac{4}{9}, \quad \begin{pmatrix} \frac{5}{9} & 0 & 0 \\ 0 & \frac{2}{3} & \frac{2}{3} \\ 0 & \frac{2}{3} & \frac{2}{3} \end{pmatrix} \mathbf{L}^{(3)} = \mathbf{0} \quad \text{gives} \quad \mathbf{L}^{(3)} = \pm \begin{pmatrix} 0 \\ \frac{1}{\sqrt{2}} \\ \frac{-1}{\sqrt{2}} \end{pmatrix}.$$

7.2 Polar Decomposition

At *any* point in a deformed body, the deformation gradient matrix $\mathbf{F}$ may be written as

$$\mathbf{F} = \mathbf{RU}, \tag{7.9}$$

where $\mathbf{U}$ is a *symmetric* matrix having the *principal stretches* λ_1, λ_2 and λ_3 as eigenvalues, and $\mathbf{R}$ is a *proper orthogonal* matrix (i.e. $\mathbf{R}^T\mathbf{R} = \mathbf{I}$, with $\det\mathbf{R} = +1$). Locally, the state of deformation may be regarded as a state of *tri-axial stretching* (with principal stretches λ_1, λ_2 and λ_3 along material directions identified by the eigenvectors $\mathbf{L}^{(1)}$, $\mathbf{L}^{(2)}$ and $\mathbf{L}^{(3)}$) followed by a rotation described by the matrix $\mathbf{R}$ (see Figure 7.2). This statement is proved as follows:

The three orthonormal eigenvectors $\mathbf{L}^{(1)}$, $\mathbf{L}^{(2)}$ and $\mathbf{L}^{(3)}$ (labelled in suitable order so that $\det\mathbf{H} = +1$) are used as columns of the matrix

$$\mathbf{H} \equiv (\mathbf{L}^{(1)} \,|\, \mathbf{L}^{(2)} \,|\, \mathbf{L}^{(3)}),$$

which is proper orthogonal. Then, since for each $k = 1, 2, 3$, pre-multiplying $\mathbf{L}^{(k)}$ by the Cauchy–Green strain matrix $\mathbf{G} \equiv \mathbf{F}^T\mathbf{F}$ gives $\mathbf{GL}^{(k)} = \lambda_k^2\mathbf{L}^{(k)}$ (without summation), the product $\mathbf{GH}$ has the (partitioned) form

$$\mathbf{GH} = (\mathbf{GL}^{(1)} \,|\, \mathbf{GL}^{(2)} \,|\, \mathbf{GL}^{(3)}) = (\lambda_1^2\mathbf{L}^{(1)} \,|\, \lambda_2^2\mathbf{L}^{(2)} \,|\, \lambda_3^2\mathbf{L}^{(3)}) .$$

The property that $\mathbf{L}^{(j)T}\mathbf{L}^{(k)} = 1$ if $j = k$, but is zero otherwise, implies that

$$\mathbf{H}^T\mathbf{GH} = \mathrm{diag}\,(\lambda_1^2 \quad \lambda_2^2 \quad \lambda_3^2) = \mathbf{D}^2, \quad \text{where} \quad \mathbf{D} \equiv \begin{pmatrix} \lambda_1 & 0 & 0 \\ 0 & \lambda_2 & 0 \\ 0 & 0 & \lambda_3 \end{pmatrix}.$$

Equivalently (since $\mathbf{H}^T\mathbf{H} = \mathbf{I}$), writing $\mathbf{F}^T\mathbf{F} = \mathbf{G} = \mathbf{HD}^2\mathbf{H}^T$ yields a diagonalization of the symmetric matrix $\mathbf{G}$.

Now, *define* the symmetric matrix $\mathbf{U}$ as

$$\mathbf{U} = \mathbf{HDH}^T, \qquad \text{so that} \quad \mathbf{U}^2 = \mathbf{HD}^2\mathbf{H}^T = \mathbf{G}, \tag{7.10}$$

($\mathbf{U}$ is a *square-root* of $\mathbf{G}$) and write $\mathbf{F} = \mathbf{RU}$. It turns out that $\mathbf{R}$ is *proper orthogonal* and so describes a pure rotation. This is shown by using

$$\mathbf{R} = \mathbf{FU}^{-1} \qquad \text{so that} \qquad \mathbf{R}^T = (\mathbf{U}^{-1})^T\mathbf{F}^T = \mathbf{U}^{-1}\mathbf{F}^T,$$

thus giving

$$\mathbf{R}^T\mathbf{R} = \mathbf{U}^{-1}\mathbf{F}^T\mathbf{F}\mathbf{U}^{-1} = \mathbf{U}^{-1}\mathbf{G}\mathbf{U}^{-1} = \mathbf{U}^{-1}\mathbf{U}^2\mathbf{U}^{-1} = \mathbf{I}.$$

This shows that $\mathbf{R}$ is orthogonal. Moreover, since $\det \mathbf{F} = (\det \mathbf{R}) \det \mathbf{U}$ while $\det \mathbf{U} = \lambda_1\lambda_2\lambda_3 > 0$ and $\det \mathbf{F} > 0$, it follows that $\det \mathbf{R} > 0$, so that $\det \mathbf{R} = +1$. Hence, $\mathbf{R}$ is a proper orthogonal matrix and so describes a rotation (without a reflection).

Example 7.4

For $\mathbf{F}$ as in Example 7.3, find the matrices $\mathbf{U}$ and $\mathbf{R}$ in the polar decomposition.

One suitable choice, from Example 7.3, of column vectors $\mathbf{L}^{(k)}$ which gives a *proper* orthogonal matrix $\mathbf{H}$ is

$$\mathbf{H} = \begin{pmatrix} 1 & 0 & 0 \\ 0 & \frac{1}{\sqrt{2}} & \frac{-1}{\sqrt{2}} \\ 0 & \frac{1}{\sqrt{2}} & \frac{1}{\sqrt{2}} \end{pmatrix} \qquad \text{associated with} \qquad \mathbf{D} = \begin{pmatrix} 1 & 0 & 0 \\ 0 & \frac{4}{3} & 0 \\ 0 & 0 & \frac{2}{3} \end{pmatrix}.$$

This allows calculation of $\mathbf{U}$ as

$$\mathbf{U} = \mathbf{HDH}^T = \mathbf{H}\begin{pmatrix} 1 & 0 & 0 \\ 0 & \frac{2\sqrt{2}}{3} & \frac{2\sqrt{2}}{3} \\ 0 & \frac{-\sqrt{2}}{3} & \frac{\sqrt{2}}{3} \end{pmatrix} = \begin{pmatrix} 1 & 0 & 0 \\ 0 & 1 & \frac{1}{3} \\ 0 & \frac{1}{3} & 1 \end{pmatrix}$$

(for which it is readily verified that $\mathbf{U}^2 = \mathbf{G}$). To find $\mathbf{U}^{-1}$, either apply a standard matrix inversion procedure to $\mathbf{U}$ or, more simply, substitute $\mathbf{D}^{-1} = \operatorname{diag}(1 \;\; \frac{3}{4} \;\; \frac{3}{2})$ into $\mathbf{U}^{-1} = \mathbf{HD}^{-1}\boldsymbol{H}^T$, to give

$$\mathbf{U}^{-1} = \begin{pmatrix} 1 & 0 & 0 \\ 0 & \frac{9}{8} & \frac{-3}{8} \\ 0 & \frac{-3}{8} & \frac{9}{8} \end{pmatrix}, \text{ so that } \mathbf{R} = \mathbf{FU}^{-1} = \begin{pmatrix} \cos\alpha & -\sin\alpha & 0 \\ \sin\alpha & \cos\alpha & 0 \\ 0 & 0 & 1 \end{pmatrix}.$$

The matrix $\mathbf{R}$ describes an anticlockwise rotation through angle α, about $\mathrm{O}X_3$, so that the deformation gradient $\mathbf{F}$ describes the change of shape in which the principal stretches $1, \frac{4}{3}$ and $\frac{2}{3}$ occur along material directions identified by $(1\;0\;0)^T$, $(0\;\frac{1}{\sqrt{2}}\;\frac{1}{\sqrt{2}})^T$ and $(0\;\frac{-1}{\sqrt{2}}\;\frac{1}{\sqrt{2}})^T$ respectively, *followed by* the anticlockwise rotation about $\mathrm{O}X_3$. It may be verified (see Exercise 7.9) that, for any deformation, there is an alternative polar decomposition

$$\mathbf{F} = \mathbf{VR} \quad , \qquad \text{where } \mathbf{V} \equiv \mathbf{RUR}^T \;(= \mathbf{RHDH}^T\mathbf{R}^T\,).$$

Notice that $\mathbf{RH}$ is another proper orthogonal matrix and that its columns are the unit vectors $\mathbf{RL}^{(k)}$ for $k = 1, 2, 3$. Moreover, since $\mathbf{V}(\mathbf{RL}^{(k)}) = (\mathbf{RUR}^T)\mathbf{RL}^{(k)} = \mathbf{RUL}^{(k)} = \mathbf{R}(\lambda_k\mathbf{L}^{(k)}) = \lambda_k(\mathbf{RL}^{(k)})$, these vectors identify the current orientations of the material line elements which have undergone the principal stretches λ_k $(k = 1, 2, 3)$. In Example 7.4, these three directions are $(\cos\alpha \;\; \sin\alpha \;\; 0)^T$, $(\frac{-1}{\sqrt{2}}\sin\alpha \;\; \frac{1}{\sqrt{2}}\cos\alpha \;\; \frac{1}{\sqrt{2}})^T$ and $(\frac{1}{\sqrt{2}}\sin\alpha \;\; \frac{-1}{\sqrt{2}}\cos\alpha \;\; \frac{1}{\sqrt{2}})^T$. The deformation gradient $\mathbf{F}$ may be envisaged as a rotation described by $\mathbf{R}$ *preceding* the tri-axial stretching along the directions now aligned along $\mathbf{RL}^{(1)}$, $\mathbf{RL}^{(2)}$ and $\mathbf{RL}^{(3)}$.

EXERCISES

7.1. For each of the following deformations $\boldsymbol{X} \to \boldsymbol{x}$, obtain the elements F_{ij} of the deformation gradient matrix:
(a) $x_1 = AX_1 + DX_1X_3\,,\;\; x_2 = BX_2 + DX_2X_3\,,$
$x_3 = 1 + CX_3 - DX_3^2$ (A, B, C and D each constant);
(b) $x_1 = aX_1 \cos\kappa X_3 - aX_2 \sin\kappa X_3\,,$
$x_2 = aX_1 \sin\kappa X_3 + aX_2 \cos\kappa X_3\,,\;\; x_3 = bX_3$ (a, b, κ constant).

7.2. The deformation $x_1 = X_1 + aX_3$, $x_2 = X_2$, $x_3 = X_3$ is an example of a *simple shear* (in this case, each plane $X_3 =$ constant undergoes a rigid body displacement through distance aX_3 parallel to $\mathrm{O}X_1$). Sketch the reference and deformed configurations of the block defined by $0 \le X_1 \le 1$, $0 \le X_2 \le 1$ and $0 \le X_3 \le 1$, when $a \ge 0$. Determine the deformation gradient matrix $\mathbf{F}$. Confirm that $(0 \;\; 1 \;\; 0)^T$ is an eigenvector of $\mathbf{G} \equiv \mathbf{F}^T\mathbf{F}$. What is the associated eigenvalue?

7.3. Confirm that each of the following describes a *simple shear:*
(a) $x_1 = X_1 + aX_3\,,\quad x_2 = X_2 + bX_3\,,\quad x_3 = X_3\,;$
(b) $x_1 = (1 + a)X_1 - aX_2\,,\;\; x_2 = aX_1 + (1 - a)X_2\,,$
$x_3 = bX_1 - bX_2 + X_3\,.$
In each case, identify planes which undergo a rigid body translation within their own plane. Determine the direction of that translation.

7.4. For each simple shear of Exercise 7.3, determine the Cauchy–Green strain matrix $\mathbf{G}$. Confirm that 1 is one of its eigenvalues and that the corresponding eigenvector is parallel to the planes which undergo rigid translations (the *shear planes*).

Find the other two principal stretches. Confirm that their product equals 1 (there is no change in volume).

7.5. For $a = b = 1$ in Exercise 7.1(b), interpret the deformation as a *torsion* about the $\mathrm{O}X_3$ axis, in which a typical plane $X_3 =$ constant undergoes a rotation. What is the angle through which it is rotated?

Show that, in this case, it is possible to write $\mathbf{F} = \mathbf{RS}$, where $\mathbf{R}$ is the orthogonal matrix

$$\mathbf{R} = \begin{pmatrix} \cos\kappa X_3 & -\sin\kappa X_3 & 0 \\ \sin\kappa X_3 & \cos\kappa X_3 & 0 \\ 0 & 0 & 1 \end{pmatrix}$$

and $\mathbf{S}$ is upper triangular. Show that, at chosen (X_1, X_2), the matrix $\mathbf{S}$ corresponds to a simple shear of amount $\kappa\sqrt{X_1^2 + X_2^2}$. Interpret this as a shear of magnitude proportional to radial distance from the OX_3 axis and tangential to the cylindrical surface $X_1^2 + X_2^2 =$ constant.

7.3 Stress

In an elastic material, a change of shape can occur only if a system of forces is maintained. Consider first a simple case, in which $\boldsymbol{x} = \lambda_1 X_1 \boldsymbol{i} + \lambda_2 X_2 \boldsymbol{j} + \lambda_3 X_3 \boldsymbol{k}$, so that the material undergoes *principal stretches* λ_1, λ_2 and λ_3 along the coordinate axes OX_1, OX_2 and OX_3, respectively. This is a *homogeneous deformation*, in which each material element has the same deformation gradient.

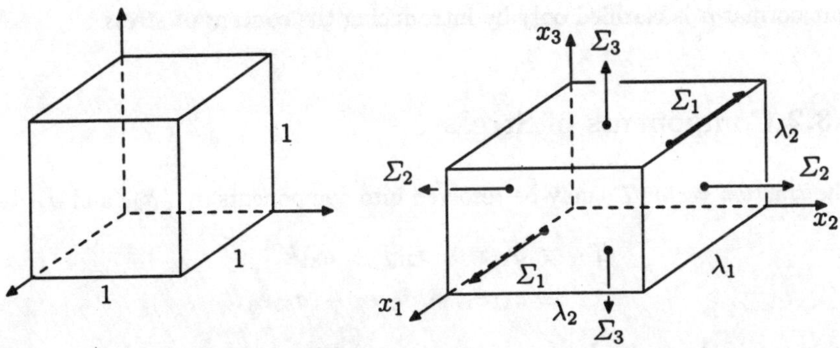

Figure 7.3 Normal tractions Σ_i arising when a unit cube is stretched in the ratios λ_1, λ_2 and λ_3 in directions parallel to the coordinate axes.

Imagine a unit cube of material in the reference configuration, with faces parallel to the coordinate planes. When this is deformed into a block with dimensions λ_1, λ_2 and λ_3, each face must carry a normal force distributed evenly over its area. The (outward) load over the faces normal to the Ox_1 axis are denoted by Σ_1, while those over the other faces are Σ_2 and Σ_3. Equivalently, the force per unit of *current area* normal to a coordinate axis will be one of

$$\frac{\Sigma_1}{\lambda_2\lambda_3}, \quad \frac{\Sigma_2}{\lambda_3\lambda_1} \quad \text{or} \quad \frac{\Sigma_3}{\lambda_1\lambda_2}.$$

7.3.1 Traction Vectors

On *any* element of surface, the force *acting over any unit of area* is known as the *traction*. For the rectangular block shown in Figure 7.3, the traction on each portion of its bounding surface is normal to that surface. However, on an element of surface within the block having general orientation denoted by a unit normal $\boldsymbol{n}$, the corresponding *traction* vector need not be parallel to $\boldsymbol{n}$.

More generally, suppose that a deformed material has deformation gradient matrix $\mathbf{F}$ at the *current position*. Imagine an element of surface at $\boldsymbol{x}$ currently having unit normal $\boldsymbol{n} = +\boldsymbol{i}$ and consider the traction (the force per unit area) exerted across this *by* the material on the side to which $\boldsymbol{i}$ is pointing *upon* the material on the opposite side of the surface. This traction is denoted by $\boldsymbol{T}_1$. It is associated with elements of surface having unit normal $\boldsymbol{i}$, but its direction is arbitrary. Clearly, by Newton's third law, the *reaction* is $-\boldsymbol{T}_1$ and so the traction vector associated with $-\boldsymbol{i}$ is $-\boldsymbol{T}_1$. On a surface element with any other current orientation $\boldsymbol{n}$, the traction vector can have a different magnitude and orientation. It may be denoted by $\boldsymbol{T} = \boldsymbol{T}(\boldsymbol{n})$, but its dependence upon the unit normal $\boldsymbol{n}$ is clarified only by introducing the concept of *stress.*

7.3.2 Components of Stress

The *traction vector* $\boldsymbol{T}_1$ may be resolved into components σ_{11}, σ_{21} and σ_{31} as

$$\begin{aligned} \boldsymbol{T}_1 &\equiv \sigma_{11}\boldsymbol{i} + \sigma_{21}\boldsymbol{j} + \sigma_{31}\boldsymbol{k} \\ &\equiv \sigma_{11}\boldsymbol{e}_1 + \sigma_{21}\boldsymbol{e}_2 + \sigma_{31}\boldsymbol{e}_3 \,. \end{aligned} \tag{7.11}$$

Here, σ_{11} is the *normal stress* in the $\boldsymbol{i} = \boldsymbol{e}_1$ direction, while σ_{21} and σ_{31} are *shear stresses* acting tangentially to a surface $x_1 =$ constant. Notice how a relabelling of the basis vectors as $\boldsymbol{i} \equiv \boldsymbol{e}_1$, $\boldsymbol{j} \equiv \boldsymbol{e}_2$ and $\boldsymbol{k} \equiv \boldsymbol{e}_3$ allows simpler identification of the *stress components* σ_{i1} with their direction $\boldsymbol{e}_i$ of action.

Similarly, on all planes $x_2 =$ const. and $x_3 =$ const., the tractions may be denoted by $\boldsymbol{T}_2$ and $\boldsymbol{T}_3$ respectively and resolved into components as

$$\begin{aligned} \boldsymbol{T}_2 &= \sigma_{12}\boldsymbol{e}_1 + \sigma_{22}\boldsymbol{e}_2 + \sigma_{32}\boldsymbol{e}_3 \,, \\ \boldsymbol{T}_3 &= \sigma_{13}\boldsymbol{e}_1 + \sigma_{23}\boldsymbol{e}_2 + \sigma_{33}\boldsymbol{e}_3 \,, \end{aligned}$$

or, in a more compact notation,

$$T_j = \sigma_{1j}e_1 + \sigma_{2j}e_2 + \sigma_{3j}e_3 = \sigma_{ij}e_i. \tag{7.12}$$

Notice again how both the summation convention and the relabelling of the unit vectors helps us by producing a compact statement.

Observe that, in a typical state of deformation, there are nine stress components σ_{ij} $(i, j = 1, 2, 3)$ at each point within the material. On elements of planes parallel to the Ox_i axes, the components σ_{11}, σ_{22} and σ_{33} are *normal stresses*, while all other stress components σ_{ij} for $i \neq j$ are *shear stresses.*

Although nine stress components exist (relative to the chosen basis directions), they are not independent. The 3×3 matrix $\boldsymbol{\sigma}$ of stress components is always *symmetric*, with

$$\sigma_{ji} = \sigma_{ij} \qquad \text{for all } i, j = 1, 2, 3. \tag{7.13}$$

This is indicated by the following simple example.

Example 7.5

By examining the moments of the tractions about the Ox_3 axis, show that, in a homogeneous deformation state, the cube

$$-1 < x_1 < 1, \quad -1 < x_2 < 1, \quad -1 < x_3 < 1$$

can remain in equilibrium only if $\sigma_{21} = \sigma_{12}$.

By symmetry about O, none of the normal stresses can exert a moment about the Ox_3 axis. Similarly, the shear tractions $\pm\sigma_{13}e_1$ and $\pm\sigma_{23}e_2$ acting over the faces $x_3 = \pm 1$ have zero resultant moment about this axis. Also, on the faces $x_1 = \pm 1$ and $x_2 = \pm 1$, the stress components σ_{3j} act parallel to the Ox_3 axis and so have no moment about it. Consequently, we need consider only the effects of the shear stress components $\pm\sigma_{12}e_1$ and $\pm\sigma_{21}e_2$. The analysis is essentially two-dimensional, as shown in Figure 7.4.

Since each face of the cube has area 4, the uniform shear stress σ_{21} acting over $x_1 = 1$ is equivalent to a resultant force $4\sigma_{21}$ acting at its midpoint P. As this face is at unit distance from the origin, this resultant force exerts a moment $4\sigma_{21}$ acting anti-clockwise about Ox_3. Likewise, the resultant $-4\sigma_{21}$ of the shear traction acting over the opposite face $x_1 = -1$ exerts an anticlockwise moment $4\sigma_{21}$. Also, the shear stresses acting over $x_2 = \pm 1$ each have moment $4\sigma_{12}$ acting clockwise. The cube can be in equilibrium only if the total moment $8\sigma_{21} - 8\sigma_{12}$ about the Ox_3 axis is zero. Hence, $\sigma_{21} = \sigma_{12}$. By taking moments about the other coordinate axes, Equation (7.13) is deduced.

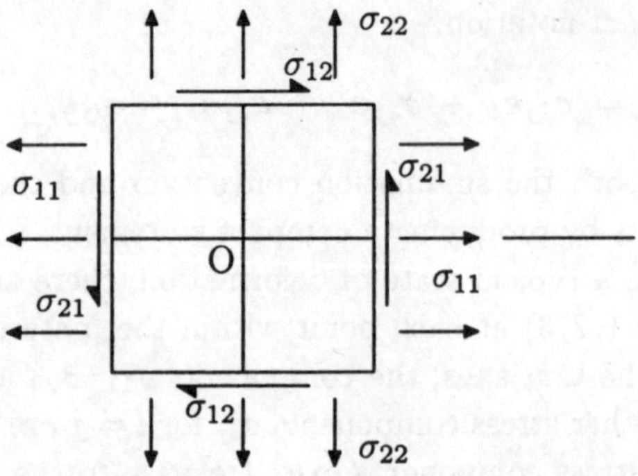

Figure 7.4 Traction components parallel to the Ox_1x_2 plane acting on the faces $x_1 = \pm 1$ and $x_2 = \pm 1$ of the cube of Example 7.5.

7.3.3 Traction on a General Surface

Knowing the components of stress σ_{ij} associated with coordinate surfaces $x_j =$ constant is sufficient to define the traction $\boldsymbol{T}$ on a surface element having *arbitrary* unit normal $\boldsymbol{n}$. The result (again using summation convention) is

$$\boldsymbol{T} = \boldsymbol{T}_j n_j = \boldsymbol{e}_1 \sigma_{1j} n_j + \boldsymbol{e}_2 \sigma_{2j} n_j + \boldsymbol{e}_3 \sigma_{3j} n_j = \boldsymbol{e}_i \sigma_{ij} n_j \,. \tag{7.14}$$

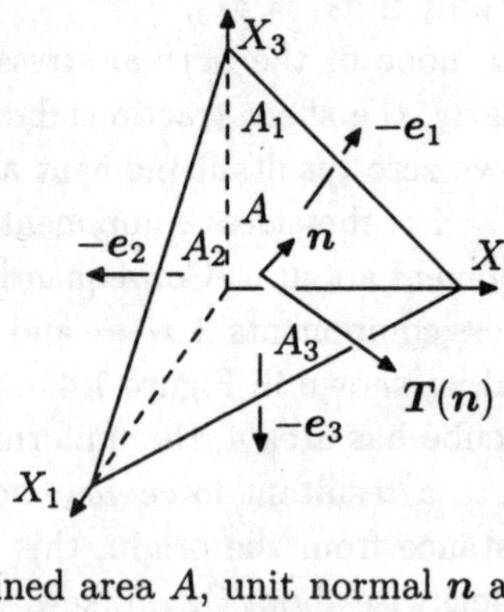

Figure 7.5 Inclined area A, unit normal $\boldsymbol{n}$ and traction $\boldsymbol{T} = \boldsymbol{T}(\boldsymbol{n})$.

This is demonstrated by considering the material tetrahedron lying between a plane passing through $\boldsymbol{x}$ with unit normal $\boldsymbol{n}$ and any three coordinate planes. Suppose that the tetrahedron is small, so that all points of its surface are sufficiently close to $\boldsymbol{x}$ that variations of σ_{ij} with position may be ignored. This amounts to considering again a portion of homogeneously deformed material. If A denotes the area of its sloping triangular face and (for simplicity) each component n_j of $\boldsymbol{n}$ is positive, then the faces lying in the coordinate planes have areas $A_j = A\boldsymbol{n}\cdot\boldsymbol{e}_j = An_j$ and outward normals $-\boldsymbol{e}_j$, respectively. Accordingly, the force on each plane face x_j = constant is $-An_j\boldsymbol{T}_j = -A_j\boldsymbol{T}_j$ (without summation). The (vector) sum of these three forces must exactly balance the force $A\boldsymbol{T}$ acting on the sloping plane. Hence, $A\boldsymbol{T} - A_1\boldsymbol{T}_1 - A_2\boldsymbol{T}_2 - A_3\boldsymbol{T}_3 = 0$, which yields Equation (7.14).

The conclusion is that, on any element of surface imagined within an elastic body and having unit normal $\boldsymbol{n}$, the component of the traction $\boldsymbol{T}$ in the direction $\boldsymbol{e}_i$ is $\sigma_{ij}n_j$. Thus it is a linear combination of the stress components acting over elements of coordinate surfaces at the same point, with the multipliers being just the components of $\boldsymbol{n}$.

7.4 Isotropic Linear Elasticity

Equation (7.14) describes how the traction components at a point depend upon the orientation of a surface element. It is also necessary to know how they depend on the state of strain. Specifically, it is necessary to know how the stress components σ_{ij} depend upon the components F_{ij} of the deformation gradient matrix.

Many different behaviours are found in practice, depending upon the material. However, there is substantial simplification if attention is restricted to *small displacements* $\boldsymbol{u} \equiv \boldsymbol{x} - \boldsymbol{X} = \boldsymbol{u}(\boldsymbol{X}, t)$, as occurs widely in the deformation of structures. The resulting theory retains in the analysis only terms linear in the displacement components $u_i = x_i - X_i$ and their derivatives.

An important concept is the *displacement gradient*. This has components

$$\frac{\partial u_i}{\partial X_j} = F_{ij} - \delta_{ij}$$

where δ_{ij} denotes the *Kronecker delta*, defined by

$$\delta_{ij} = \begin{cases} 1 & \text{if } i = j\,, \\ 0 & \text{otherwise}\,. \end{cases}$$

(Notice that the displacement gradient matrix is just $\mathbf{F} - \mathbf{I}$, where $\mathbf{I}$ is the 3×3 unit matrix.)

Since the stress components depend on the components F_{ij} of deformation gradient and are zero in the reference (stress-free) state ($\mathbf{F} = \mathbf{I}$), it is natural to take the stress components σ_{ij} to be linear in the nine elements $\partial u_i/\partial X_j$. This yields the *linear theory of elasticity*, which is a widely used approximation, valid when the *displacement gradient* is small. However, it may still involve a large number of material coefficients (or *elastic moduli*), particularly in some crystalline materials having complicated crystal symmetries.

Fortunately, the theory is considerably simplified for materials which have no inherent preferred directions, unlike wood or slate which have obvious *grain* or *layering*. Such a material is said to be *isotropic*. Important practical examples include rubber and many metals (unless they are inspected on the microscopic scale).

7.4.1 The Constitutive Law

In isotropic linear elasticity, *only two* elastic moduli need to be introduced (they are readily determined for a particular material by experiment). The rule relating stress to strain (the *stress–strain law*) is

$$\sigma_{ij} = \lambda \Delta \delta_{ij} + \mu \left(\frac{\partial u_i}{\partial X_j} + \frac{\partial u_j}{\partial X_i} \right) . \tag{7.15}$$

Here, the quantity Δ is the *dilatation*

$$\Delta \equiv \frac{\partial u_1}{\partial X_1} + \frac{\partial u_2}{\partial X_2} + \frac{\partial u_3}{\partial X_3} = \frac{\partial u_p}{\partial X_p} , \tag{7.16}$$

which measures the incremental change in volume per unit of reference volume. The material constants (or *elastic moduli*) λ and μ are known as the Lamé coefficients – μ is the *shear modulus*, while λ is the *bulk modulus*. Their rôles and significance will be shown through examples. However, first the concept of the *linearized strain* matrix **e** is introduced.

The components of **e** (a *symmetric* matrix) are evaluated as

$$e_{ij} = \frac{1}{2} \left(\frac{\partial u_i}{\partial X_j} + \frac{\partial u_j}{\partial X_i} \right) .$$

Its diagonal elements are $e_{11} = \partial u_1 / \partial X_1$, $e_{22} = \partial u_2 / \partial X_2$ and $e_{33} = \partial u_3 / \partial X_3$, so that Δ is its *trace* $\mathrm{tr}\,\mathbf{e} \equiv e_{11} + e_{22} + e_{33} = e_{kk} = \Delta$ (the sum of the diagonal elements). In terms of **e**, Equation (7.15) may be compactly written as

$$\sigma_{ij} = \lambda (\mathrm{tr}\,\mathbf{e}) \delta_{ij} + 2\mu e_{ij} = \lambda e_{kk} \delta_{ij} + 2\mu e_{ij} . \tag{7.17}$$

Clearly, σ_{ij} is always symmetric, as required by (7.13).

The name *linearized strain* results from the fact that, when terms quadratic in the elements of $\mathbf{e}$ are neglected, it follows that

$$\mathbf{G} = \begin{pmatrix} 1+\frac{\partial u_1}{\partial X_1} & \frac{\partial u_2}{\partial X_1} & \frac{\partial u_3}{\partial X_1} \\ \frac{\partial u_1}{\partial X_2} & 1+\frac{\partial u_2}{\partial X_2} & \frac{\partial u_3}{\partial X_2} \\ \frac{\partial u_1}{\partial X_3} & \frac{\partial u_2}{\partial X_3} & 1+\frac{\partial u_3}{\partial X_3} \end{pmatrix} \begin{pmatrix} 1+\frac{\partial u_1}{\partial X_1} & \frac{\partial u_1}{\partial X_2} & \frac{\partial u_1}{\partial X_3} \\ \frac{\partial u_2}{\partial X_1} & 1+\frac{\partial u_2}{\partial X_2} & \frac{\partial u_2}{\partial X_3} \\ \frac{\partial u_3}{\partial X_1} & \frac{\partial u_3}{\partial X_2} & 1+\frac{\partial u_3}{\partial X_3} \end{pmatrix}$$

$$\approx \begin{pmatrix} 1+2\frac{\partial u_1}{\partial X_1} & \frac{\partial u_2}{\partial X_1}+\frac{\partial u_1}{\partial X_2} & \frac{\partial u_3}{\partial X_1}+\frac{\partial u_1}{\partial X_3} \\ \frac{\partial u_1}{\partial X_2}+\frac{\partial u_2}{\partial X_1} & 1+2\frac{\partial u_2}{\partial X_2} & \frac{\partial u_3}{\partial X_2}+\frac{\partial u_2}{\partial X_3} \\ \frac{\partial u_1}{\partial X_3}+\frac{\partial u_3}{\partial X_1} & \frac{\partial u_2}{\partial X_3}+\frac{\partial u_3}{\partial X_2} & 1+2\frac{\partial u_3}{\partial X_3} \end{pmatrix},$$

so that, to this accuracy $\mathbf{G} \approx \mathbf{I} + 2\mathbf{e} \approx (\mathbf{I}+\mathbf{e})^2$. Consequently, the deformed length $|\boldsymbol{l}|$ of a line element represented by $\mathbf{L}$ is $|\boldsymbol{l}| = (\boldsymbol{l}^T\boldsymbol{l})^{1/2} = (\mathbf{L}^T\mathbf{G}\mathbf{L})^{1/2} \approx \{\mathbf{L}^T(\mathbf{I}+\mathbf{e})^T(\mathbf{I}+\mathbf{e})\mathbf{L}\}^{1/2} = |(\mathbf{I}+\mathbf{e})\mathbf{L}|$. Thus, the symmetric matrix $\mathbf{e}$ (with all elements much smaller than unity) allows calculation of the relative change of length for all line elements at a point $\boldsymbol{x}$. Moreover, to the same accuracy, the ratio of reference volume to deformed volume is $\det \mathbf{F} \approx 1 + e_{11} + e_{22} + e_{33}$, so confirming Δ as the incremental change in volume.

It can be shown (see Exercise 7.11) that, to similar accuracy, the local rotation matrix $\mathbf{R}$ is approximately $\mathbf{I}+\mathbf{S}$, where the skew-symmetric (or anti-symmetric) matrix $\mathbf{S}$ has elements $S_{ij} = \frac{1}{2}(\partial u_i/\partial X_j - \partial u_j/\partial X_i)$.

7.4.2 Stretching, Shear and Torsion

Example 7.6

Find the components of displacement gradient and of stress when $u_1 = aX_1$, $u_2 = u_3 = 0$.

Here $\dfrac{\partial u_1}{\partial X_1} = a\,,\quad \dfrac{\partial u_i}{\partial X_j} = 0$ otherwise. Thus, there is an extension $e_{11} = a$ in the $\mathrm{O}X_1$ direction, without any lateral change of dimension. The dilatation is $\Delta = a + 0 + 0 = a$, so that

$$\begin{pmatrix} \sigma_{11} & \sigma_{12} & \sigma_{13} \\ \sigma_{21} & \sigma_{22} & \sigma_{23} \\ \sigma_{31} & \sigma_{32} & \sigma_{33} \end{pmatrix} = \lambda a \begin{pmatrix} 1 & 0 & 0 \\ 0 & 1 & 0 \\ 0 & 0 & 1 \end{pmatrix} + 2\mu \begin{pmatrix} a & 0 & 0 \\ 0 & 0 & 0 \\ 0 & 0 & 0 \end{pmatrix}.$$

Reading off components gives $\sigma_{11} = (\lambda + 2\mu)a$ (the *axial tensile stress*, which is proportional to the *axial extension* a). However, this is not the only non-zero stress component, since $\sigma_{22} = \sigma_{33} = \lambda a$. Thus, a one-dimensional extension without lateral contraction must be accompanied by a *lateral stress* which is a *positive* multiple of a.

Also, $\sigma_{ij} = 0$ for all $i \neq j$.

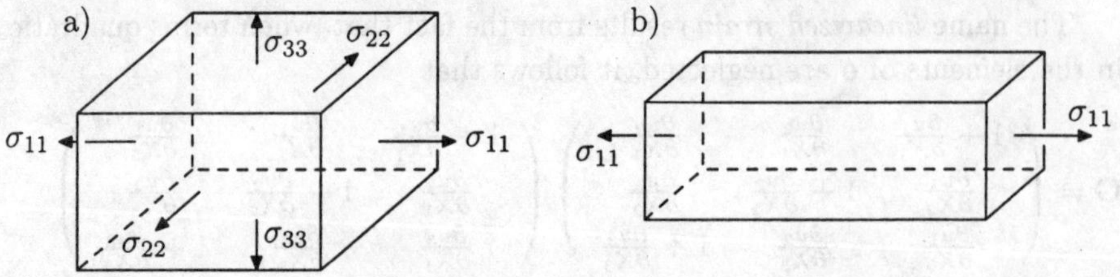

Figure 7.6 Normal components of stress on the faces of a block stretched parallel to Ox_1 a) as in Example 7.6 without lateral displacements and b) with lateral contraction as for the bar as in Example 7.7.

The calculation of Example 7.6 is readily modified to allow for arbitrary lateral expansion or contraction:

Example 7.7

Find the components of displacement gradient and stress when $u_1 = aX_1$, $u_2 = bX_2$, $u_3 = bX_3$.

Here $\dfrac{\partial u_1}{\partial X_1} = a = e_{11}$, $e_{22} = \dfrac{\partial u_2}{\partial X_2} = b = \dfrac{\partial u_3}{\partial X_3} = e_{33}$ and $e_{ij} = 0$ otherwise. Since the dilatation is $\Delta = a + b + b = a + 2b$, the stress components are

$$\begin{pmatrix} \sigma_{11} & \sigma_{12} & \sigma_{13} \\ \sigma_{21} & \sigma_{22} & \sigma_{23} \\ \sigma_{31} & \sigma_{32} & \sigma_{33} \end{pmatrix} = \lambda(a+2b)\begin{pmatrix} 1 & 0 & 0 \\ 0 & 1 & 0 \\ 0 & 0 & 1 \end{pmatrix} + 2\mu \begin{pmatrix} a & 0 & 0 \\ 0 & b & 0 \\ 0 & 0 & b \end{pmatrix},$$

so giving the *axial tension* $\sigma_{11} = (a+2b)\lambda + 2a\mu$, the *lateral stresses* $\sigma_{22} = \sigma_{33} = (a+2b)\lambda + 2b\mu$ and with all other stress components vanishing ($\sigma_{ij} = 0$ for $i \neq j$).

Example 7.7 allows determination of b such that a bar, or rod, may be stretched with extension a so that all lateral boundaries remain traction-free (i.e. $\sigma_{i2} = 0 = \sigma_{i3}$). Clearly, $\sigma_{22} = \sigma_{33} = \lambda a + 2(\lambda+\mu)b = 0$ is the required choice, so leading to

$$b = \frac{-\lambda}{2(\lambda+\mu)}a.$$

In this deformation, $\sigma_{11} \propto a$ and $\sigma_{ij} = 0$ otherwise.

Definition: The *Poisson ratio* ν is defined as

$$\nu = \frac{\lambda}{2(\lambda+\mu)}.$$

It is the ratio of the *lateral contraction* $-b$ to the extension a when a bar or rod (of any uniform cross-sectional shape) is in *uniaxial tension* (with lateral surfaces free of traction). Thus a principal stretch $1+a$ along the bar is accompanied by change of all lateral dimensions by the ratio $1+b=1-\nu a$. The phenomenon is known as *Poisson contraction.*
Definition. Young's modulus is given by

$$E \equiv \frac{\mu(3\lambda+2\mu)}{\lambda+\mu}.$$

It describes the ratio of tension to extension when an elastic rod is stretched in uniaxial tension (cf. Example 7.7). This agrees with Hooke's Law, which states that tension is proportional to extension ($\sigma_{11}=Ea=E\partial u_1/\partial X_1$).

Example 7.8 (Simple shear)

The displacements $u_1=\gamma X_2$, $u_2=u_3=0$ involve no change in volume (i.e. $\Delta=0$). Find the corresponding stress distribution when γ is small.

Since $u_3=0$, (i.e. $x_3=X_3$) a two-dimensional sketch illustrates the deformation. Each plane $X_2=$ constant slides in the Ox_1 direction, through a distance γX_2, which is proportional to X_2. The parameter γ is the *shear*, such that $\tan^{-1}\gamma$ is the *angle of shear.* Since $\dfrac{\partial u_1}{\partial X_2}=\gamma$, but $\dfrac{\partial u_i}{\partial X_j}=0$ otherwise, then $e_{12}=\frac{1}{2}\gamma=e_{21}$ with all other components of $\mathbf{e}$ equal to zero. Provided $|\gamma|$ is small, the stress components are given by linear elasticity as

$$\sigma_{12}=\mu\gamma=\sigma_{21} \quad \text{but} \quad \sigma_{ij}=0 \quad \text{otherwise.}$$

(Moreover, for a slab defined by $0<X_2<b$, the *displacements* are everywhere small when $b|\gamma|$ is small.)

Consider the traction over any plane $X_2=c$ (with unit normal $+\mathbf{e}_2$). Then, since $\sigma_{22}=\sigma_{32}=0$, the only non-vanishing component of traction is σ_{12}. The traction vector is then in the direction of the shear displacements. Moreover, the ratio of shear stress to shear equals μ, which explains the name *shear modulus.* Note also that, since $\sigma_{21}=\sigma_{12}=\mu\gamma\neq 0$ there are shear stress components acting on the planes $x_1=$ constant (this result is not so intuitive, but is consistent with the analysis in Example 7.5).

It is possible also to consider dynamic (unsteady) shear deformations, in which $u_1=u_1(X_2,t)$, while $u_2=0=u_3$ (see Example 7.10 and also Section 8.4.1). However, these *shear waves*, like any static deformations with non-uniform deformation gradient, require that the balance of forces is analysed. For static problems, the *equilibrium equations* of linear elasticity are obtained,

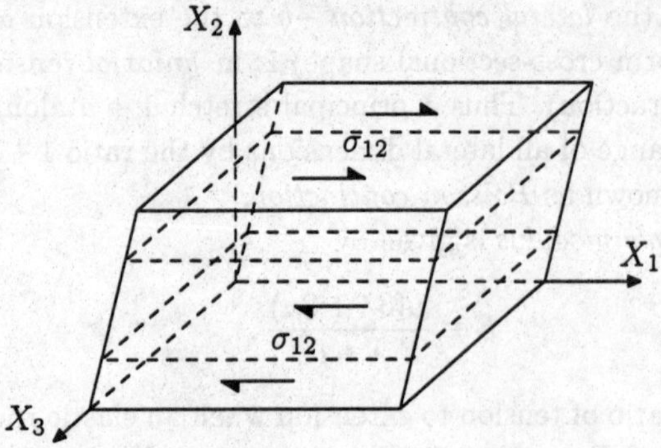

Figure 7.7 Simple shear of planes $X_2 =$ constant in the direction parallel to OX_1, showing associated shear tractions σ_{12}.

rather like the balance laws of Chapter 4, by considering the total (or resultant) force acting on the material currently occupying the region $\mathcal{R}$. At a typical point within $\mathcal{R}$, there may be a body force per unit volume $\rho \boldsymbol{b}$ (where ρ is the density and the body force $\boldsymbol{b}$ acting on unit mass has cartesian components b_i in the directions of $\boldsymbol{e}_i$). At a typical point of the surface $\partial\mathcal{R}$ with outward unit normal $\boldsymbol{n}$ and area δS, the $\boldsymbol{e}_i$ component of the tractive force is $\sigma_{ij}n_j\delta S$. Thus, the total force in the $\boldsymbol{e}_i$ direction $(i = 1, 2, 3)$ is the aggregate of the body and surface forces and must vanish, if the material within the region $\mathcal{R}$ is to remain at rest. This requires that

$$\iint_{\partial\mathcal{R}} \sigma_{ij}n_j \mathrm{d}S + \iiint_{\mathcal{R}} \rho b_i \mathrm{d}V = 0 .$$

Now, the divergence theorem (4.4) may be applied to the first term since, for each choice $i = 1, 2, 3$ a vector $\boldsymbol{\sigma}_i$ may be defined by $\boldsymbol{\sigma}_i = \sigma_{i1}\boldsymbol{e}_1 + \sigma_{i2}\boldsymbol{e}_2 + \sigma_{i3}\boldsymbol{e}_3$, so that $\sigma_{ij}n_j = \boldsymbol{\sigma}_i \cdot \boldsymbol{n}$. Thus, the law of static balance of forces becomes

$$\iiint_{\mathcal{R}} \{\operatorname{div}\boldsymbol{\sigma}_i + \rho b_i\} \mathrm{d}V = \iiint_{\mathcal{R}} \left(\frac{\partial \sigma_{ij}}{\partial x_j} + \rho b_i\right) \mathrm{d}V = 0 . \qquad (7.18)$$

As previously, it is legitimate to argue that a statement such as (7.18) can hold for *all* regions $\mathcal{R}$ only if the integrand vanishes everywhere. Hence, the *equilibrium equation* for static linear elasticity is

$$\frac{\partial \sigma_{ij}}{\partial x_j} + \rho b_i = 0 \qquad (i = 1, 2, 3). \qquad (7.19)$$

In situations in which the body force (typically due to gravity) is unimportant when compared with forces arising from the stresses, Equation (7.19) simplifies to

$$\frac{\partial \sigma_{ij}}{\partial x_j} = 0 \qquad (i = 1, 2, 3). \qquad (7.20)$$

This may be recognized as the statement that each of the three vector fields $\sigma_i(x_1, x_2, x_3)$ has zero divergence. Clearly, in each of Examples 7.6–7.8 the stress is uniform, so that Equation (7.20) is trivially satisfied for each $i = 1, 2, 3$.[2] For the next important example, static simple torsion, the equilibrium equation (7.20) is satisfied, but less trivially.

Example 7.9 (Simple torsion of a circular rod)

Suppose that each plane X_3 = constant is rotated about the Ox_3 axis through a small angle proportional to X_3. If all displacements are small, approximate the configuration of Example 7.5 to show that $u_1 \approx -\kappa X_3 X_2$, $u_2 = \kappa X_3 X_1$ and $u_3 = 0$. Find the corresponding stress components σ_{ij}. After replacing $\boldsymbol{X}_k$ by $\boldsymbol{x}_k$ in the expressions for σ_{ij}, verify that these expressions satisfy the equilibrium equation (7.20).

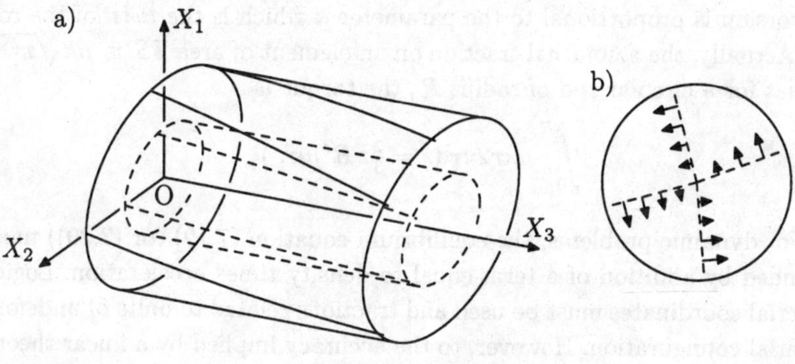

Figure 7.8 a) Rotation of planes X_3 = constant through an angle κX_3 around the OX_3 axis. b) Azimuthal displacements within a plane X_3 = const., when κX_3 is small.

Use the approximations $\cos \kappa X_3 \approx 1$, $\sin \kappa X_3 \approx \kappa X_3$ in Exercise 7.1(b), so making u_1 and $u_2 \propto X_3$. Differentiating then gives

$$\frac{\partial u_i}{\partial X_j} = \begin{pmatrix} 0 & -\kappa X_3 & -\kappa X_2 \\ \kappa X_3 & 0 & \kappa X_1 \\ 0 & 0 & 0 \end{pmatrix} \qquad \text{which yields} \quad \Delta = 0\,.$$

[2] The discerning reader will have noted that, in (7.18) and (7.19), the stress components σ_{ij} are regarded as functions of the *current location* $\boldsymbol{x}$ rather than functions of $\boldsymbol{X}$, as in (7.15). However, in considering small displacements, it is commonplace to associate stresses and indeed displacements (u_1, u_2, u_3) with the Eulerian coordinates x_1, x_2 and x_3 (often written as the familar coordinates (x, y, z)). The error is consistent with the approximations used in deriving the stress–strain law (7.15).

Hence $\mathbf{e}$ is found and inserted into (7.17) to show that

$$\begin{pmatrix} \sigma_{11} & \sigma_{12} & \sigma_{13} \\ \sigma_{21} & \sigma_{22} & \sigma_{23} \\ \sigma_{31} & \sigma_{32} & \sigma_{33} \end{pmatrix} = \mu \begin{pmatrix} 0 & 0 & -\kappa X_2 \\ 0 & 0 & \kappa X_1 \\ -\kappa X_2 & \kappa X_1 & 0 \end{pmatrix}$$

$$\approx \mu \begin{pmatrix} 0 & 0 & -\kappa x_2 \\ 0 & 0 & \kappa x_1 \\ -\kappa x_2 & \kappa x_1 & 0 \end{pmatrix}.$$

To check that the stresses satisfy (7.20), compute $\partial\sigma_{11}/\partial x_1 + \partial\sigma_{12}/\partial x_2 + \partial\sigma_{13}/\partial x_3 = 0 + 0 + \partial(-\kappa x_2)/\partial x_3 = 0$ (for $i = 1$), $0 + 0 + \partial(\kappa x_1)/\partial x_3 = 0$ (for $i = 2$) and $\partial(-\kappa x_2)/\partial x_1 + \partial(\kappa x_1)/\partial x_2 + 0 = 0$ (for $i = 3$).

Observe that on *any* plane $X_3 =$ constant, the traction components are $\sigma_{13} = -\mu\kappa x_2$, $\sigma_{23} = \mu\kappa x_1$ and $\sigma_{33} = 0$. Thus the stress is a shear stress in planes containing the axial and azimuthal direction, with magnitude $\mu\kappa$ times the radius. The *torque* (i.e. the total moment about the axis) required to cause the torsion is proportional to the parameter κ which is the *twist* of the rod.

(Actually, the azimuthal traction on an element of area δS is $\mu\kappa\sqrt{x_1^2 + x_2^2}$ so that for a circular rod of radius R, the torque is

$$\int_0^R \mu\kappa r 2\pi r \mathrm{d}r = \tfrac{2}{3}\pi R^3 \mu\kappa\,.)$$

For dynamic problems, the equilibrium equation (7.19) (or (7.20)) must be amended by addition of a term equal to density times acceleration. Logically, material coordinates must be used and tractions related to units of undeformed material configuration. However, to the accuracy implied by a linear theory, in either equation (7.19) or (7.20) the correct term to include is $-\rho\partial^2 u_i/\partial t^2$. When body forces are negligible, this leads to the elastodynamic equation

$$\rho\frac{\partial^2 u_i}{\partial t^2} = \frac{\partial \sigma_{ij}}{\partial x_j} \qquad (i = 1, 2, 3)\,. \tag{7.21}$$

A simple example of elastic motions is the dynamic version of simple shear, which illustrates the phenomenon of shear waves.

Example 7.10 (Shear waves)

Consider shearing motions purely parallel to the OX_2 axis, but depending on only x_1 and t. They are described by $u_2 = u(x_1, t), u_1 = u_3 = 0$. Show that $u(x_1, t)$ satisfies the wave equation. What is the wavespeed?

The acceleration is in the $j = \mathbf{e}_2$ direction, with magnitude $\dfrac{\partial^2 u}{\partial t^2}$. Since the only non-zero components of stress are $\sigma_{12} = \sigma_{21} = \mu\dfrac{\partial u}{\partial x_1}$, it transpires that the shear stress, like the acceleration is independent of position on *each* plane $x_1 =$ constant. The elastodynamic equation (7.21) is thus satisfied trivially for $i = 1$ and $i = 3$. For $i = 2$, it becomes

$$\rho\frac{\partial^2 u}{\partial t^2} = \frac{\partial}{\partial x_1}\left(\mu\frac{\partial u}{\partial x_1}\right)$$

so that $u(x_2, t)$ satisfies the *wave equation* (5.11), in which the wavespeed is $c \equiv \sqrt{\mu/\rho}$. This quantity is known as the *shear speed* $c = c_S$ of the elastic material. It is important in seismology. Clearly, by analogy with the solutions of (5.11), waves with $u_2 = f(x_1 - c_S t)$ and having *any* waveform may travel at speed c_S in the positive direction parallel to Ox_1, while other disturbances $u_2 = g(x_1 + c_S t)$ may travel in the opposite direction. In fact, shear waves may propagate in *any direction*, while the material motion may be in *any* direction orthogonal to that *propagation direction.*

(There also are other important elastic wavespeeds. For example, *longitudinal waves* travel faster – at wavespeed $\sqrt{(\lambda + 2\mu)/\rho}$, known as the *dilatational speed* c_L – notice how $\lambda + 2\mu$ is the ratio of longitudinal stress to extension when there is no Poisson contraction (see Example 7.7). These, and other, elastic waves are discussed in Chapter 8.)

EXERCISES

7.6. Use the relation $\partial u_i/\partial X_j = F_{ij} - \delta_{ij}$ between the components of the displacement gradient and those of the deformation gradient matrix $\mathbf{F}$ to find the displacement gradient corresponding to each of the deformations of Exercise 7.3(a) and (b). Confirm these results by first identifying the components $u_i(\boldsymbol{X})$ and then differentiating (observe how the elements $\partial u_i/\partial X_j$ are linear in a and b). Relate the corresponding matrices with elements $\partial u_i/\partial X_j$ to the shearing deformations found in Exercise 7.3.

7.7. For the Cauchy–Green matrices $\mathbf{G}$ computed in Exercise 7.4 for the deformations in Exercise 7.3(a) and (b), relate the linearizations (for small a and b) to e_{ij}.

7.8. Use the *stress–strain* law (7.17) to determine the stress components σ_{ij} in each of the following cases (it may help first to calculate the dilatation $\Delta = \operatorname{tr}\mathbf{e}$):
(a) $u_1 = AX_1\,,\ u_2 = BX_2\,,\ u_3 = 0\,;$
(b) $u_1 = 0\,,\ u_2 = \gamma X_1\,,\ u_3 = 0\,;$

(c) $u_1 = AX_3\,,\ \ u_2 = BX_3\,,\ \ u_3 = 0$;
(d) the matrix $\mathbf{e}$ of Exercise 7.7(b).

7.9. If the displacement components (u_1, u_2, u_3) are given by $u_1 = aX_1 + A\cos(X_2 - \omega t)$, $u_2 = bX_2 + cX_3$, $u_3 = dX_3 - A\sin(X_2 - \omega t)$, determine the velocity components $v_i = \partial u_i/\partial t$ and the components $\partial u_i/\partial X_j$ of the displacement gradient.

Calculate the corresponding components of $\mathbf{e}$, the dilatation Δ and the stress components σ_{ij} (using (7.17)).

Confirm that these displacements satisfy the elastodynamic equation $\rho\partial v_i/\partial t = \partial\sigma_{ij}/\partial X_j$ when $\rho\omega^2 = \mu$.

7.10. Derive from the result $\mathbf{F} = \mathbf{RU}$, where $\mathbf{R}^T\mathbf{R} = \mathbf{I}$ and $\mathbf{U} = \mathbf{U}^T$, the left-polar decomposition $\mathbf{F} = \mathbf{VR}$. Confirm that $\mathbf{V}$ is symmetric and has the same eigenvalues as does $\mathbf{U}$.

7.11. Use the polar decomposition $\mathbf{F} = \mathbf{RU}$ and the approximation $\mathbf{U} \approx \mathbf{I} + \mathbf{e}$ to show that, correct to terms linear in the elements $\partial u_i/\partial X_j$, the elements of $\mathbf{R}$ are $\delta_{ij} + \frac{1}{2}(\partial u_i/\partial X_j - \partial u_j/\partial X_i)$. (This confirms the statement of Section 7.4.1 that $\mathbf{R} \approx \mathbf{I} + \mathbf{S}$, where $\mathbf{S}$ is skew-symmetric.)

8 Vibrations and Waves

8.1 Wave Reflection and Refraction

In previous chapters, wavelike motion of strings, of gas within tubes and of elastic bodies has been encountered. Waves are, of course, all around us – as surface motions on seas and lakes and in radio, light and numerous other *electromagnetic* phenomena. This chapter aims to illustrate how the mathematics of the complex exponential $e^{\imath\omega t} = \cos\omega t + \imath\sin\omega t$ (where $\imath^2 = -1$) is particularly useful for analysing many features of waves, such as reflection, refraction and guiding by surfaces or within layers or ducts. Through its use, many features common to acoustic, elastic and electromagnetic waves will be revealed.

8.1.1 Use of the Complex Exponential

The simplest ODE which describes oscillations is the *simple harmonic (SHM) equation* (cf. (2.5))

$$\ddot{x}(t) + \omega^2 x = \frac{\mathrm{d}^2 x}{\mathrm{d}t^2} + \omega^2 x = 0 \tag{8.1}$$

in which dots denote time derivatives, and for which the general solution may be written in a number of equivalent ways as

$$x = a\cos\omega t + b\sin\omega t = A\cos(\omega t - \phi) = A\cos\omega(t - t_0) = \mathrm{Re}\left(Ae^{\imath(\phi-\omega t)}\right).$$

Here, $A = \sqrt{a^2+b^2}$ is the *amplitude* of the oscillation $x(t)$, ω is its *frequency* (measured in radians/second) and ϕ is its *phase lag* (where $\cos\phi = a/\sqrt{a^2+b^2}$ and $\sin\phi = b/\sqrt{a^2+b^2}$). Observe that this general solution to (8.1) may be written as $x = \text{Re}\,(\mathcal{A}e^{-\imath\omega t})$, where the *complex amplitude* $\mathcal{A}$ records both the amplitude $A = \text{Re}\,\mathcal{A}$ and the phase $\phi = \arg\mathcal{A}$. This notation is readily extended to describe *travelling waves* such as $u = A\cos k(x-ct-x_0)$ to be written as $u(x,t) = \text{Re}\,(\mathcal{A}e^{\imath k(x-ct)})$, where $\arg\mathcal{A} = -kx_0$. In many calculations, it is common to omit the symbol Re, to work using complex algebra but to recognise that only the real part of the final answer has physical significance.

Example 8.1

By superposing the two special solutions $e^{\pm\imath\omega t}$ of (8.1), show that the general *real* solution has the form $x(t) = \text{Re}(\mathcal{A}e^{-\imath\omega t})$ in which $\mathcal{A}$ is an arbitrary complex constant.

Write $x(t) = a_+e^{\imath\omega t} + a_-e^{-\imath\omega t}$, where a_+ and a_- are complex constants. This is the most general complex solution of (8.1) (since it allows arbitrary choice of $x(0)$ and $\dot{x}(0)$). A solution $x(t)$ is real for all t only if $\bar{x}(t) = x(t)$, where a bar denotes the complex conjugate. Thus, rearranging the statement $x(t) - \bar{x}(t) = 0$ gives $a_+e^{\imath\omega t} + a_-e^{-\imath\omega t} - \bar{a}_+e^{-\imath\omega t} - \bar{a}_-e^{\imath\omega t} = 0$, so requiring that $a_+ - \bar{a}_- = 0$ and $a_- - \bar{a}_+ = 0$. Both these requirements are satisfied if $a_- = \bar{a}_+$. Write $a_- = \bar{a}_+ = \frac{1}{2}\mathcal{A}$, then the general real solution becomes

$$x(t) = \tfrac{1}{2}\{\mathcal{A}e^{-\imath\omega t} + \bar{\mathcal{A}}e^{\imath\omega t}\} = \text{Re}\{\mathcal{A}e^{-\imath\omega t}\}\,.$$

(N.B. The choice $x(t) = \text{Re}\{\bar{\mathcal{A}}e^{\imath\omega t}\}$ is equally valid, with $\bar{\mathcal{A}}$ arbitrary.)

Example 8.2 (Forced oscillations)

Find the solution $x(t)$ to the ODE governing undamped forced oscillations

$$\ddot{x}(t) + \omega^2 x = F\cos\Omega t \qquad \text{with} \qquad x(0) = 0 = \dot{x}(0)\,.$$

Observe that $F\cos\Omega t = \text{Re}(Fe^{-\imath\Omega t})$, so that if $z(t)$ satisfies $\ddot{z}(t) + \omega^2 z = Fe^{\imath\Omega t}$, then $x(t) = \text{Re}\,z$ satisfies $\ddot{x}(t) + \omega^2 x = F\cos\Omega t$. The function $x_C(t) = \text{Re}(\mathcal{A}e^{-\imath\omega t})$ is the *complementary function* (a solution to the homogeneous problem in which $F = 0$). Add to this a *particular solution* $x_P(t)$ found as the real part of a function $z_P = Be^{-\imath\Omega t}$.

Substituting $z_P = Be^{-\imath\Omega t}$ and $\ddot{z}_P(t) = -\Omega^2 Be^{-\imath\Omega t}$ gives

$$\{-\Omega^2 B + \omega^2 B\}e^{-\imath\Omega t} = Fe^{-\imath\Omega t}\,.$$

Thus, $B = F/(\omega^2 - \Omega^2)$ so that $z_P(t) = e^{-\imath\Omega t}F/(\omega^2 - \Omega^2)$. This gives $x_P(t) = \text{Re}(e^{-\imath\Omega t}F/(\omega^2 - \Omega^2))$, which is added to $x_C(t)$ to yield

$$x(t) = x_C(t) + x_P(t) = \mathrm{Re}(\mathcal{A}\mathrm{e}^{-\imath\omega t}) + F\cos\Omega t/(\omega^2 - \Omega^2)\,.$$

Inserting the initial conditions yields $\mathrm{Re}\,\mathcal{A} + F/(\omega^2 - \Omega^2) = x(0) = 0$ and $\mathrm{Re}(\imath\omega\mathcal{A}) + 0 = 0$. Hence $\mathcal{A} = -F/(\omega^2 - \Omega^2)$, so giving the required solution

$$x(t) = \frac{F}{\omega^2 - \Omega^2}\{\cos\Omega t - \cos\omega t\}\,.$$

The complex exponential is useful also in describing many wave motions. Recall from Chapter 5 that transverse motions $v(x,t)$ of a stretched string satisfy $v_{xx} - c^{-2}v_{tt} = 0$. This has solutions $v = \mathrm{Re}\{\mathcal{V}(x)\mathrm{e}^{-\imath\omega t}\}$ for which $\mathcal{V}''(x) + \omega^2 c^{-2}\mathcal{V} = 0$. As this ODE has general solution $\mathcal{V} = \mathcal{A}\mathrm{e}^{\imath kx} + \mathcal{B}\mathrm{e}^{-\imath kx}$, the expression for general *time-harmonic* motions of the string is

$$v(x,t) = \mathrm{Re}\left\{\mathcal{A}\mathrm{e}^{\imath(kx-\omega t)} + \mathcal{B}\mathrm{e}^{-\imath(kx+\omega t)}\right\}, \tag{8.2}$$

where $k = \omega/c$ with ω being the angular frequency. Writing the constants $\mathcal{A}$ and $\mathcal{B}$ as $\mathcal{A} = a\mathrm{e}^{\imath\phi_1}$, $\mathcal{B} = b\mathrm{e}^{\imath\phi_2}$ with a, b, ϕ_1 and ϕ_2 real then yields

$$v(x,t) = a\cos(kx - \omega t + \phi_1) + b\cos(kx + \omega t - \phi_2)\,,$$

which may be recognised as the special case of D'Alembert's solution (5.15) in which $v(x,t)$ is time-harmonic. The *complex amplitude* $\mathcal{A}$ is associated with a right-propagating wave, of amplitude a and phase ϕ_1, while $\mathcal{B}$ is the complex amplitude of the left-propagating wave of amplitude b and phase ϕ_2.

It is clear also how standing waves (cf. Section 5.3.1) arise when $|\mathcal{B}| = |\mathcal{A}|$. Indeed, for $\mathcal{A} = -\imath a\mathrm{e}^{\imath\alpha}, \mathcal{B} = \imath a\mathrm{e}^{\imath\alpha}$, (8.2) yields $v = \mathrm{Re}\left\{2a\sin kx\,\mathrm{e}^{-\imath(\omega t-\alpha)}\right\}$, which gives $v = 2a\sin kx\,\cos(\omega t - \alpha)$ when a is taken as real.

8.1.2 Plane Waves

In Chapter 5, various wave phenomena depending on just one spatial coordinate x and time t were considered. We live in a three-dimensional world, so it is frequently necessary to consider waves in three dimensions. A familiar example is acoustics, in which small changes $\bar{\rho}$ in gas density are associated with pressure changes. As in Section 5.5.1, the pressure is expressed as $p = p_0 + c^2\bar{\rho}$. To the appropriate approximation, the momentum within a region $\mathcal{R}$ is $\iiint_{\mathcal{R}} \rho_0\boldsymbol{u}(\boldsymbol{x},t)\,\mathrm{d}V$, where ρ_0 is the density at pressure p_0 and $\boldsymbol{u}(\boldsymbol{x},t)$ is the gas velocity. This leads to the momentum equation (in the *linearized approximation*)

$$\frac{\mathrm{d}}{\mathrm{d}t}\iiint_{\mathcal{R}} \rho_0\boldsymbol{u}\mathrm{d}V = -\iint_{\partial\mathcal{R}} c^2\bar{\rho}\boldsymbol{n}\mathrm{d}S = -\iiint_{\mathcal{R}} c^2\boldsymbol{\nabla}\bar{\rho}\,\mathrm{d}V\,. \tag{8.3}$$

This yields, by methods familiar from earlier chapters,

$$\rho_0 \frac{\partial \boldsymbol{u}}{\partial t} + c^2 \nabla \bar{\rho} = 0\,, \tag{8.4}$$

while the corresponding approximation to the mass conservation equation is

$$\frac{\partial \bar{\rho}}{\partial t} + \rho_0 \nabla \cdot \boldsymbol{u} = 0\,. \tag{8.5}$$

As in Section 6.3.2 the velocity is irrotational, so that there is a velocity potential $\phi(\boldsymbol{x}, t)$ such that $\boldsymbol{u} = \nabla\phi$. Since substitution into Equation (8.4) gives $\nabla(\rho_0 \partial\phi/\partial t + c^2\bar{\rho}) = 0$, it is possible to write, without loss of generality,

$$\bar{\rho} = -\rho_0 c^{-2} \frac{\partial \phi}{\partial t} \quad , \qquad \boldsymbol{u} = \nabla\phi\,, \tag{8.6}$$

where ϕ is the *acoustic potential.* Then, substitution into (8.5) gives

$$\frac{\partial^2 \phi}{\partial t^2} = c^2 \nabla^2 \phi = c^2 \left(\frac{\partial^2 \phi}{\partial x^2} + \frac{\partial^2 \phi}{\partial y^2} + \frac{\partial^2 \phi}{\partial z^2} \right)\,, \tag{8.7}$$

which is the *three-dimensional wave equation* (a generalization to three dimensions of Equation (5.10) or (5.23)).

Equation (8.7) clearly possesses solutions like (8.2) describing sound waves travelling parallel to Ox. Similarly it allows disturbances which are functions of $y \pm ct$, or of $z \pm ct$. In fact it allows *plane waves* of the form

$$\phi = \mathrm{Re}\{\mathcal{A} e^{\imath(k_1 x + k_2 y + k_3 z - \omega t)}\} \tag{8.8}$$

whenever the constants k_1, k_2, k_3, ω are related through $\omega^2 = c^2(k_1^2 + k_2^2 + k_3^2)$. The interpretation is as follows:

The density perturbation $\bar{\rho}$, pressure perturbation $\bar{p} = c^2\bar{\rho}$ and gas velocity $\boldsymbol{u} = (k_1\boldsymbol{e}_1 + k_2\boldsymbol{e}_2 + k_3\boldsymbol{e}_3)\mathrm{Re}\{\imath\mathcal{A} \exp \imath(k_1 x + k_2 y + k_3 z - \omega t)\}$ each are constant on moving planes $k_1 x + k_2 y + k_3 z - \omega t =$ constant, which have $k_1\boldsymbol{e}_1 + k_2\boldsymbol{e}_2 + k_3\boldsymbol{e}_3$ as a constant normal vector. Hence, the solution describes sinusoidal pressure disturbances travelling in the direction $k_1\boldsymbol{e}_1 + k_2\boldsymbol{e}_2 + k_3\boldsymbol{e}_3$ at the *acoustic speed* $\omega/\sqrt{(k_1^2 + k_2^2 + k_3^2)} = c = \sqrt{p'(\rho)}$. The vector $k_1\boldsymbol{e}_1 + k_2\boldsymbol{e}_2 + k_3\boldsymbol{e}_3$ is known as the *wave vector*, often denoted by $\boldsymbol{k}$. It may be chosen *arbitrarily* (it does not denote the unit vector parallel to Oz). Its direction defines the direction in which disturbances advance, its magnitude $k = |\boldsymbol{k}| = \sqrt{(k_1^2 + k_2^2 + k_3^2)}$ is the *wavenumber* which equals $2\pi/(\text{wavelength})$. The (radian) *frequency* is ω. Note that the speed c at which plane disturbances advance through the gas is independent of both direction and wavelength. Note also that these planes $k_1 x + k_2 y + k_3 z - \omega t =$ constant (on which the pressure is constant) are the loci on which $\arg\phi =$ constant. They are frequently referred to as *planes of constant phase.*

8.1.3 Reflection at a Rigid Wall

Suppose that an acoustic plane wave having (complex) acoustic potential $\phi_I = \mathcal{A}\exp\imath(\boldsymbol{k}\cdot\boldsymbol{x}-\omega t)$ approaches a rigid surface at $z=0$ through a region of gas occupying $z>0$. (Throughout this analysis, it is assumed that only the *real part* has physical significance.) The wave vector $\boldsymbol{k}$ must point towards $z=0$, so that $k_3<0$. There is no loss of generality in choosing axes so that $k_2=0$ (i.e. the wave vector is parallel to the Oxz plane as in Figure 8.1). It is anticipated that a wave will be reflected back into $z>0$. The motion within $z\geq 0$ must satisfy (8.7) and the condition

$$\boldsymbol{u}\cdot\boldsymbol{e}_3 = \frac{\partial\phi}{\partial z} = 0 \qquad \text{on} \quad z=0\,,$$

which states that the normal component of velocity vanishes at the rigid wall. The required solution of (8.7) is a superposition of the *incident wave* with complex velocity potential ϕ_I and a suitable *reflected wave* with potential $\phi_R = \mathcal{B}\exp\imath(\tilde{\boldsymbol{k}}\cdot\boldsymbol{x}-\tilde{\omega}t)$. It is readily verified (by substitution) that $\phi(\boldsymbol{x},t)\equiv\phi_I+\phi_R$ satisfies (8.7) whenever $|\boldsymbol{k}|=\omega/c$ and $|\tilde{\boldsymbol{k}}|=\tilde{\omega}/c$ (with $\mathcal{A}$ and $\mathcal{B}$ arbitrary). The condition at $z=0$ then becomes

$$0=\frac{\partial\phi_I}{\partial z}+\frac{\partial\phi_R}{\partial z}=\imath\mathcal{A}k_3\mathrm{e}^{\imath(k_1x-\omega t)}+\imath\mathcal{B}\tilde{k}_3\mathrm{e}^{\imath(\tilde{k}_1x-\tilde{\omega}t)}\,. \tag{8.9}$$

Since this condition must hold as either x or t vary, it is necessary that $\tilde{\omega}=\omega$ and $\tilde{k}_1=k_1$. Then, from $|\tilde{\boldsymbol{k}}|^2=\omega^2/c^2=k_1^2+k_3^2=|\boldsymbol{k}|^2$, it follows that $\tilde{k}_3=-k_3$. Finally, condition (8.9) reduces to $\imath\{\mathcal{A}-\mathcal{B}\}=0$. Thus, the potential describing an incident plane wave and its plane wave reflection is

$$\phi=\mathrm{Re}\left\{\mathcal{A}\left(\mathrm{e}^{\imath(k_1x+k_3z-\omega t)}+\mathrm{e}^{\imath(k_1x-k_3z-\omega t)}\right)\right\}, \tag{8.10}$$

where, for $k_3<0$, the first term describes the incident wave.

The reflected and incident waves have *wave normals* $\boldsymbol{k}/|\boldsymbol{k}|$ and $\tilde{\boldsymbol{k}}/|\tilde{\boldsymbol{k}}|$ which are equally inclined to the normal $\boldsymbol{e}_3$ to the rigid wall. They also have equal frequencies, equal wavelengths and equal amplitudes $|\mathcal{A}|$. In fact, since all these properties are independent of frequency, a more general version of (8.10) holds, describing the reflection of an incident plane wave of *arbitrary* waveform. Thus, if the incident wave has velocity potential

$$\phi_I=f(\boldsymbol{k}\cdot\boldsymbol{x}-c|\boldsymbol{k}|t)\,,$$

where f is any twice-differentiable function, the potential

$$\phi(\boldsymbol{x},t)=f(k_1x+k_3z-c|\boldsymbol{k}|t)+f(k_1x-k_3z-c|\boldsymbol{k}|t) \tag{8.11}$$

describes the superposition of incident and reflected plane waves.

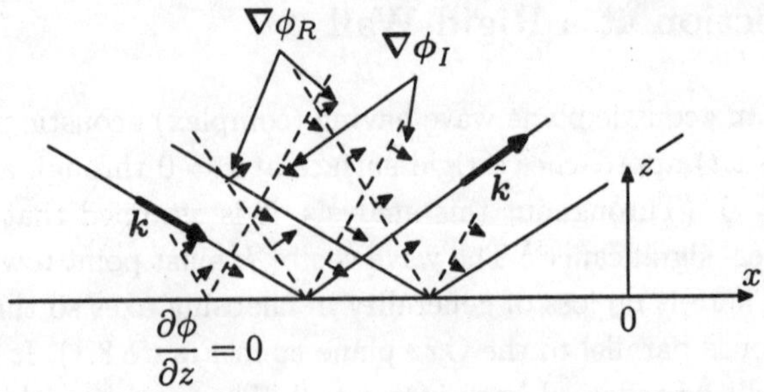

Figure 8.1 An incident plane wave with wave normal parallel to $\boldsymbol{k} = k_1\boldsymbol{e}_1 + k_3\boldsymbol{e}_3$ (with $k_3 < 0$) and its reflection with wave vector $\tilde{\boldsymbol{k}} = k_1\boldsymbol{e}_1 - k_3\boldsymbol{e}_3$ at the rigid wall $z = 0$.

Example 8.3

A plane wave travelling in the direction of the vector $2\boldsymbol{e}_1 - \boldsymbol{e}_3$ is incident from $z > 0$ upon the rigid wall $z = 0$. As it passes through the point $(x, y, z) = (0, 0, c)$ it causes a pressure disturbance which is a pulse $\bar{p}(0, 0, c, t) = t(1-t)$ for $0 \le t \le 1$ but is zero for all other t. Find the total pressure disturbance at $(x, y, z) = (2c, 0, 2c)$, identifying both the incident and reflected parts.

The potential $\phi_I(x, y, z, t) = f(2x - z - \sqrt{5}ct)$ describes a plane wave having the correct orientation and travelling at speed c. The corresponding pressure perturbation is $-\rho_0\partial\phi_I/\partial t = -\sqrt{5}\rho_0 c f'(2x - z - \sqrt{5}ct)$. Thus, for $0 \le t \le 1$, $-\sqrt{5}\rho_0 c f'(-c - \sqrt{5}ct) = t(1-t)$, with f = constant in each of $t < 0$ and $t > 1$.

The acoustic potential $\phi = f(2x - z - \sqrt{5}ct) + f(2x + z - \sqrt{5}ct)$ satisfies the condition $\phi_z = 0$ at $z = 0$. It predicts that at $(x, y, z) = (2c, 0, 2c)$, the pressure disturbance is

$$\begin{aligned} -\rho_0\partial\phi/\partial t &= -\sqrt{5}\rho_0 c\{f'(2c - \sqrt{5}ct) + f'(6c - \sqrt{5}ct)\} \\ &= \begin{cases} (t - \frac{3}{\sqrt{5}})(1 + \frac{3}{\sqrt{5}} - t) & \frac{3}{\sqrt{5}} < t < \frac{3}{\sqrt{5}} + 1 \\ (t - \frac{7}{\sqrt{5}})(1 + \frac{7}{\sqrt{5}} - t) & \frac{7}{\sqrt{5}} < t < \frac{7}{\sqrt{5}} + 1 \\ 0 & \text{for all other } t. \end{cases} \end{aligned}$$

The *incident wave* is the parabolic pulse arriving at $t = \frac{3}{\sqrt{5}}$ and passing the point $(2c, 0, 2c)$ by $t = 1 + \frac{3}{\sqrt{5}}$. The *reflected wave* is the pulse which arrives at $t = \frac{7}{\sqrt{5}}$. At this location (x, y, z), the reflected pulse does not arrive until after the incident pulse has completely passed. However, it is easily checked that in $z < \frac{\sqrt{5}}{2}c$ the two pulses overlap.

8.1.4 Refraction at an Interface

When a plane wave encounters an interface between two regions of different densities and sound speeds, it is converted into both a reflected and a transmitted wave. As before, the reflected wave and incident wave have equal inclinations to the normal to the interface, but the transmitted wave is *refracted*; it has a wave normal which differs from that of the incident wave.

Suppose that region $z > 0$ is occupied by a gas having ambient density ρ_1 and sound speed c_1, while the region $z < 0$ is occupied by a fluid of ambient density ρ_2 through which sound can travel at speed c_2. In the two regions, the acoustic potentials are written as $\phi = \phi_i(\boldsymbol{x}, t)$ $(i = 1, 2)$ which satisfy appropriate versions of the wave equation $\partial^2\phi_i/\partial t^2 = c_i^2\nabla^2\phi_i$. At the interface, the pressure and normal component of velocity must each be continuous, so giving

$$\frac{\partial\phi_1}{\partial z} = \boldsymbol{u}\cdot\boldsymbol{e}_3 = \frac{\partial\phi_2}{\partial z} \quad , \qquad \rho_1\frac{\partial\phi_1}{\partial t} = -\bar{p} = \rho_2\frac{\partial\phi_2}{\partial t} \qquad \text{on} \quad z = 0\,.$$

When potentials are assumed to be superpositions of plane waves, it is clear that all three waves must have the same dependence on x and t, namely through the factor $\mathrm{e}^{-\imath(k_1x-\omega t)}$. Thus, the potentials are written as (the real parts of)

$$\phi_1 = \mathcal{A}\exp\imath(k_1x + k_3z - \omega t) + \mathcal{R}\exp\imath(k_1x + \tilde{k}_3z - \omega t) \ \text{ in } z \ge 0\,, \qquad (8.12)$$

$$\phi_2 = \mathcal{T}\exp\imath(k_1x + \bar{k}_3z - \omega t) \qquad \text{in } z \le 0\,; \qquad (8.13)$$

all potentials are periodic in x with period $2\pi/k_1$. Substitution into $\partial^2\phi_1/\partial t^2 = c_1^2\nabla^2\phi_1$ in $z > 0$ and into $\partial^2\phi_2/\partial t^2 = c_2^2\nabla^2\phi_2$ in $z < 0$ gives

$$\{k_1^2 + k_3^2 - c_1^{-2}\omega^2\}\mathcal{A}\mathrm{e}^{\imath(k_1x+k_3z-\omega t)}$$
$$+\{k_1^2 + \tilde{k}_3^2 - c_1^{-2}\omega^2\}\mathcal{R}\mathrm{e}^{\imath(k_1x+\tilde{k}_3z-\omega t)} = 0\,,$$
$$\{k_1^2 + \bar{k}_3^2 - c_2^{-2}\omega^2\}\mathcal{T}\mathrm{e}^{\imath(k_1x+\bar{k}_3z-\omega t)} = 0\,,$$

so that $\tilde{k}_3 = -k_3$ and $c_1^2\{k_1^2 + k_3^2\} = \omega^2 = c_2^2\{k_1^2 + \bar{k}_3^2\}$. This has a geometrical interpretation, since writing $k = \{k_1^2 + k_3^2\}^{1/2}$ and $\bar{k} = \{k_1^2 + \bar{k}_3^2\}^{1/2}$ gives $c_1k = c_2\bar{k}$ while $k\sin\psi_1 = k_1 = \bar{k}\sin\psi_2$. Here ψ_1 and ψ_2 are the inclinations to the normal $-\boldsymbol{e}_3$ to the interface of the wave vectors $\boldsymbol{k}$ and $\bar{\boldsymbol{k}}$ of the *incident wave* and *transmitted wave*, respectively. This yields the relation $(\sin\psi_1)/c_1 = (\sin\psi_2)/c_2$ which is analogous to *Snell's law* describing how in optics a *ray of light* (normal to the planes of constant phase) is bent, or *refracted*, at an interface (see also Section 9.1.3).

At $z = 0$, the matching conditions become

$$\imath k_3(\mathcal{A} - \mathcal{R})\exp\imath(k_1x - \omega t) = \frac{\partial\phi_1}{\partial z} = \frac{\partial\phi_2}{\partial z} = \imath\bar{k}_3\mathcal{T}\exp\imath(k_1x - \omega t)\,,$$
$$\imath\omega\rho_1(\mathcal{A} + \mathcal{R})\exp\imath(k_1x - \omega t) = \bar{p} = \imath\omega\rho_2\mathcal{T}\exp\imath(k_1x - \omega t)\,.$$

After simplification as $k_3(\mathcal{A}-\mathcal{R}) = \bar{k}_3\mathcal{T}$ and $\rho_1(\mathcal{A}+\mathcal{R}) = \rho_2\mathcal{T}$, these simultaneous equations may be solved to yield the (complex) transmitted amplitude

$$\mathcal{T} = \frac{2}{\rho_2/\rho_1 + \bar{k}_3/k_3}\mathcal{A} \tag{8.14}$$

and the (complex) reflected amplitude

$$\mathcal{R} = \frac{\rho_2/\rho_1 - \bar{k}_3/k_3}{\rho_2/\rho_1 + \bar{k}_3/k_3}\mathcal{A}. \tag{8.15}$$

Equation (8.15) shows that the ratio $\mathcal{R}/\mathcal{A}$ depends upon the ratio $\rho_1\bar{k}_3/(\rho_2 k_3)$. Equation (8.14) gives the corresponding ratio $\mathcal{T}/\mathcal{A}$ of amplitudes of the transmitted and incident waves. Both ratios are real. Thus, for specified incident wave parameters $\mathcal{A}$, k_1 and ω, all remaining parameters in the expressions (8.12) and (8.13) are determined using formulae $\tilde{k}_3 = -k_3 = \{\omega^2 c_1^{-2} - k_1^2\}^{1/2}$, $\bar{k}_3 = -\{\omega^2 c_2^2 - k_1^2\}^{1/2}$ which are equivalent to Snell's law. Notice in particular that there is a special case $\bar{k}_3/k_3 = \rho_2/\rho_1$ in which there is no reflected wave.

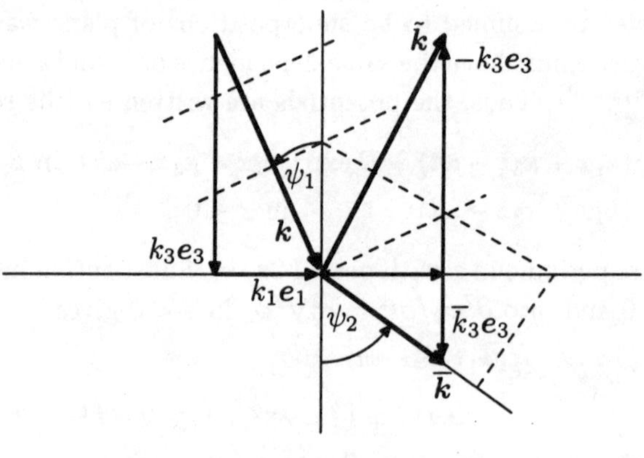

Figure 8.2 Reflected and refracted plane waves on two sides of the interface $z = 0$.

Example 8.4

Although water is often treated as incompressible, in many fluid flows (see Chapter 6) it is sufficiently compressible that *underwater sound* may propagate. Suppose that water occupies the region $z < 0$ with air occupying $z > 0$. Typical values for the parameters are $\rho_1 = 0.0013$ kg/m^3, $\rho_2 = 1.0$ kg/m^3, $c_1 = 333$ m/sec and $c_2 = 1,500$ m/sec. Show that, when a plane wave is incident at angle ψ_1 upon the water surface from the air, there is a transmitted wave in $z < 0$ only for $\psi_1 < \sin^{-1} 0.222$. Find this transmitted wave.

Use $k_3 = -k\cos\psi_1$, $\bar{k}_3 = -\bar{k}\cos\psi_2$ with the above results and with $\rho_2/\rho_1 \approx 769$, $\sin\psi_1 = (c_1/c_2)\sin\psi_2 = 0.222\sin\psi_2$ to deduce that

$$\bar{k}_3/k_3 = \frac{\tan\psi_1}{\tan\psi_2} = \frac{\sqrt{(0.222)^2 - \sin^2\psi_1}}{\cos\psi_1} = \sqrt{1 - 0.9507\sec^2\psi_1} \equiv F(\psi_1) \quad \text{(say)}.$$

Hence, provided that $\sin\psi_1 < 0.222$, there is a transmitted wave with potential

$$\phi = \frac{2}{769 + F(\psi_1)}\mathrm{Re}\left(\mathcal{A}\exp\imath\left\{k\left[x\sin\psi_1 - z\sqrt{0.0493 - \sin^2\psi_1}\right] - \omega t\right\}\right).$$

Hardly surprisingly, in view of the great disparity in densities, the speeds $|\boldsymbol{u}|$ are much smaller in the water than in the air, even though the pressure disturbances are comparable in magnitude.

8.1.5 Total Internal Reflection

Snell's law $\sin\psi_1/c_1 = \sin\psi_2/c_2$ shows that the representation (8.12), (8.13) cannot be valid when $\sin\psi_1 > c_1/c_2$. As in Example 8.4, there cannot be a transmitted wave for which $\bar{k}_3$ is real. However, replacing (8.13) by

$$\phi_2 = \mathcal{T}\mathrm{e}^{\gamma z}\exp\imath(k_1 x - \omega t) \qquad \text{with } \gamma > 0 \tag{8.16}$$

gives a potential which remains bounded throughout $z \leq 0$ and which gives $\omega^2 = c_2^2(k_1^2 - \gamma^2) = c_1^2 k^2$, with a real root $\gamma = k\{\sin^2\psi_1 - (c_1/c_2)^2\}^{-1/2}$. Here, and in the matching conditions across $z = 0$, the only change from Section 8.1.4 is to replace $i\bar{k}_3$ by γ. Thus, it is readily checked that (8.14) and (8.15) are replaced by

$$\mathcal{T} = \frac{2}{\rho_2/\rho_1 - i\gamma/k_3}\mathcal{A}\,, \quad \mathcal{R} = \frac{\rho_2/\rho_1 + i\gamma/k_3}{\rho_2/\rho_1 - i\gamma/k_3}\mathcal{A}\,. \tag{8.17}$$

Notice that the amplitude $|\mathcal{R}|$ of the reflected wave equals the amplitude $|\mathcal{A}|$ of the incident wave. This is the phenomenon known classically (in ray optics) as *total internal reflection.* While it is true that the plane reflected wave has the same amplitude as the incident wave, it is not true that all the disturbance is confined to the region $z > 0$. Within the 'faster' medium occupying $z < 0$, there is a time-varying acoustic disturbance with potential

$$\begin{aligned}\phi_2 &= \mathrm{Re}\left\{\frac{\rho_2/\rho_1 + i\gamma/k_3}{\rho_2/\rho_1 - i\gamma/k_3}A\mathrm{e}^{\gamma z}\mathrm{e}^{\imath(k_1 x - \omega t)}\right\} \\ &= |\mathcal{A}|\mathrm{e}^{\gamma z}\cos(k_1 x - \omega t + 2\tan^{-1}(\gamma\rho_1/k_3\rho_2) + \arg\mathcal{A})\,,\end{aligned}$$

so that the amplitude decays exponentially with distance from the interface.

Example 8.5

For the parameters of Example 8.4, find the ratio γ/k. Hence show that as $\psi_1 \to \pi/2$, the attenuation parameter γ approaches $1.025k$.

From the formula $\gamma = k\{\sin^2\psi_1 - (c_1/c_2)^2\}^{-1/2}$, it follows that $\gamma/k = \{\sin^2\psi_1 - (0.222)^2\}^{-1/2}$. Hence, as $\psi_1 \to \pi/2$, $\gamma \to k(0.9507)^{-1/2} \approx 1.025k$.

Note that, since the incident wave has wavelength $2\pi/k$ and the disturbance amplitude decays by a factor e as γz varies by one unit, the typical *penetration depth* γ^{-1} is substantially less than one wavelength.

EXERCISES

8.1. Confirm that the potential (8.10) describing wave reflection has the form $\phi(\boldsymbol{x},t) = \mathrm{Re}\{\Phi(z)\exp \imath(k_1x - \omega t)\}$.

Obtain (8.10) by the following alternative process: Substitute $\phi(\boldsymbol{x},t) = \mathrm{Re}\{\Phi(z)\exp \imath(k_1x - \omega t)\}$ into both the wave equation (8.7) and the boundary conditions to show that $\Phi''(z) \propto \Phi$, with $\Phi'(0) = 0$. Then find the most general solution $\phi(\boldsymbol{x},t)$.

8.2. Show that a function $\phi = \mathrm{e}^{\imath(k_1x+k_2y-\omega t)}\Phi(z)$ satisfies $\phi_{tt} = c^2\nabla^2\phi$ if, and only if,

$$\Phi''(z) = \left(k_1^2 + k_2^2 - \frac{\omega^2}{c^2}\right)\Phi(z)\,.$$

Obtain the general solution of this equation in a region where $c = c_1$ and where $\omega^2 > c_1^2(k_1^2+k_2^2)$. Interpret this solution as a superposition of two plane waves.

If, in the region $z < 0$ the wavespeed is $c = c_2$ where $\omega^2 < c_2^2(k_1^2+k_2^2)$, show that the only functions $\Phi(z)$ which are bounded as $z \to -\infty$ are $\Phi(z) = \mathcal{C}\mathrm{e}^{\gamma z}$ for an appropriate choice of γ. Determine this choice.

Use the above results to construct $\phi = \mathrm{Re}\{\Phi(z)\mathrm{e}^{\imath(k_1x+k_2y-\omega t)}\}$ describing total internal reflection at $z = 0$ (i.e. use the fact that both $\partial\phi/\partial z$ and $\rho\,\partial\phi/\partial t$ should be continuous at $z = 0$).

8.3. *Impedance matching.* If the reference density and acoustic speed are ρ_1 and c_1 in $z > 0$, while they are ρ_2 and c_2 in $z < 0$, show that a plane wave with potential $\phi_I = \mathrm{Re}\{\mathcal{A}_1\mathrm{e}^{\imath(kz-\omega t)}\}$ incident normally upon the interface $z = 0$ from $z > 0$ produces no reflected wave if $\rho_2c_2 = \rho_1c_1$. Show also that in this case $\mathcal{A}_2/\mathcal{A}_1 = \rho_1/\rho_2$. [The media in $z > 0$ and $z < 0$ are said to have *matched impedances*.]

Generalize the result to deduce that for certain sequences of layered media with parameters $\{\rho_n, c_n\}$ there are no reflected waves. Determine the corresponding ratio $|\mathcal{A}_n/\mathcal{A}_1|$ of amplitudes.

8.4. *Radial symmetry.* Show that radially symmetric solutions $\phi = \Phi(r,t)$ of the wave equation $\phi_{tt} = c^2\nabla^2\phi$ satisfy $(r\Phi)_{tt} = c^2(r\Phi)_{rr}$, where $r^2 = x^2 + y^2 + z^2$. [Hint: An analogy with Section 2.3 may help.] Deduce that $\phi = Ar^{-1}\sin k(r - ct - \alpha)$ is the solution describing propagation of acoustic waves of frequency kc radially outward from a time-harmonic *point source* at $r = 0$. Write down the corresponding pressure perturbation $-\rho_0\partial\phi/\partial t$. Determine also the corresponding acoustic velocity $\boldsymbol{\nabla}\phi$.

8.5. Adapt the solution of Exercise 8.4 so as to describe the potential due to an oscillatory source at $\boldsymbol{x} = a\boldsymbol{e}_3$. Determine the corresponding acoustic velocity $\boldsymbol{\nabla}\phi$ everywhere on the plane $z = 0$. Show that, if an *image* oscillatory source is located at the *image point* $\boldsymbol{x} = -a\boldsymbol{e}_3$ and the potentials are added, the corresponding velocity component $u_3 = \boldsymbol{u}\cdot\boldsymbol{e}_3$ vanishes over the plane $z = 0$. [This solution describes reflection of radiation from a point source by a rigid wall.]

8.2 Guided Waves

Many disturbances are strongly *guided* by boundaries or interfaces, so that they travel in a particular direction or parallel to certain surfaces. Typically, they are more complicated than plane waves, but may be analysed as ***travelling waves*** involving a factor such as $e^{\imath(kx-\omega t)}$, where the *phase speed* ω/k depends upon the frequency ω. The simplest example is of acoustic waves confined between two rigid, plane walls.

8.2.1 Acoustic Waves in a Layer

Acoustic disturbances confined between two rigid walls at $z = 0$ and $z = a$ are governed by the wave equation $\phi_{tt} = c^2\nabla^2\phi$ (8.7), with the potential ϕ satisfying the boundary conditions $\phi_z = 0$ at both $z = 0$ and $z = a$ (cf. Section 8.1.3). All these equations allow a search for solutions of the form

$$\phi(\boldsymbol{x}, t) = \mathrm{Re}\{\Phi(z)\exp\imath(k_1x + k_2y - \omega t)\},$$

describing a pressure disturbance $-\rho_0\phi_t$ which advances in the direction parallel to the vector $k_1\boldsymbol{e}_1 + k_2\boldsymbol{e}_2$. Writing the phase factor as $k_1x + k_2y - \omega t \equiv \psi$ and substituting into (8.7) yields

$$\nabla^2\phi = -(k_1^2+k_2^2)\Phi(z)e^{\imath\psi} + \Phi''(z)e^{\imath\psi} = c^{-2}\phi_{tt} = (\omega/c)^2\Phi e^{\imath\psi},$$

so that, with the boundary conditions $\phi_z = \Phi'(z)e^{\imath\psi} = 0$ on $z = 0, a$, this gives

$$\Phi''(z) = (\beta^2 - \omega^2/c^2)\Phi \quad \text{in } 0 < z < a\,, \quad \Phi'(0) = 0\,, \quad \Phi'(a) = 0\,, \tag{8.18}$$

where $\beta \equiv \{k_1^2 + k_2^2\}^{1/2}$. Since this boundary-value problem for $\Phi(z)$ is closely related to that in Section 5.3.1 for standing waves $Y(x)$, the most general solution is seen to be $\Phi = \cos(n\pi z/a)$ for some integer $n = 0, 1, 2, \ldots$. Moreover, this solution requires the choice $\beta^2 - \omega^2/c^2 = -n^2\pi^2/a^2$. The interpretation is as follows:

For each choice of unit vector $\hat{\boldsymbol{k}} = (\boldsymbol{e}_1\cos\theta + \boldsymbol{e}_2\sin\theta)$ parallel to the layer and for each wavelength $2\pi/\beta$, there is a family of *guided modes* having potential

$$\phi = \mathrm{Re}\left[\mathcal{A}\exp\{\imath\,(\beta(x\cos\theta + y\sin\theta) - \omega t)\}\right]\cos\frac{n\pi z}{a} \tag{8.19}$$

with frequency related to the *propagation constant* β through

$$\omega^2 = c^2\left(\beta^2 + \frac{n^2\pi^2}{a^2}\right), \quad n = 0, 1, 2, \ldots\,. \tag{8.20}$$

The pressure perturbation $\bar{p} = -\rho_0\phi_t$ and the components $u_1 \equiv \phi_x$ and $u_2 \equiv \phi_y$ of the velocity $\boldsymbol{u} = u_1\boldsymbol{e}_1 + u_2\boldsymbol{e}_2$ each depend on z through the factor $\cos(n\pi z/a)$ and are periodic in time with (radian) frequency ω. The disturbance is sinusoidal in $x\cos\theta + y\sin\theta$ and travels parallel to $\hat{\boldsymbol{k}}$ with speed $c_p = \omega/\beta = c\{1 + (n\pi/(a\beta))^2\}^{1/2}$ which increases with both wavelength $\lambda = 2\pi/\beta$ and with *mode number* n.

The *phase speed* c_p is independent of the angle of propagation relative to the Ox axis, but depends upon both n and β. It is the speed at which an observer would need to travel parallel to $\hat{\boldsymbol{k}}$ in order to observe a steady pattern of disturbances. This pattern is illustrated by considering the special case $k_2 = 0$, which corresponds to choosing the Ox axis to be parallel to $\hat{\boldsymbol{k}}$.

Example 8.6

Show that guided acoustic modes (for $n > 0$) travelling parallel to Ox in the layer having rigid walls $z = 0, a$ may be regarded as a superposition of plane waves undergoing repeated reflections at the two walls. Use this to explain geometrically how the speed c_p is related to β. Determine the velocity field within the mode.

When $k_2 = 0$, the potential $\phi = \text{Re}\{Ae^{\imath(\beta x - \omega t)} \cos(n\pi z/a)\}$ describes the guided modes. Writing $A = 2Ae^{\imath\alpha}$, then gives

$$\begin{aligned}\phi(x,z,t) &= 2A\cos(\beta x - \omega t + \alpha)\cos(n\pi z/a) \\ &= A\cos(\beta x + (n\pi/a)z - \omega t + \alpha) + A\cos(\beta x - (n\pi/a)z - \omega t + \alpha).\end{aligned}$$

This is the superposition of two plane waves, each of amplitude A and having respective wave vectors $\beta\mathbf{e}_1 \pm (n\pi/a)\mathbf{e}_3$ as in (8.10). Since all plane acoustic waves have the same speed c, it is necessary that $|\beta\mathbf{e}_1 \pm (n\pi/a)\mathbf{e}_3| = \omega/c$. This yields the *dispersion relation* (8.20) connecting β to ω and n.

In the special case $n = 0$, the wave is simply a plane wave travelling in the direction of Ox. Otherwise, the two plane waves have normals inclined at angles $\pm\delta$ to Ox, where $\tan\delta = n\pi/(a\beta)$. Each wave describes a sinusoidal pattern with time periodicity $2\pi/\omega$ and with spatial periodicity $2\pi c/\omega$ in the direction of the wave normal. Hence, the x–periodicity is $2\pi c/(\omega\cos\delta) = 2\pi/\beta$, so giving $c_p = c/\cos\delta$. Eliminating δ between $\tan\delta = n\pi/(a\beta)$ and $\cos\delta = c\beta/\omega$, readily yields the relation (8.20).

Within the guided mode, the velocity components are $u_1 = -2A\beta\sin(\beta x - \omega t + \alpha)\cos(n\pi z/a)$, $u_2 = 0$ and $u_3 = -2(n\pi A/a)\cos(\beta x - \omega t + \alpha)\sin(n\pi z/a)$. (As required, the velocity component u_3 vanishes at both $z = 0$ and $z = a$.)

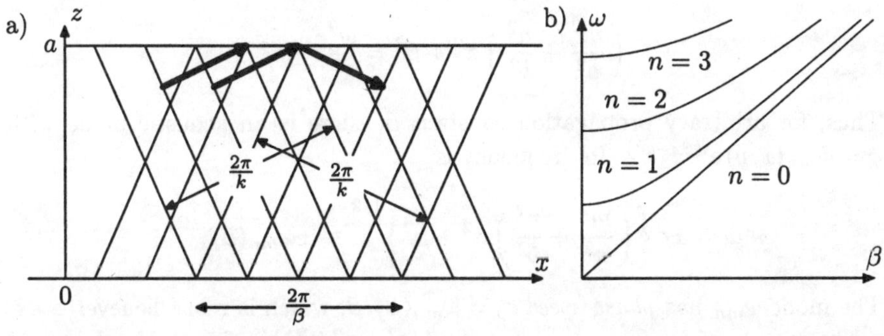

Figure 8.3 a) Reflected plane waves making up the guided mode between two rigid planes $z = 0, a$. b) The corresponding dispersion relation.

8.2.2 Waveguides and Dispersion

In any system possessing guided modes involving a factor $e^{\imath(\beta x - \omega t)}$, the governing equations and boundary conditions will cause β to be related to ω through a *dispersion relation*, such as (8.20). If the dispersion relation is presented graphically as in Figure 8.3, the graph shows how the *phase speed* $c_p = \omega/\beta$ depends upon the frequency ω. In many cases (as indeed for (8.20)), it reveals that there may be many distinct modes having the same propagation constant β (and consequently the same periodicity in x). For the waves considered in Section 8.2.1, there is an infinite number of modes for *each* value of β, but for each mode there is a minimum frequency (the *cut-off frequency*) $\bar{\omega} = n\pi c/a$ below which that mode cannot propagate. Similar features arise for many guided waves.

Example 8.7 (The rectangular duct)

Find the acoustic modes which can propagate within the rectangular duct having rigid walls at $x = 0, a$ and at $y = 0, b$. For each mode, find the cut-off frequency and find how the phase speed depends on frequency.

Within the duct, the velocity is $\boldsymbol{u} = \nabla\phi$ where $\phi_{tt} = c^2\nabla^2\phi$, with $u_1 = \phi_x = 0$ on $x = 0, a$ and $u_2 = \phi_y = 0$ on $y = 0, b$. Seek propagating modes $\phi = \mathrm{Re}\{\Phi(x,y)\mathrm{e}^{\imath(\beta z - \omega t)}\}$. This gives

$$\Phi_{xx} + \Phi_{yy} = (\beta^2 - c^{-2}\omega^2)\Phi\,, \quad \Phi_x(0,y) = \Phi_x(a,y) = 0 = \Phi_y(x,0) = \Phi_y(x,b)\,.$$

Separable solutions $\Phi = X(x)\,Y(y) = \cos\dfrac{m\pi x}{a}\cos\dfrac{n\pi y}{b} \equiv \Phi_{mn}(x,y)$ exist for each pair of integers $m, n = 0, 1, 2, \ldots$ and give the *dispersion relation*

$$\left(\frac{m^2}{a^2} + \frac{n^2}{b^2}\right)\pi^2 + \beta^2 = \frac{\omega^2}{c^2}\,.$$

Thus, for arbitrary propagation constant β, there is an acoustic mode with $\phi = \Phi_{mn}(x,y)\,\mathrm{e}^{\imath(\beta z - \omega t)}$. Its frequency is

$$\omega = \pm c\left\{\left(\frac{m^2}{a^2} + \frac{n^2}{b^2}\right)\pi^2 + \beta^2\right\}^{1/2} \equiv \pm\omega_{mn}(\beta)\,.$$

The mode Φ_{mn} has *phase speed* $c_p = \omega_{mn}(\beta)/\beta$, which is real whenever ω exceeds the *cut-off frequency* $\bar{\omega}_{mn} \equiv \pi c\{m^2/a^2 + n^2/b^2\}^{1/2}$. For the fundamental plane wave mode Φ_{00} the phase speed takes the constant value c and the cut-off frequency vanishes. For all other modes, the curve in the β-ω plane is part of a hyperbola. In the limit of exceedingly short wavelengths ($\beta \to \infty$), the phase speed approaches c, but as the wavelength $2\pi/\beta$ increases, so also does the phase speed. Moreover, it increases without bound (see Figure 8.4).

Other types of waveguide are formed when a layer having one wavespeed c is sandwiched between regions having a higher wavespeed. A naturally occurring example is the *underwater acoustic waveguide*. A simplified model is to imagine

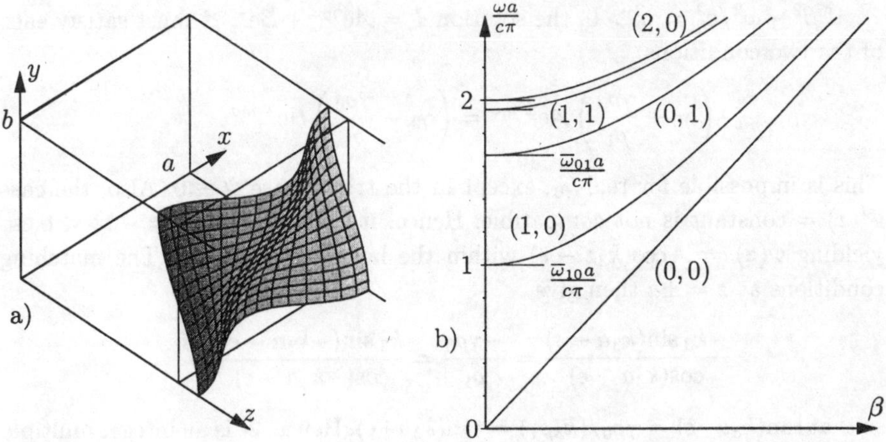

Figure 8.4 a) Guided (2, 1) mode within the rectangular duct of Example 8.7, having rigid walls at $x = 0, a$, $y = 0, b$. b) Some branches of the dispersion relation, showing the cut-off frequency $\bar{\omega}_{mn}$ for each (m, n), when $a = 1.6\,b$.

that a fluid layer of density ρ_0 and acoustic speed c_0 occupies the region $|z| < a$ and is surrounded by fluid of unlimited extent having density ρ_1 and acoustic speed c_1. In this case, waves travelling parallel to Ox are sought in the form $\phi = \Phi(z)e^{i(\beta x - \omega t)}$, so that substitution into (8.7) with appropriate acoustic speeds $c = c_0, c_1$ gives

$$\Phi''(z) = \begin{cases} (\beta^2 - \omega^2/c_1^2)\Phi & \text{for } |z| > a\,, \\ (\beta^2 - \omega^2/c_0^2)\Phi & \text{for } |z| < a\,. \end{cases}$$

At each of $z = \pm a$, the velocity component $u_3 = \phi_z$ and the pressure perturbation $-\rho\phi_t$ must be continuous, so that Φ' and $\rho\Phi$ must be continuous.

Since $\Phi(z)$ satisfies an equation $\Phi''(z) = \alpha\Phi$ both within ($|z| < a$) and outside ($|z| > a$) the guiding layer, the dependence of Φ on z in each region is either sinusoidal, linear or a sum of two exponentials. The disturbance will be concentrated near the guiding layer only if the outer layers allow exponential decay of Φ, which requires that $\beta^2 - \omega^2/c_1^2 > 0$. Then, possible forms for $\Phi(z)$ in $|z| > a$ are found, by writing $\sqrt{\beta^2 - \omega^2/c_1^2} \equiv \gamma$, as

$$\Phi = \mathcal{B}_+ e^{-\gamma z} \quad \text{for } z \geq a\,; \qquad \Phi = \mathcal{B}_- e^{\gamma z} \quad \text{for } z \leq -a\,.$$

The continuity conditions at $z = a$ require that the solution to $\Phi''(z) = \alpha_0\Phi$ in $-a < z < a$, where $\alpha_0 \equiv \beta^2 - \omega^2/c_0^2$, must satisfy $\Phi'(a) = -\gamma\mathcal{B}_+e^{-\gamma a}$ and $\rho_0\Phi(a) = \rho_1\mathcal{B}_+e^{-\gamma a}$. This gives $\Phi'(a)/\Phi(a) = -\gamma\rho_0/\rho_1$, while similarly at $z = -a$ it is necessary that $\Phi'(-a)/\Phi(-a) = \gamma\rho_0/\rho_1$.

If $\beta^2 - \omega^2/c_0^2 = \gamma_0^2 > 0$, the solution $\Phi = \mathcal{A}e^{\gamma_0 z} + \mathcal{B}e^{-\gamma_0 z}$ must satisfy each of the two conditions

$$\left(\gamma_0 + \frac{\gamma\rho_0}{\rho_1}\right)\mathcal{A}e^{\pm\gamma_0 a} = \left(\gamma_0 - \frac{\gamma\rho_0}{\rho_1}\right)\mathcal{B}e^{\mp\gamma_0 a}.$$

This is impossible for real γ_0, except in the trivial case $\Phi \equiv 0$. Also, the case $\Phi'(z) = $ constant is not compatible. Hence, it follows that $\alpha_0 = -k_3^2 < 0$, so yielding $\Phi(z) = \mathcal{A}\cos(k_3 z - \epsilon)$ within the layer $-a \le z \le a$. The matching conditions at $z = \pm a$ then give

$$\frac{-k_3\sin(k_3 a - \epsilon)}{\cos(k_3 a - \epsilon)} = \frac{-\gamma\rho_0}{\rho_1} = \frac{k_3\sin(-k_3 a - \epsilon)}{\cos(-k_3 a - \epsilon)},$$

so that $\tan(k_3 a - \epsilon) = \gamma\rho_0/(k_3\rho_1) = \tan(k_3 a + \epsilon)$. Hence, 2ϵ is an integer multiple of π, so giving just two possibilities $\Phi(z) = \mathcal{A}\cos k_3 z$ and $\Phi(z) = \mathcal{A}\sin k_3 z$. The first possibility gives *symmetric modes* with

$$\Phi(z) = \begin{cases} \mathcal{A}\cos k_3 z & \text{for } |z| \le a, \\ (\rho_0/\rho_1)\mathcal{A}e^{\gamma(a-|z|)}\cos k_3 a & \text{for } |z| \ge a, \end{cases} \tag{8.21}$$

where γ and k_3 are related through $\gamma\rho_0\cos k_3 a = k_3\rho_1\sin k_3 a$. This not only determines a countable set of possibilities for ω at each value of the propagation constant β, but also it gives the dispersion relation

$$\frac{\rho_1}{\rho_0}\tan\left(a\sqrt{\omega^2 c_0^{-2} - \beta^2}\right) = \left\{\frac{\beta^2 - \omega^2/c_1^2}{\omega^2 c_0^{-2} - \beta^2}\right\}^{1/2}.$$

For the symmetric modes, this has the parametric form

$$\omega/c_0 = (1 - c_0^2/c_1^2)^{-1/2}k_3\left[1 + \left(\rho_1^2/\rho_0^2\right)^2\tan^2 k_3 a\right]^{1/2},$$

$$\beta = (1 - c_0^2/c_1^2)^{-1/2}k_3\left[\left(c_0^2/c_1^2\right)^2 + \left(\rho_1^2/\rho_0^2\right)^2\tan^2 k_3 a\right]^{1/2}$$

with $k_3 a \in [\frac{1}{2}n\pi, \frac{1}{2}(n+1)\pi)$, $n = 0, 2, 4, \ldots$. The second possibility ($\epsilon = \frac{1}{2}\pi$) gives *anti-symmetric modes* with

$$\Phi(z) = \begin{cases} \mathcal{A}\sin k_3 z & \text{for } |z| \le a, \\ (\rho_0/\rho_1)\mathcal{A}e^{\gamma(a-z)}\sin k_3 a & \text{for } z \ge a, \\ -(\rho_0/\rho_1)\mathcal{A}e^{\gamma(z+a)}\sin k_3 a & \text{for } z \le -a, \end{cases} \tag{8.22}$$

where $\gamma\rho_0\sin k_3 a = -k_3\rho_1\cos k_3 a$ so that the dispersion relation is

$$\frac{\rho_0}{\rho_1}\tan\left(a\sqrt{\omega^2 c_0^{-2} - \beta^2}\right) = -\left\{\frac{\omega^2 c_0^{-2} - \beta^2}{\beta^2 - \omega^2/c_1^2}\right\}^{1/2}.$$

Again, for a typical mode, there is a parametric representation, with $\tan^2 k_3 a$ replaced by $\cot^2 k_3 a$, with $k_3 a \in [\frac{1}{2}n\pi, \frac{1}{2}(n+1)\pi)$, $n = 1, 3, 5, \ldots$. These formulae, though complicated, show that guided waves exist only for $c_0 < |\omega/\beta| < c_1$, but that for typical frequency ω there are many possibilities for β. Indeed, since the n-th mode has cut-off frequency $\bar{\omega}_n = (c_0/a)(1 - c_0^2/c_1^2)^{-1/2}(\frac{1}{2}n\pi)$, $n = 0, 1, 2, \ldots$, there exists n propagating modes when $\bar{\omega}_{n-1} \le \omega < \bar{\omega}_n$. This illustrates some features arising widely for guided modes. In particular, the waves are trapped or guided by layers in which the propagation speed of plane waves is slower than it is in the surrounding regions, and also there may be many such guided modes. For the underwater acoustic waveguide, the function $\Phi(z)$ describing the symmetric mode $n = 0$ and anti-symmetric mode $n = 1$ are shown in Figure 8.5, together with dispersion curves for $n = 0, 1$ and 2.

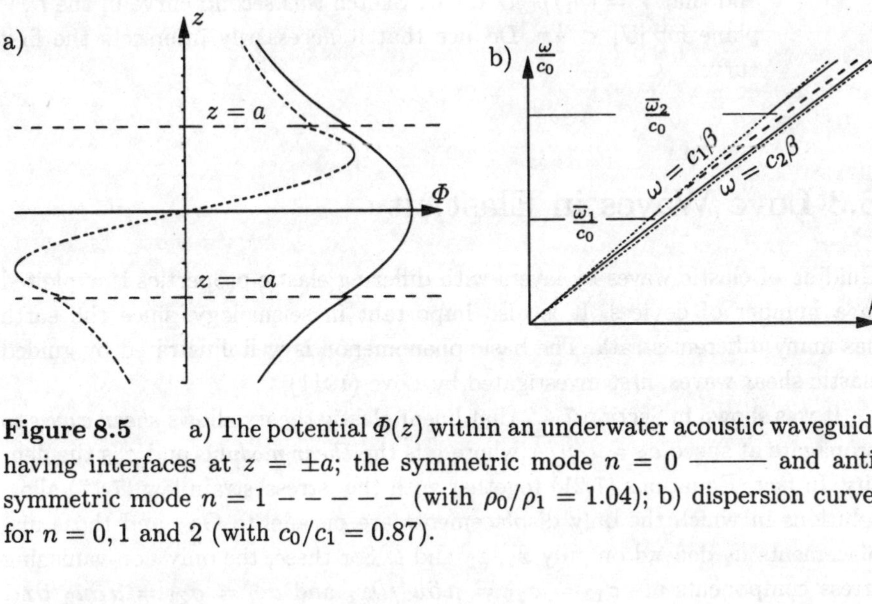

Figure 8.5 a) The potential $\Phi(z)$ within an underwater acoustic waveguide having interfaces at $z = \pm a$; the symmetric mode $n = 0$ ——— and anti-symmetric mode $n = 1$ – – – – – (with $\rho_0/\rho_1 = 1.04$); b) dispersion curves for $n = 0, 1$ and 2 (with $c_0/c_1 = 0.87$).

EXERCISES

8.6. Within a layer $x \ge 0$, $0 \le z \le a$, two-dimensional acoustic modes satisfy $\phi_{tt} = c^2(\phi_{xx} + \phi_{zz})$ with $\phi_z = 0$ on $z = 0, a$. Seek waves of frequency ω in the separable form $\phi = \mathrm{Re}\{e^{-\imath\omega t}X(x)Z(z)\}$. Show that $Z = \cos(n\pi z/a)$ and that, provided that $\omega > n\pi c/a$, waves travelling *away from* the end $x = 0$ may exist, with $X(x) \propto e^{\imath\beta_n x}$ and $\beta_n = +(c^{-2}\omega^2 - n^2\pi^2/a^2)^{1/2}$. When $0 < \omega < n\pi c/a$, find the corresponding solutions $\phi(x, z, t)$ which are bounded as $x \to \infty$.

8.7. In Exercise 8.6, if $N\pi c/a < \omega < (N+1)\pi c/a$, obtain the expression

$$\phi = \mathrm{Re}\left\{ \mathrm{e}^{-\imath\omega t} \left[\sum_{n=0}^{N} A_n \mathrm{e}^{\imath\beta_n x} \cos\frac{n\pi z}{a} + \sum_{n=N+1}^{\infty} A_n \mathrm{e}^{-\gamma_n x} \cos\frac{n\pi z}{a} \right] \right\}$$

for disturbances excited at the end $x = 0$ by specifying the velocity component as $\phi_x(0, z, t) = U(z)\cos\omega t$.

Deduce that $U(z)$ has an expansion as a Fourier cosine series. Hence, determine the coefficients A_n in terms of $U(z)$.

8.8. For the *symmetric modes* in the underwater acoustic waveguide, confirm as follows that, for $c_0 < c_1$, at least one mode exists: Write $k_3 a = U$ and $\gamma a = V$. Show that $U^2 + V^2 = a^2\omega^2(c_0^{-2} - c_1^{-2})$ and that $V = (\rho_1/\rho_0)U \tan U$. Sketch this second curve in the U, V plane for $|U| < \frac{1}{2}\pi$. Deduce that it necessarily intersects the first curve.

8.3 Love Waves in Elasticity

Guiding of elastic waves by layers with differing elastic properties is exploited in a number of devices. It is also important in seismology, since the earth has many different strata. The basic phenomenon is well illustrated by guided elastic shear waves, first investigated by Love (1911).

It was shown in Section 7.4.2 that linear elastic theory allows *shear waves* to propagate at speed $c_S = \sqrt{\mu/\rho}$, where μ is the *shear modulus* and ρ is the density. In fact, Equations (7.21) together with the stress–strain law (7.17) allow solutions in which the only displacements are parallel to $\mathrm{O}x_2$ and those displacements u_2 depend on only x_1, x_3 and t. For these, the only non-vanishing stress components are $\sigma_{12} = \sigma_{21} = \mu\,\partial u_2/\partial x_1$ and $\sigma_{32} = \sigma_{23} = \mu\,\partial u_2/\partial x_3$. Since none of these depend upon x_2, it is seen that the elastodynamic equation (7.21) is automatically satisfied for $i = 1, 3$. Only the equation governing motions parallel to $\mathrm{O}x_2$ needs consideration. It is

$$\frac{\partial}{\partial x_1}\left(\mu\frac{\partial u_2}{\partial x_1}\right) + \frac{\partial}{\partial x_3}\left(\mu\frac{\partial u_2}{\partial x_3}\right) = \rho\frac{\partial^2 u_2}{\partial t^2}.$$

After simplification of notation ($x_1 = x$, $x_3 = z$ and $u_2(x_1, x_3, t) = u(x, z, t)$), this becomes

$$u_{tt} = c^2(u_{xx} + u_{zz}), \qquad \text{where } c = \sqrt{\mu/\rho}. \tag{8.23}$$

This is the two-dimensional wave equation for functions $u(x, z, t)$.

For shear disturbances in a solid occupying $z \leq 0$ and having density ρ_0 and shear modulus μ_0 within the layer $-a < z < 0$, but having density ρ_1 and shear modulus μ_1 in the substrate $z < -a$, Equation (8.23) should be analysed with both the displacement u and the shear traction $\sigma_{23} = \mu\, \partial u/\partial z$ being continuous across $z = -a$. If, in addition, the disturbances are consistent with a traction-free surface $z = 0$ and are concentrated close to the layer $-a < z < 0$, then solutions to (8.23) are sought with

$$u_z(x, 0, t) = 0; \quad u \text{ and } \mu u_z \text{ continuous at } z = -a\,; \qquad u \to 0 \text{ as } z \to -\infty\,. \tag{8.24}$$

The relevant solutions are similar to those of the underwater acoustic waveguide.

Seek $u(x, z, t) = \mathrm{Re}\{U(z)\mathrm{e}^{\imath(\beta x - \omega t)}\}$, so that

$$U''(z) = \begin{cases} (\beta^2 - \omega^2/c_0^2)U & \text{for } -a < z < 0\,, \\ (\beta^2 - \omega^2/c_1^2)U & \text{for } z < -a\,, \end{cases}$$

where $c_0 = \sqrt{\mu_0/\rho_0}$ and $c_1 = \sqrt{\mu_1/\rho_1}$. The required solution must have exponential decay as $z \to -\infty$, so that $U \propto \mathrm{e}^{\gamma z}$ in $z \leq -a$, with $\gamma = \sqrt{\beta^2 - \omega^2/c_1^2}$ as before. Also, since $U'(0) = 0$, the solution within $-a \leq z \leq 0$ must have $U(z) \propto \cos k_3 z$ for $k_3 = \sqrt{\omega^2 c_0^{-2} - \beta^2}$. Continuity of U at $z = -a$ then gives the *modal displacements*

$$U(z) = \begin{cases} A \cos k_3 z & \text{in } -a \leq z \leq 0 \\ A(\cos k_3 a)\, \mathrm{e}^{\gamma(z+a)} & \text{for } z \leq -a\,, \end{cases} \tag{8.25}$$

while continuity of traction gives $-\mu_0 k_3 \sin k_3(-a) = \mu_1 \gamma \sin k_3 a$. These mode shapes are analogous to the portion $z \leq 0$ of those for $\rho\Phi(z)$ in the symmetric case (8.22) of underwater acoustic waves. The slight change is that the dispersion relation for Love waves is

$$\frac{\mu_0}{\mu_1} \tan a\sqrt{\omega^2 c_0^{-2} - \beta^2} = \left\{\frac{\beta^2 - \omega^2/c_1^2}{\omega^2 c_0^{-2} - \beta^2}\right\}^{1/2}.$$

Again, modes guided by the surface layer exist only for $c_0 < c_1$. However, at any value β, there may be many modes having periodicity $2\pi/\beta$ in x (see Exercise 8.9). Each has a distinct phase speed ω/β lying in $c_0 < \omega/\beta < c_1$.

Example 8.8

Rewrite the expression $U(z)\mathrm{e}^{\imath(\beta x - \omega t)}$ arising in Love wave displacements so as to show that in $-a < z < 0$ it is the sum of two plane waves of equal amplitude.

Show also that these may be regarded as the incident and reflected waves for total internal reflection at the interface $z = -a$.

Solution (8.25) in $-a < z < 0$ gives $U(z)e^{\imath(\beta x-\omega t)} = \mathcal{B}(\cos k_3 z)e^{\imath(\beta x-\omega t)} = \frac{1}{2}\mathcal{B}\{e^{\imath(\beta x-k_3 z-\omega t)}+e^{\imath(\beta x+k_3 z-\omega t)}\}$. Thus, the real part describes two plane waves of equal amplitude and having wave normals $\beta e_1 \pm k_3 e_3$ (where $k_3 > 0$). To investigate conditions near $z = -a$, write $\hat{z} = z + a$. Then, for $\hat{z} < 0$ the solution has $U = \mathcal{B}(\cos k_3 a)e^{\gamma\hat{z}}$, while within the layer $0 < z < a$ the solution is $U = \frac{1}{2}\mathcal{B}\{e^{-\imath k_3(\hat{z}-a)} + e^{\imath k_3(\hat{z}-a)}\}$. This corresponds to an incident wave of complex amplitude $\mathcal{A} = \frac{1}{2}\mathcal{B}\exp \imath k_3 a$ and a reflected wave of complex amplitude $\mathcal{R} = \frac{1}{2}\mathcal{B}\exp -\imath k_3 a$ which together excite in $z < -a$ an exponentially decaying disturbance with amplitude $\mathcal{T} = \mathcal{B}\cos k_3 a$. As in (8.17), this has $|\mathcal{R}| = |\mathcal{A}|$ and $\arg(\mathcal{R}/\mathcal{T}) = \arg(\mathcal{T}/\mathcal{A})$.

8.4 Elastic Plane Waves

Besides the elastic shear waves of Example 7.10 and the Love waves of Section 8.3, other wave motions are illustrated by considering plane waves. Example 7.10 shows that displacements $\boldsymbol{u} = u_2(x_1, t)\boldsymbol{e}_2$ travelling in the $\mathrm{O}x_1$ direction have wavespeed $c_S = \sqrt{\mu/\rho}$. Analogously, shear displacements $\boldsymbol{u} = u_3(x_1, t)\boldsymbol{e}_3$ may travel parallel to $\mathrm{O}x_1$ with velocity $\pm c_S$. Since an isotropic material defines no preferred directions, it should be unsurprising that plane waves may travel in *any* direction $\hat{\boldsymbol{k}}$ at speed c_S, provided that the induced displacements are shear displacements having some *polarization* $\boldsymbol{a}$ which is real and perpendicular to $\hat{\boldsymbol{k}}$. These waves are found by seeking plane waves with displacement (the real part of)

$$\boldsymbol{u}(\boldsymbol{x}, t) = \boldsymbol{a}e^{\imath(k_1 x_1+k_2 x_2+k_3 x_3-\omega t)} = \boldsymbol{a}\exp \imath(\boldsymbol{k}\cdot\boldsymbol{x} - \omega t) \quad (\boldsymbol{a} = \text{ const.}) \quad (8.26)$$

in which *each* component of displacement oscillates at the same frequency ω and, at each point, the motion is parallel to the polarization vector $\boldsymbol{a}$. Throughout this section, complex displacements are treated, although only real parts have physical relevance.

8.4.1 Elastic Shear Waves

In an isotropic elastic material, with stress components σ_{ij} related to the deformation gradient $\partial u_i/\partial x_j$ through (7.21), it transpires that two types of plane wave exist for each choice of the vector $\boldsymbol{k}$. One type is a shear wave generalizing that of Example 7.10, the other is a *longitudinal* (or *dilatational*) wave.

In component form, (8.26) gives $u_i = a_i \exp\imath(\boldsymbol{k}\cdot\boldsymbol{x} - \omega t)$. Since $\boldsymbol{k}\cdot\boldsymbol{x} - \omega t$ has partial derivatives $\partial(\boldsymbol{k}\cdot\boldsymbol{x} - \omega t)/\partial x_j = k_j$ (for $j = 1, 2, 3$), it follows that

$$\frac{\partial u_i}{\partial x_j} = \imath k_j a_i \exp\imath(\boldsymbol{k}\cdot\boldsymbol{x} - \omega t)\,, \qquad \frac{\partial u_j}{\partial x_i} = \imath k_i a_j \exp\imath(\boldsymbol{k}\cdot\boldsymbol{x} - \omega t)$$

and the dilatation is $\Delta = \imath k_i a_i \exp\imath(\boldsymbol{k}\cdot\boldsymbol{x} - \omega t)$. In the special case $\boldsymbol{a}\cdot\boldsymbol{k} = a_i k_i = 0$ (remember the summation convention), the disturbance induces no dilatation so that the stress components required in the elastodynamic equation (7.21) become $\sigma_{ij} = \mu\imath(a_i k_j + a_j k_i)\exp\imath(\boldsymbol{k}\cdot\boldsymbol{x} - \omega t)$. Then, since

$$\frac{\partial \sigma_{ij}}{\partial x_j} = \mu(\imath^2)k_j(a_i k_j + a_j k_i)\exp\imath(\boldsymbol{k}\cdot\boldsymbol{x} - \omega t) = -\mu\boldsymbol{k}\cdot\boldsymbol{k}a_i\exp\imath(\boldsymbol{k}\cdot\boldsymbol{x} - \omega t)$$

(where the relation $a_j k_j = \boldsymbol{a}\cdot\boldsymbol{k} = 0$ has been used) and since $\partial^2 u_i/\partial t^2 = -\omega^2 a_i \exp\imath(\boldsymbol{k}\cdot\boldsymbol{x} - \omega t)$, Equation (7.21) gives

$$-\rho\omega^2 a_i \exp\imath(\boldsymbol{k}\cdot\boldsymbol{x} - \omega t) = -\mu\boldsymbol{k}\cdot\boldsymbol{k}a_i\exp\imath(\boldsymbol{k}\cdot\boldsymbol{x} - \omega t)\,.$$

This confirms that, for *arbitrary* wave normal $\hat{\boldsymbol{k}} = \boldsymbol{k}/|\boldsymbol{k}|$ and arbitrary frequency ω a plane shear wave exists whenever $\boldsymbol{a}\cdot\boldsymbol{k} = 0$. It has wavelength $2\pi/|\boldsymbol{k}| = 2\pi c_S\omega^{-1}$ and advances in the direction of $\hat{\boldsymbol{k}}$ at speed $c_S = \omega/|\boldsymbol{k}| = \sqrt{\mu/\rho}$. The *polarization direction* denoted by $\boldsymbol{a}$ is the direction in which motions take place. It may be chosen as *any* direction orthogonal to the wave normal $\hat{\boldsymbol{k}}$, so showing that *shear waves* can travel in any direction and may be *polarized* in any orthogonal direction, but have phase speed c_S which is entirely independent of orientation.

Example 8.9

By superposing two shear waves $\boldsymbol{u} = a\boldsymbol{e}_2 e^{\imath(k_1x_1 - \omega t)}$ and $\boldsymbol{u} = \imath b\boldsymbol{e}_3 e^{\imath(k_1x_1 - \omega t)}$ travelling parallel to Ox_1 and having orthogonal polarizations $\boldsymbol{e}_2$ and $\boldsymbol{e}_3$, with both a and b real, show that shear waves travelling at speed c_S may be (i) *circularly polarized* (for $b = \pm a$), or (ii) *elliptically polarized* ($b \neq \pm a$) [the cases $a = 0$ and $b = 0$ give linear polarization].

In the real expression $\boldsymbol{u} = \mathrm{Re}\{(a\boldsymbol{e}_2 + \imath b\boldsymbol{e}_3)e^{\imath(k_1x_1 - \omega t)}\}$, the displacement components are $u_2 = a\cos(k_1x_1 - \omega t)$ and $u_3 = -b\sin(k_1x_1 - \omega t)$. Writing this as $\boldsymbol{u} = a\boldsymbol{e}_2\cos\theta + b\boldsymbol{e}_3\sin\theta$ shows (i) that, for $b = \pm a$ the displacement amplitude $|\boldsymbol{u}| = a$ remains constant through the wave, while the direction of displacement is $\theta = \mp(k_1x_1 - \omega t)$. Thus, as the wave advances in the $\boldsymbol{e}_1$ direction, the shear displacement at each chosen position $\boldsymbol{x}$ is a vector of constant magnitude a which rotates at constant angular speed ω around the $\boldsymbol{e}_1$ direction. The cases $b = \pm a$ describe right-handed and left-handed *circularly polarized* shear waves, respectively.

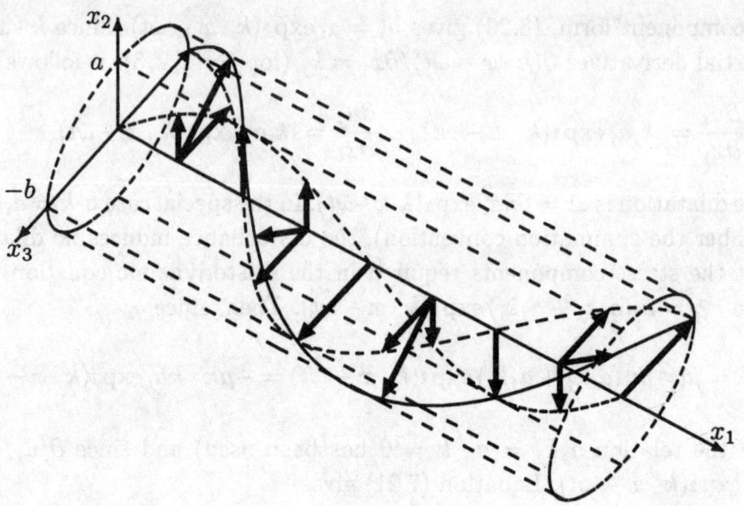

Figure 8.6 Two linearly polarized plane shear waves and their superposition as an elliptically polarized wave.

If $b \neq \pm a$, (ii) the direction of shear polarization still rotates around the e_1 direction, but since the components satisfy

$$\frac{u_2^2}{a^2} + \frac{u_3^2}{b^2} = 1\,,$$

the shear wave is known as elliptically polarized. The polarization ellipse has its principal axes aligned along e_2 and e_3. In the limiting cases $b = 0$ and $a = 0$, the ellipse degenerates to portions of straight line aligned along e_2 and e_3, respectively, so describing *linearly polarized* shear waves.

The above example shows how circularly and elliptically polarized shear waves arise as superpositions of two orthogonally polarized plane waves (8.26) having phases differing by a quarter-period. In fact, superposition of *any two* linearly polarized plane waves having a common wave vector $\boldsymbol{k}$ describes an elliptically polarized shear wave (or one of its limiting cases a circularly polarized or a linearly polarized wave). Similar features are shared by electromagnetic waves (see Section 9.1.2).

8.4.2 Dilatational Waves

To show that plane waves must be *either* shear waves *or* dilatational waves, in (8.26) write the vector $\boldsymbol{a}$ as $\boldsymbol{a} = \tilde{\boldsymbol{a}} + b\boldsymbol{k}$ with $\tilde{\boldsymbol{a}} \cdot \boldsymbol{k} = \tilde{a}_j k_j = 0$ as in shear waves, so that b is the displacement component parallel to $\boldsymbol{k}$ (the longitudinal component). Then, since $\boldsymbol{a}$ has components $a_i = \tilde{a}_i + bk_i$, the associated (complex) components of displacement gradient become

$$\frac{\partial u_i}{\partial x_j} = \imath(\tilde{a}_i k_j + bk_i k_j) \exp \imath(\boldsymbol{k} \cdot \boldsymbol{x} - \omega t),$$

so giving dilatation $\Delta = \imath b\boldsymbol{k} \cdot \boldsymbol{k} \exp \imath(\boldsymbol{k} \cdot \boldsymbol{x} - \omega t)$. The stress components then become

$$\sigma_{ij} = \imath \{\lambda \delta_{ij} b\boldsymbol{k} \cdot \boldsymbol{k} + \mu(\tilde{a}_i k_j + \tilde{a}_j k_i + 2bk_i k_j)\} \exp \imath(\boldsymbol{k} \cdot \boldsymbol{x} - \omega t),$$

so that (7.21) yields

$$\begin{aligned}
\frac{\partial \sigma_{ij}}{\partial x_j} &= (\imath)^2 \{\lambda b\boldsymbol{k} \cdot \boldsymbol{k} k_i + \mu \tilde{a}_i k_j k_j + \mu \tilde{a}_j k_j k_i + 2\mu b k_i k_j k_j\} \exp \imath(\boldsymbol{k} \cdot \boldsymbol{x} - \omega t) \\
&= -\{\mu \boldsymbol{k} \cdot \boldsymbol{k} \tilde{a}_i + (\lambda + 2\mu)\boldsymbol{k} \cdot \boldsymbol{k} b k_i\} \exp \imath(\boldsymbol{k} \cdot \boldsymbol{x} - \omega t) \\
&= \rho \frac{\partial^2 u_i}{\partial t^2} = -\omega^2 \rho(\tilde{a}_i + bk_i) \exp \imath(\boldsymbol{k} \cdot \boldsymbol{x} - \omega t). \qquad (8.27)
\end{aligned}$$

By considering the contributions to this equation which are perpendicular to $\boldsymbol{k}$, it is seen that $(\mu \boldsymbol{k} \cdot \boldsymbol{k} - \rho\omega^2)\tilde{\boldsymbol{a}} = \boldsymbol{0}$. This gives either $\tilde{\boldsymbol{a}} = \boldsymbol{0}$, or else $\rho\omega^2 = \mu\, \boldsymbol{k} \cdot \boldsymbol{k}$. The choice $\omega/|\boldsymbol{k}| = \pm c_S = \pm\sqrt{\mu/\rho}$ describes the shear waves of Section 8.4.1, since it requires that $b = 0$. If $b \neq 0$, the only way to balance the $b\boldsymbol{k}$ terms is to choose $\rho\omega^2 = (\lambda + 2\mu)\boldsymbol{k} \cdot \boldsymbol{k}$, with $\tilde{\boldsymbol{a}} = \boldsymbol{0}$. This describes *longitudinal* waves that travel at speed $c_L \equiv \sqrt{(\lambda + 2\mu)/\rho}$, with all displacements and motions being parallel to the (arbitrarily chosen) wave normal $\hat{\boldsymbol{k}} = \boldsymbol{k}/|\boldsymbol{k}|$. Thus, plane longitudinal waves may travel in *any* direction through an isotropic elastic solid, with squared wavespeed c_L^2, which is the extensional modulus (in the absence of lateral contraction) divided by density. In fact, c_L has an even greater significance, since in *any* disturbance, the dilatation Δ obeys the three-dimensional wave equation (8.7) with $c = c_L$. This is shown by first differentiating (7.21) to give

$$\rho \frac{\partial^2 \Delta}{\partial t^2} = \rho \frac{\partial^2}{\partial t^2} \left(\frac{\partial u_i}{\partial x_i} \right) = \frac{\partial}{\partial x_i} \left(\rho \frac{\partial^2 u_i}{\partial t^2} \right) = \frac{\partial}{\partial x_i} \left(\frac{\partial \sigma_{ij}}{\partial x_j} \right), \qquad (8.28)$$

while using the summation convention for repeated indices i and j.

Now, from $\sigma_{ij} = \lambda \Delta \delta_{ij} + \mu(\partial u_i/\partial x_j + \partial u_j/\partial x_i)$ it readily follows that

$$\begin{aligned}
\frac{\partial^2}{\partial x_i \partial x_j}(\lambda \Delta \delta_{ij}) &= \lambda \delta_{ij} \frac{\partial^2 \Delta}{\partial x_i \partial x_j} = \lambda \left(\frac{\partial^2 \Delta}{\partial x_1^2} + \frac{\partial^2 \Delta}{\partial x_2^2} + \frac{\partial^2 \Delta}{\partial x_3^2} \right) = \lambda \nabla^2 \Delta, \\
\frac{\partial^2}{\partial x_i \partial x_j} \left(\mu \frac{\partial u_i}{\partial x_j} \right) &= \mu \frac{\partial^2}{\partial x_j \partial x_j} \left(\frac{\partial u_i}{\partial x_i} \right) = \mu \frac{\partial^2 \Delta}{\partial x_j \partial x_j} = \mu \nabla^2 \Delta, \\
\frac{\partial^2}{\partial x_i \partial x_j} \left(\mu \frac{\partial u_j}{\partial x_i} \right) &= \mu \frac{\partial^2}{\partial x_i \partial x_i} \left(\frac{\partial u_j}{\partial x_j} \right) = \mu \nabla^2 \left(\frac{\partial u_j}{\partial x_j} \right) = \mu \nabla^2 \Delta.
\end{aligned}$$

Thus, combining these results with (8.28) gives

$$\rho\frac{\partial^2\Delta}{\partial t^2} = (\lambda+2\mu)\nabla^2\Delta \quad \text{or} \quad \frac{\partial^2\Delta}{\partial t^2} = c_L^2\left(\frac{\partial^2\Delta}{\partial x_1^2}+\frac{\partial^2\Delta}{\partial x_2^2}+\frac{\partial^2\Delta}{\partial x_3^2}\right). \tag{8.29}$$

Equation (8.29) shows that, in any elastic disturbance, changes in volume travel at the *dilatational speed* c_L. These disturbances are closely analogous to the pressure waves in acoustics. However, since elastic materials support both shear waves and dilatational waves, reflections and refractions of elastic waves at boundaries are much more complicated to analyse than are acoustic reflections and refractions. Typically reflections and refractions involve conversion between the two types of disturbance. In seismology, dilatational waves are evident since they are the fastest waves which can travel through a solid. Hence, after a major seismic event, the leading part of the signal received at a distant seismic station is interpreted as due to P-waves (pressure waves), while later parts (typically having much greater amplitude) are due to shear waves (S-waves) and more complicated modes such as Love waves and Rayleigh waves (see Exercise 8.13).

Example 8.10

Show that plane elastic waves of the form $\boldsymbol{u} = \boldsymbol{a}e^{\imath(k_1x+k_3z-\omega t)}$ exist when either $\rho\omega^2 = (\lambda+2\mu)(k_1^2+k_3^2)$, or $\rho\omega^2 = \mu(k_1^2+k_3^2)$. Show that, in the first case, $\boldsymbol{a} = \alpha_1(k_1\boldsymbol{e}_1+k_3\boldsymbol{e}_3)$, while in the second case $\boldsymbol{a} = \alpha_2\boldsymbol{e}_2+\alpha_3(k_3\boldsymbol{e}_1-k_1\boldsymbol{e}_3)$, for some choices of constants α_1, α_2 and α_3.

The plane wave has wave vector $\boldsymbol{k} = k_1\boldsymbol{e}_1+k_3\boldsymbol{e}_3$. Two orthogonal vectors are $\boldsymbol{e}_2$ and $k_3\boldsymbol{e}_1-k_1\boldsymbol{e}_3 = \boldsymbol{b}$ (say). Write $\boldsymbol{a} = \alpha_1\boldsymbol{k}+\alpha_2\boldsymbol{e}_2+\alpha_3\boldsymbol{b}$, so that

$$u_1 = (\alpha_1k_1+\alpha_3k_3)e^{\imath(k_1x+k_3z-\omega t)}, \qquad u_2 = \alpha_2e^{\imath(k_1x+k_3z-\omega t)},$$
$$u_3 = (\alpha_1k_3-\alpha_3k_1)e^{\imath(k_1x+k_3z-\omega t)}.$$

The (complex) displacement-gradient components $\partial u_i/\partial u_j$ then become

$$\frac{\partial u_i}{\partial x_j} = \begin{pmatrix} k_1(\alpha_1k_1+\alpha_3k_3) & 0 & k_3(\alpha_1k_1+\alpha_3k_3) \\ k_1\alpha_2 & 0 & k_3\alpha_2 \\ k_1(\alpha_1k_3-\alpha_3k_1) & 0 & k_3(\alpha_1k_3-\alpha_3k_1) \end{pmatrix} \imath e^{\imath(k_1x+k_3z-\omega t)},$$

with $x_1 = x$ and $x_3 = z$. Then, $\Delta = \alpha_1(k_1^2+k_3^2)e^{\imath(k_1x+k_3z-\omega t)}$ and the (complex) stress components are found to be

$$\sigma_{11} = \{\alpha_1[(\lambda + 2\mu)k_1^2 + \lambda k_3^2] + 2\alpha_3\mu k_1 k_3\}e^{i(k_1x+k_3z-\omega t)},$$
$$\sigma_{12} = \sigma_{21} = \alpha_2\mu k_1 e^{i(k_1x+k_3z-\omega t)}, \; \sigma_{22} = \alpha_1\lambda(k_1^2 + k_3^2)e^{i(k_1x+k_3z-\omega t)},$$
$$\sigma_{13} = \sigma_{31} = \{2\alpha_1\mu k_1 k_3 + \alpha_3\mu(k_3^2 - k_1^2) + 2\alpha_3\}e^{i(k_1x+k_3z-\omega t)},$$
$$\sigma_{23} = \sigma_{32} = \alpha_2\mu k_3 e^{i(k_1x+k_3z-\omega t)},$$
$$\sigma_{33} = \{\alpha_1[\lambda k_1^2 + (\lambda + 2\mu)k_3^2] - 2\alpha_3\mu k_1 k_3\}e^{i(k_1x+k_3z-\omega t)}.$$

Substituting into $\rho\partial^2 u_i/\partial t^2 = \partial\sigma_{i1}/\partial x + \partial\sigma_{i3}/\partial z$ for each of $i = 1, 2, 3$ and then cancelling the complex exponentials gives

$$-\rho\omega^2(\alpha_1 k_1 + \alpha_3 k_3) = -(k_1^2 + k_3^2)[(\lambda + 2\mu)\alpha_1 k_1 + \mu\alpha_3 k_3],$$
$$-\rho\omega^2\alpha_2 = -(k_1^2 + k_3^2)\mu\alpha_2,$$
$$-\rho\omega^2(\alpha_1 k_3 - \alpha_3 k_1) = -(k_1^2 + k_3^2)[(\lambda + 2\mu)\alpha_1 k_3 - \mu\alpha_3 k_1].$$

Clearly, for $\alpha_2 \neq 0$ this requires that $\rho\omega^2 = \mu(k_1^2 + k_3^2)$, so requiring that $\alpha_1 = 0$ but allowing α_3 to be arbitrary. If $\rho\omega^2 \neq \mu(k_1^2 + k_3^2)$, then $\alpha_2 = \alpha_3 = 0$. In this case, either $\rho\omega^2 \neq (\lambda + 2\mu)(k_1^2 + k_3^2)$ (so giving $\boldsymbol{u} \equiv \boldsymbol{0}$) or $\rho\omega^2 = (\lambda + 2\mu)(k_1^2 + k_3^2)$, with α_1 arbitrary. The possibilities $\boldsymbol{a} = \alpha_1(k_1\boldsymbol{e}_1 + k_3\boldsymbol{e}_3)$ and $\boldsymbol{a} = \alpha_2\boldsymbol{e}_2 + \alpha_3(k_3\boldsymbol{e}_1 - k_1\boldsymbol{e}_3)$ thus arise for waves travelling at the dilatational speed c_L and the shear speed c_S, respectively.

EXERCISES

8.9. In the dispersion relation for Love waves (see Section 8.3), put $ak_3 = X$ and $a\gamma = Y$ and so demonstrate that possible modes correspond to intersections of the curve $Y = (\mu_0/\mu_1)X \tan X$ and the circle $X^2 + Y^2 = a^2\omega^2(\mu_0^{-1}\rho_0 - \mu_1^{-1}\rho_1)$.

Deduce that, for $a\omega < \frac{1}{2}\pi(c_0^{-2} - c_1^{-2})^{-1/2}$, only a single mode (having $ak_3 < \frac{1}{2}\pi$) exists (in this mode, there is no depth at which $U = 0$). Deduce also that, for $\pi(c_0^{-2} - c_1^{-2})^{-1/2} < a\omega < \frac{3}{2}\pi(c_0^{-2} - c_1^{-2})^{-1/2}$ three modes exist (one in each of $0 < ak_3 < \frac{1}{2}\pi$, $\frac{1}{2}\pi < ak_3 < \pi$ and $\pi < ak_3 < \frac{3}{2}\pi$).

8.10. Confirm that, for all choices of *complex* $\mathcal{A}_2$ and $\mathcal{A}_3$, the displacements

$$\boldsymbol{u} = \mathrm{Re}\{(\mathcal{A}_2\boldsymbol{e}_2 + \mathcal{A}_3\boldsymbol{e}_3)e^{i(kx-\omega t)}\} \qquad \text{with} \qquad \rho\omega^2 = \mu k^2$$

describe elastic shear waves travelling parallel to Ox. Confirm also that

(a) if $\mathcal{A}_3/\mathcal{A}_2$ is real, the wave is a linearly polarized shear wave,
(b) if $\mathcal{A}_3/\mathcal{A}_2 = \pm\imath$, the wave is a circularly polarized shear wave.

By writing $\mathcal{A}_2 = b + c$ and $\mathcal{A}_3 = \imath(b - c)$, show that this general shear wave may be regarded as the superposition of a right-handed circularly polarized shear wave of amplitude $|b|$ and a left-handed circularly polarized shear wave of amplitude $|c|$.
Express $|b|$ and $|c|$ in terms of $\mathcal{A}_2$ and $\mathcal{A}_3$.

8.11. Confirm that the displacements in Exercise 8.10 generalize those of Example 8.9 (since $\boldsymbol{a}$ is *any* constant *complex* vector orthogonal to $\boldsymbol{e}_1$).

Split $\boldsymbol{a}$ into its real and imaginary parts $\boldsymbol{a} = \boldsymbol{b} + \imath\boldsymbol{c}$. Show that displacements at a typical point may be written parametrically as $\boldsymbol{u} = \boldsymbol{b}\cos\alpha + \boldsymbol{c}\sin\alpha$, with α varying periodically in time. Explain why, in general, this describes motion around an ellipse.

8.12. In an elastic material occupying $y \equiv x_2 \leq 0$ and having density ρ and Lamé coefficients λ and μ, seek two-dimensional displacements $\boldsymbol{u} = u_1\boldsymbol{e}_1 + u_2\boldsymbol{e}_2$ in the form of travelling disturbances $\boldsymbol{u} = \mathrm{Re}\{\boldsymbol{A}\mathrm{e}^{\gamma y}\mathrm{e}^{\imath k(x-ct)}\}$. Show, by writing $\lambda + 2\mu = \rho c_L^2$, $\mu = \rho c_S^2$, $\gamma/k = \Gamma$ and $\boldsymbol{A} = A\boldsymbol{e}_1 + B\boldsymbol{e}_2$, that A and B must satisfy

$$(c^2 - c_L^2 + \Gamma^2 c_S^2)A + \imath\Gamma(c_L^2 - c_S^2)B = 0\,,$$
$$\imath\Gamma(c_L^2 - c_S^2)A + (c^2 + \Gamma^2 c_L^2 - c_S^2)B = 0\,.$$

Hence, for $0 < c < c_S < c_L$, show that there are two positive choices for γ/k, namely $\Gamma_1 = \sqrt{1 - c^2/c_S^2}$, $\Gamma_2 = \sqrt{1 - c^2/c_L^2}$, with corresponding ratios $\imath B_1/A_1 = \Gamma_1^{-1}$ and $\imath B_2/A_2 = \Gamma_2$ for $\imath B/A$.

8.13. By superposing the two solutions $\boldsymbol{u}$ found in Exercise 8.12 and by taking A_1 and A_2 to be real, deduce that the displacements

$$u_1 = (A_1\mathrm{e}^{k\Gamma_1 y} + A_2\mathrm{e}^{k\Gamma_2 y})\cos k(x - ct)\,,$$
$$u_2 = (A_1\Gamma_1^{-1}\mathrm{e}^{k\Gamma_1 y} + A_2\Gamma_2\mathrm{e}^{k\Gamma_2 y})\sin k(x - ct)\,, \quad u_3 = 0$$

describe elastic disturbances travelling at speed c parallel to Ox and decaying as $y \to -\infty$. In these disturbances, show that the components of traction acting over $y = 0$ are

$$\sigma_{12} = \rho k c_S^2[(\Gamma_1 + \Gamma_1^{-1})A_1 + 2\Gamma_2 A_2]\cos k(x - ct),$$
$$\sigma_{22} = \rho k[2c_S^2 A_1 + (\Gamma_2^2 c_L^2 - c_L^2 + 2c_S^2)A_2]\sin k(x - ct), \quad \sigma_{32} = 0.$$

Hence deduce that all traction components σ_{i2} vanish on $y = 0$ when $(\Gamma_1^2 + 1)^2 = 4\Gamma_1\Gamma_2$. Show also that this condition yields a cubic equation for c^2, namely

$$f(c^2) \equiv c^6 - 8c_S^2c^4 + 8c_S^4(3 - 2c_S^2/c_L^2)c^2 - 16c_S^6(1 - c_S^2/c_L^2) = 0.$$

Confirm that $f(0) < 0$ and that $f(c_S^2) = c_S^6 > 0$, so that there exists a speed $c = c_R$ at which elastic waves may travel adjacent to a traction-free surface.

[These *Rayleigh waves*, obtained by putting $c = c_R$ in Γ_1, Γ_2, u_1 and u_2, were discovered by Lord Rayleigh in 1885. For *all* wavelengths $2\pi/k$, they travel at the Rayleigh speed c_R ($< c_S < c_L$). Rayleigh waves are thus an unusual type of guided wave, since they are *non-dispersive* – there is no tendency for waves of different wavelengths to separate out.
Despite the layered structure of the earth, Rayleigh waves, like Love waves, are significant carriers of seismic disturbances.]

9
Electromagnetic Waves and Light

9.1 Physical Background

Historically, much of the motivation for the study of waves, together with much of the terminology such as reflection, refraction and polarization, came from observation and interpretation of the behaviour of light. Awareness, by the early nineteenth century, that electricity and magnetism are interrelated phenomena, together with the mathematical formulation of observed 'laws' named after Ampère, Faraday, Gauss and others, led finally to a description using a set of equations which predicts the propagation of waves. Moreover, those waves are *transverse* in nature, which is consistent with the observation that as light propagates in one direction, it is possible with some crystalline materials to detect that light is associated with preferred directions orthogonal to the propagation direction. A brief motivation for the set of equations known as Maxwell's equations follows in Section 9.1.1, but more extensive accounts may be found in Lorrain, Corson and Lorrain (2000). The reader prepared to accept Maxwell's equations as the basis of electromagnetism may proceed directly to Section 9.1.2.

9.1.1 The Origin of Maxwell's Equations

In Section 3.4, it was seen that electric charge density ρ_e acts as a source for static *electric fields* $\boldsymbol{E}$, so that for any region $\mathcal{R}$ with boundary surface $\partial\mathcal{R}$

having outward unit normal $\boldsymbol{n}$

$$\iiint_{\mathcal{R}} \rho_e \,\mathrm{d}V = \text{ charge contained within } \mathcal{R} = \epsilon_0 \iint_{\partial\mathcal{R}} \boldsymbol{E}\cdot\boldsymbol{n}\,\mathrm{d}S\,. \tag{9.1}$$

In Section 4.4.2, the rate of charge flow per unit of area was measured by the *electric current density* $\boldsymbol{J}$ (an example of a flux vector). Since charge is neither created nor destroyed, there is a *law of charge conservation*

$$\iiint_{\mathcal{R}} \frac{\partial \rho_e}{\partial t}\,\mathrm{d}V + \iint_{\partial\mathcal{R}} \boldsymbol{J}\cdot\boldsymbol{n}\,\mathrm{d}S = 0\,, \tag{9.2}$$

or (in local form) $\mathrm{div}\boldsymbol{J} + \partial\rho_e/\partial t = \boldsymbol{\nabla}\cdot\boldsymbol{J} + \partial\rho_e/\partial t = 0$.

It was known from Ampère's experiments that current flowing along a long thin straight wire creates a *magnetic field* in the azimuthal direction (i.e. tangential to circles centred on the wire and lying in planes orthogonal to the wire) with strength inversely proportional to distance from the wire. Thus, the magnetic loop tension (the line integral of the tangential component of the field around each such circular loop) is independent of radius. From this, Ampère's law was deduced, stating that there exists a vector field known as *magnetic field intensity* $\boldsymbol{H}$ such that, if $\mathcal{C}$ is *any* fixed closed loop in three-dimensional space and $\mathcal{S}$ is any surface spanning $\mathcal{C}$ with $\boldsymbol{n}$ as unit normal, then

$$\oint_{\mathcal{C}} \boldsymbol{H}\cdot\mathrm{d}\boldsymbol{r} = \text{ rate of charge flow through } \mathcal{C} = \iint_{\mathcal{S}} \boldsymbol{J}\cdot\boldsymbol{n}\,\mathrm{d}S\,. \tag{9.3}$$

However, with the benefit of vector calculus, it is readily seen that Equations (9.1)–(9.3) are inconsistent. Indeed, elimination of integrals involving $\partial\rho_e/\partial t$ between (9.1) and (9.2) leads to

$$\iint_{\partial\mathcal{R}} \left(\boldsymbol{J} + \frac{\partial(\epsilon_0\boldsymbol{E})}{\partial t}\right)\cdot\boldsymbol{n}\mathrm{d}S = 0\,,$$

which shows that if $\mathcal{R}$ is the region between two *different* choices $\mathcal{S}_1$ and $\mathcal{S}_2$ of surfaces spanning a loop $\mathcal{C}$, then (9.3) gives two distinct values for $\oint_{\mathcal{C}} \boldsymbol{H}\cdot\mathrm{d}\boldsymbol{r}$ when the electric field $\boldsymbol{E}$ is *unsteady* (i.e. depends on time). This mathematical inconsistency led Maxwell to propose *on theoretical grounds* that Ampère's law (9.3) should be modified by including the term $\partial(\epsilon_0\boldsymbol{E})/\partial t$ with $\boldsymbol{J}$ in the integral on the right. Thus, for unsteady fields, Ampère's law (9.3) must be replaced by

$$\oint_{\mathcal{C}} \boldsymbol{H}\cdot\mathrm{d}\boldsymbol{r} = \iint_{\mathcal{S}} \left(\boldsymbol{J} + \frac{\partial(\epsilon_0\boldsymbol{E})}{\partial t}\right)\cdot\boldsymbol{n}\,\mathrm{d}S\,. \tag{9.4}$$

The second important physical law connecting magnetic and electric fields is Faraday's law, which results from his observation that, when a conducting loop $\mathcal{C}$ is placed in a time-varying magnetic field, then an electric current flows around that loop. The interpretation is that the loop integral $\oint_C \boldsymbol{E} \cdot \mathrm{d}\boldsymbol{r}$ is proportional to the time rate of change of the magnetic flux through the loop, so leading to

$$\oint_C \boldsymbol{E} \cdot \mathrm{d}\boldsymbol{r} = -\frac{\mathrm{d}}{\mathrm{d}t} \iint_S \mu_0 \boldsymbol{H} \cdot \boldsymbol{n} \mathrm{d}S \tag{9.5}$$

for all surfaces S spanning an arbitrary closed loop C, where the constant μ_0 is known as the *permeability of free space.*

The governing equations of electromagnetism are (9.4) and (9.5), or their local forms $\nabla \times \boldsymbol{H} = \boldsymbol{J} + \partial(\epsilon_0 \boldsymbol{E})/\partial t$ and $\nabla \times \boldsymbol{E} = -\mu_0 \partial \boldsymbol{H}/\partial t$, together with Gauss's law (9.1) and an equivalent law

$$\iint_{\partial \mathcal{R}} \mu_0 \boldsymbol{H} \cdot \boldsymbol{n} \mathrm{d}S = 0\,, \tag{9.6}$$

which states that there is no magnetic analogue of a point source of magnetism (the most elementary object is a *magnetic dipole*). These four equations, together with a law (typically generalizing Ohm's law) relating the electric current $\boldsymbol{J}$ to the electric field $\boldsymbol{E}$, describe electromagnetic fields in many situations. However, in order to describe refraction at an interface it is necessary to include one pair of complications. In most materials, the presence of an electric field causes the creation of a distribution of molecular *dipoles* (cf. Section 3.4), which may be described using an *electric polarization density* $\boldsymbol{P}$. If this polarization density depends upon $\boldsymbol{r}$ (i.e. is non-uniform), there is a contribution $-\nabla \cdot \boldsymbol{P} = -\mathrm{div}\, \boldsymbol{P}$ to the charge density ρ_e, so that the right-hand term of (9.1) should be $\iint_{\partial\mathcal{R}} (\epsilon_0 \boldsymbol{E} + \boldsymbol{P}) \cdot \boldsymbol{n} \mathrm{d}S$. The vector $\boldsymbol{D} \equiv \epsilon_0 \boldsymbol{E} + \boldsymbol{P}$ is known as the *electric displacement.* For consistency, the term $\partial(\epsilon_0 \boldsymbol{E})/\partial t$ within (9.4) should also be replaced by $\partial \boldsymbol{D}/\partial t$, which is known as Maxwell's *displacement current.*

Analogously, in *magnetic materials*, there may be a *magnetization density* $\boldsymbol{M}(\boldsymbol{r}, t)$ which gives rise to the need to modify (9.6) and (9.5) and to define the *magnetic induction* $\boldsymbol{B} \equiv \mu_0 (\boldsymbol{H} + \boldsymbol{M})$. The end result is to produce the set of equations

$$\mathrm{div}\, \boldsymbol{D} = \nabla \cdot \boldsymbol{D} = \rho_e \qquad \text{(Gauss's law)} \tag{9.7}$$

$$\mathrm{div}\, \boldsymbol{B} = \nabla \cdot \boldsymbol{B} = 0 \qquad \text{(Gauss's magnetic law)} \tag{9.8}$$

$$\mathrm{curl}\, \boldsymbol{E} = \nabla \times \boldsymbol{E} = -\frac{\partial \boldsymbol{B}}{\partial t} \qquad \text{(Faraday's law)} \tag{9.9}$$

$$\mathrm{curl}\, \boldsymbol{H} = \nabla \times \boldsymbol{H} = \frac{\partial \boldsymbol{D}}{\partial t} + \boldsymbol{J} \qquad \text{(Ampère's law)} \tag{9.10}$$

which should be supplemented by constitutive laws relating $\boldsymbol{B}$ to $\boldsymbol{H}$ and $\boldsymbol{D}$ and $\boldsymbol{J}$ to $\boldsymbol{E}$. In many materials, appropriate assumptions are that the electric polarization $\boldsymbol{P}$ is proportional to $\boldsymbol{E}$, while the magnetization $\boldsymbol{M}$ is proportional to $\boldsymbol{H}$ (or is fixed in a *permanent magnet*, or is zero in a *non-magnetic material*), so yielding the constitutive relations

$$\boldsymbol{D} = \epsilon \boldsymbol{E}\,, \qquad \boldsymbol{B} = \mu \boldsymbol{H}\,, \tag{9.11}$$

where ϵ is the *permittivity* of the material and μ is its *magnetic permeability*. If the material conducts electricity, then it is common to write $\boldsymbol{J} = \sigma_c \boldsymbol{E}$ (see Section 4.4.2), with σ_c the electrical conductivity.

9.1.2 Plane Electromagnetic Waves

Equations (9.7)–(9.10) are known collectively as *Maxwell's equations* in honour of James Clerk Maxwell (1831–79), for his rôle in completing the unification of the theories of electricity and magnetism, in particular his demonstration that they predict wave propagation – even through empty space, within which $\boldsymbol{P} = \boldsymbol{0}$, $\boldsymbol{M} = \boldsymbol{0}$ and $\boldsymbol{J} = \boldsymbol{0}$.

Important features of electromagnetic waves are revealed by considering plane waves in any non-conducting medium, with $\boldsymbol{D}$ and $\boldsymbol{B}$ given by (9.11). For simplicity, take these waves to have $\boldsymbol{E} = \boldsymbol{E}(x,t)$ and $\boldsymbol{H} = \boldsymbol{H}(x,t)$, so that the O$x$ axis is in the direction of propagation (with $x = x_1$). Then, by resolving Equations (9.7)–(9.10) into cartesian components, it is found that

$$\epsilon \frac{\partial E_1}{\partial x} = 0\,, \qquad \mu \frac{\partial H_1}{\partial x} = 0\,, \tag{9.12}$$

$$0 = -\mu \frac{\partial H_1}{\partial t}\,, \qquad 0 = \epsilon \frac{\partial E_1}{\partial t} \tag{9.13}$$

and

$$-\frac{\partial E_3}{\partial x} = -\mu \frac{\partial H_2}{\partial t}\,, \qquad -\frac{\partial H_3}{\partial x} = \epsilon \frac{\partial E_2}{\partial t}\,, \tag{9.14}$$

$$\frac{\partial E_2}{\partial x} = -\mu \frac{\partial H_3}{\partial t}\,, \qquad \frac{\partial H_2}{\partial x} = \epsilon \frac{\partial E_3}{\partial t}\,. \tag{9.15}$$

Equations (9.12) and (9.13) collectively state that both of E_1 and H_1 are independent of both x and t. Thus, the *longitudinal* components of the fields are constant and uniform. They do not describe waves and so will be omitted from the discussion.

Equations (9.14) and (9.15) may be separated into two pairs, linking E_3 with H_2 and E_2 with H_3, respectively. Moreover, each pair allows elimination of the magnetic field components as follows:

$$\frac{\partial^2 E_2}{\partial x^2} = -\mu \frac{\partial^2 H_3}{\partial x \partial t} = \epsilon\mu \frac{\partial^2 E_2}{\partial t^2}\,, \quad \frac{\partial^2 E_3}{\partial x^2} = \mu \frac{\partial^2 H_2}{\partial x \partial t} = \epsilon\mu \frac{\partial^2 E_3}{\partial t^2}\,.$$

Thus, *each* of the *transverse components* of electric field satisfies the *wave equation* with wavespeed c given by $c^2 = (\epsilon\mu)^{-1}$. Likewise, it is readily shown that the transverse magnetic field components H_2 and H_3 satisfy the wave equation with the same wavespeed $c = (\epsilon\mu)^{-1/2}$: the *speed of light*. This speed depends on both the permittivity and the permeability of the medium through which the waves travel and, in empty space, it is $(\epsilon_0\mu_0)^{-1/2} = 2.998 \times 10^8$ m/sec $\equiv c_0$, known as the speed of light *in vacuo*.

The inter-connection between the transverse electric and magnetic field components is well illustrated by linearly polarized plane waves. For example, consider $E_2 = a\cos(kx - \omega t + \alpha)$ so that

$$\frac{\partial H_3}{\partial x} = -\epsilon\frac{\partial E_2}{\partial t} = -\epsilon\omega a\sin(kx - \omega t + \alpha),$$
$$\frac{\partial H_3}{\partial t} = -\mu^{-1}\frac{\partial E_2}{\partial x} = \mu^{-1}ka\sin(kx - \omega t + \alpha),$$

giving

$$H_3 = \frac{\epsilon\omega}{k}a\cos(kx - \omega t + \alpha) = \frac{k}{\mu\omega}a\cos(kx - \omega t + \alpha) = \sqrt{\epsilon/\mu}\,E_2(x,t)$$

(after omitting a redundant constant), where the phase speed is, of course, $\omega/k = c = (\epsilon\mu)^{-1/2}$. This describes a plane wave travelling in the $\boldsymbol{e}_1$ direction, having wavenumber k and transverse electric field $\boldsymbol{E} = a\boldsymbol{e}_2\cos(kx - \omega t + \alpha)$, together with a transverse magnetic field $\boldsymbol{H} = \sqrt{\epsilon/\mu}\,a\boldsymbol{e}_3\cos(kx - \omega t + \alpha)$ which is orthogonal to both $\boldsymbol{e}_1$ and $\boldsymbol{E}$ and oscillates *in phase* with $\boldsymbol{E}$.

It is readily shown that another plane wave is given by

$$\boldsymbol{E} = b\boldsymbol{e}_3\cos(kx - \omega t + \beta), \quad \boldsymbol{H} = \sqrt{\epsilon/\mu}\,b\boldsymbol{e}_2\cos(kx - \omega t + \beta)$$

and that each of these linearly polarized plane waves has the property

$$\boldsymbol{H}(x,t) = (\omega\mu)^{-1}\boldsymbol{k} \times \boldsymbol{E}, \qquad \text{where } \boldsymbol{k} = k\boldsymbol{e}_1 \text{ is the wave vector.} \qquad (9.16)$$

Example 9.1

Show that, if $\boldsymbol{a}$ and $\boldsymbol{k}$ are any two real, orthogonal vectors, then Maxwell's equations (9.7)–(9.10) together with $\boldsymbol{J} = \boldsymbol{0}$ and the linear constitutive laws (9.11) have the solution

$$\boldsymbol{E} = \mathrm{Re}\{\mathcal{A}\boldsymbol{a}\exp\imath(\boldsymbol{k}\cdot\boldsymbol{x} - \omega t)\}, \quad \boldsymbol{H} = \mathrm{Re}\{\mathcal{A}(\omega\mu)^{-1}\boldsymbol{k}\times\boldsymbol{a}\exp\imath(\boldsymbol{k}\cdot\boldsymbol{x} - \omega t)\}.$$

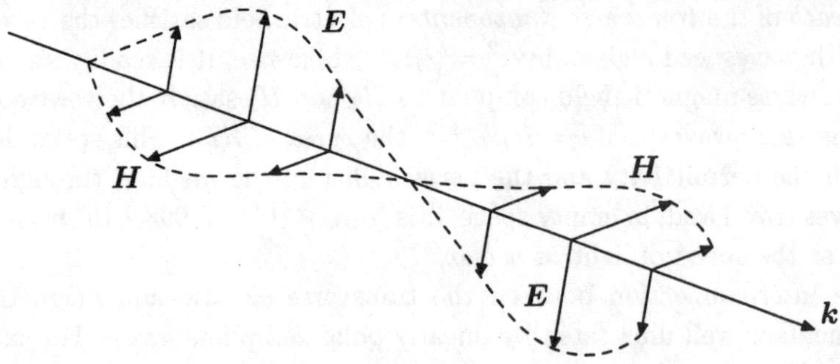

Figure 9.1 A linearly polarized electromagnetic plane wave, showing how the magnetic field $\boldsymbol{H}$ is polarized orthogonally to the electric field $\boldsymbol{E}$.

From $\boldsymbol{D} = \epsilon\boldsymbol{E}$ and from the vector identity $\nabla\cdot(\phi\boldsymbol{u}) = \phi\nabla\cdot\boldsymbol{u} + \nabla\phi\cdot\boldsymbol{u}$ (see e.g. Matthews, 1998, Chapter 4) which holds for any scalar function $\phi(\boldsymbol{x},t)$ and any vector function $\boldsymbol{u}(\boldsymbol{x},t)$, it is verified that

$$\begin{aligned}\nabla\cdot\boldsymbol{D} &= \epsilon\mathrm{Re}\{\boldsymbol{a}\cdot\mathcal{A}\nabla[\imath(\boldsymbol{k}\cdot\boldsymbol{x}-\omega t)]\exp\imath(\boldsymbol{k}\cdot\boldsymbol{x}-\omega t)\}\\ &= \epsilon\mathrm{Re}\{\imath\mathcal{A}\boldsymbol{a}\cdot\boldsymbol{k}\exp\imath(\boldsymbol{k}\cdot\boldsymbol{x}-\omega t)\} = 0\end{aligned}$$

whenever $\boldsymbol{a}$ is orthogonal to $\boldsymbol{k}$. To calculate curl $\boldsymbol{E}$, either use component form, or recall another vector identity $\mathrm{curl}(\phi\boldsymbol{u}) = \nabla\times(\phi\boldsymbol{u}) = \phi\nabla\times\boldsymbol{u} + \nabla\phi\times\boldsymbol{u}$. Then, since $\nabla(\boldsymbol{k}\cdot\boldsymbol{x}-\omega t) = \boldsymbol{k}$ and $\nabla\times\boldsymbol{a} = \boldsymbol{0}$, it is found that

$$\nabla\times\boldsymbol{E} = \mathrm{Re}\{\mathcal{A}\imath(\boldsymbol{k}\times\boldsymbol{a})\exp\imath(\boldsymbol{k}\cdot\boldsymbol{x}-\omega t)\},$$

while $$\partial\boldsymbol{E}/\partial t = \mathrm{Re}\{-\imath\omega\mathcal{A}\boldsymbol{a}\exp\imath(\boldsymbol{k}\cdot\boldsymbol{x}-\omega t)\}.$$

Similarly, since $\boldsymbol{k}\times\boldsymbol{a}$ is a constant vector orthogonal to $\boldsymbol{k}$, it follows that $\nabla\cdot(\mu\boldsymbol{H}) = 0$ and that

$$\mu\partial\boldsymbol{H}/\partial t = \mu\mathrm{Re}\{-\imath\omega(\omega\mu)^{-1}\mathcal{A}(\boldsymbol{k}\times\boldsymbol{a})\exp\imath(\boldsymbol{k}\cdot\boldsymbol{x}-\omega t)\} = -\nabla\times\boldsymbol{E}.$$

Also,

$$\begin{aligned}\nabla\times\boldsymbol{H} &= (\omega\mu)^{-1}\mathrm{Re}\{\mathcal{A}\imath\boldsymbol{k}\times(\boldsymbol{k}\times\boldsymbol{a})\exp\imath(\boldsymbol{k}\cdot\boldsymbol{x}-\omega t)\}\\ &= k^2(\omega\mu)^{-1}\mathrm{Re}\{-\imath\mathcal{A}\boldsymbol{a}\exp\imath(\boldsymbol{k}\cdot\boldsymbol{x}-\omega t)\} = \epsilon\frac{\partial\boldsymbol{E}}{\partial t}.\end{aligned}$$

Thus, all four of Maxwell's equations are satisfied.

Example 9.1 shows that linearly polarized waves may travel with fields constant on planes having *any* unit normal $\hat{\boldsymbol{k}} = \boldsymbol{k}/k$, *any* wavenumber $k = |\boldsymbol{k}|$ and having electric field $\boldsymbol{E}$ polarized in *any* direction $\boldsymbol{a}$ which is perpendicular to $\boldsymbol{k}$, provided that the frequency is $\omega = k(\epsilon\mu)^{-1/2}$. The magnetic field $\boldsymbol{H}$ is then also linearly polarized, in the direction of $\boldsymbol{k} \times \boldsymbol{a}$, so that the wave vector $\boldsymbol{k}$, the orientation $\boldsymbol{a}$ of electric polarization and the orientation of magnetic polarization form a right-handed triad of orthogonal vectors. The example leading to (9.16) is just a simple example of a configuration with these general properties.

Just as elastic shear waves in two orthogonal polarizations may be combined to give circularly, or elliptically, polarized waves (see Example 8.9 and Exercise 8.11), two solutions of the form of Example 9.1 may be superposed so as to describe circularly polarized electromagnetic waves.

Example 9.2 (Circularly polarized light)

Show that, for any wavenormal $\hat{\boldsymbol{k}}$, any orthogonal (real) unit vector $\hat{\boldsymbol{a}}$ and *any* complex amplitude $\mathcal{A}$ the electric field

$$\boldsymbol{E} = \mathrm{Re}\{\mathcal{A}(\hat{\boldsymbol{a}} + \imath\hat{\boldsymbol{a}} \times \hat{\boldsymbol{k}}) \exp \imath(\boldsymbol{k} \cdot \boldsymbol{x} - \omega t)\}\,, \qquad \text{with } \boldsymbol{k} = k\hat{\boldsymbol{k}} = |\boldsymbol{k}|\hat{\boldsymbol{k}}\,,$$

describes a right-handed circularly polarized electromagnetic wave of frequency ω and wavenumber $k = (\epsilon\mu)^{1/2}\omega$. Find the associated magnetic excitation $\boldsymbol{H}$.

Both $\hat{\boldsymbol{a}}$ and $\hat{\boldsymbol{a}} \times \hat{\boldsymbol{k}}$ are unit vectors perpendicular to the wave vector $\boldsymbol{k}$, so that $\boldsymbol{E} = \mathrm{Re}\{\mathcal{A}\hat{\boldsymbol{a}} \exp \imath(\boldsymbol{k}\cdot\boldsymbol{x}-\omega t)\}$ and $\boldsymbol{E} = \mathrm{Re}\{\imath\mathcal{A}\hat{\boldsymbol{a}}\times\hat{\boldsymbol{k}} \exp \imath(\boldsymbol{k}\cdot\boldsymbol{x}-\omega t)\}$ are each electric fields in electromagnetic plane waves. The magnetic fields corresponding to each of these cases are $\boldsymbol{H} = (\epsilon/\mu)^{1/2}\mathrm{Re}\{\mathcal{A}\hat{\boldsymbol{k}} \times \hat{\boldsymbol{a}} \exp \imath(\boldsymbol{k} \cdot \boldsymbol{x} - \omega t)\}$ and $\boldsymbol{H} = (\epsilon/\mu)^{1/2}\mathrm{Re}\{-\imath\mathcal{A}\hat{\boldsymbol{a}} \exp \imath(\boldsymbol{k}\cdot\boldsymbol{x}-\omega t)\}$, respectively. Hence, the superposition $\boldsymbol{E} = \mathrm{Re}\{\mathcal{A}(\hat{\boldsymbol{a}} + \imath\hat{\boldsymbol{a}} \times \hat{\boldsymbol{k}}) \exp \imath(\boldsymbol{k} \cdot \boldsymbol{x} - \omega t)\} = |\mathcal{A}|\{\hat{\boldsymbol{a}} \cos\phi - \hat{\boldsymbol{k}} \times \hat{\boldsymbol{a}} \sin\phi\}$, in which $\phi \equiv \boldsymbol{k}\cdot\boldsymbol{x} - \omega t + \arg\mathcal{A}$, describes an electromagnetic wave with field $\boldsymbol{E}$ of constant magnitude $|\boldsymbol{E}| = |\mathcal{A}|$ and with direction which, at each fixed $\boldsymbol{x}$, rotates around the direction $\hat{\boldsymbol{k}}$ in a clockwise, or right-handed, sense (with period $2\pi/\omega$).

The corresponding magnetic field $\boldsymbol{H} = (\epsilon/\mu)^{1/2}|\mathcal{A}|\{\hat{\boldsymbol{k}} \times \hat{\boldsymbol{a}} \cos\phi + \hat{\boldsymbol{a}} \sin\phi\}$ also rotates in a right-handed sense.

There is a corresponding left-handed circularly polarized wave (replace $\hat{\boldsymbol{a}} + \imath\hat{\boldsymbol{a}} \times \hat{\boldsymbol{k}}$ by $\hat{\boldsymbol{a}} - \imath\hat{\boldsymbol{a}} \times \hat{\boldsymbol{k}}$ in the statement for $\boldsymbol{E}$). It can be shown that, just as the general (elliptically polarized) wave of frequency ω travelling in the $\hat{\boldsymbol{k}}$ direction may be written as a superposition of linearly polarized waves in any two orthogonal polarizations, the general wave may also be expressed as a superposition of right-handed and left-handed waves of suitable (complex) amplitudes (either linearly polarized or circularly polarized waves may be used as a *basis* for the solutions). Since, in the *optical frequency range*, we perceive electromagnetic radiation as light, the above results describe useful properties of visible light. However, they apply equally to many other frequencies, such as those in the infra-red, radio and X-ray wavebands.

9.1.3 Reflection and Refraction of Electromagnetic Waves

Reflection and refraction of waves at a plane interface is best analysed using linearly polarized plane waves as incident waves. The methods used are very similar to those used in Section 8.1.3 and Section 8.1.4 for acoustics, though the boundary conditions are more complicated. For the case of a perfectly conducting boundary $z = 0$ to the region $z \geq 0$, appropriate conditions are

$$\boldsymbol{E} \times \boldsymbol{e}_3 = \boldsymbol{0} \quad \text{and} \quad \boldsymbol{B} \cdot \boldsymbol{e}_3 = 0 \qquad \text{at } z = 0 \,. \tag{9.17}$$

For a non-conducting interface $z = 0$ between a region $z > 0$ within which $(\epsilon, \mu) = (\epsilon_+, \mu_+)$ and a region $z < 0$ in which $(\epsilon, \mu) = (\epsilon_-, \mu_-)$ the appropriate conditions are *continuity conditions* (at $z = 0$)

$$\begin{gathered} \epsilon_+ E_{3+} = \epsilon_- E_{3-} \,, \quad \mu_+ H_{3+} = \mu_- H_{3-} \,, \\ \boldsymbol{E}_+ \times \boldsymbol{e}_3 = \boldsymbol{E}_- \times \boldsymbol{e}_3 \quad \text{and} \quad \boldsymbol{H}_+ \times \boldsymbol{e}_3 = \boldsymbol{H}_- \times \boldsymbol{e}_3 \,, \end{gathered} \tag{9.18}$$

(i.e. the normal components of $\boldsymbol{D}$ and $\boldsymbol{B}$ are continuous, as are the tangential components of $\boldsymbol{E}$ and $\boldsymbol{H}$). The two cases will now be considered in turn.

Reflection at a conductor. Suppose that a plane wave with wave vector $\boldsymbol{k}$ is incident from $z > 0$ upon the conductor at $z = 0$ (i.e. $k_3 < 0$). To simplify the calculations, choose axes so that $\boldsymbol{k}$ lies in the plane of $\boldsymbol{e}_1$ and $\boldsymbol{e}_3$ (i.e. take $\boldsymbol{k} = k_1\boldsymbol{e}_1 + k_3\boldsymbol{e}_3$) at angle ψ_I to $-\boldsymbol{e}_3$, so that $k_1 = k \sin\psi_I$, $k_3 = -k\cos\psi_I$. Since the boundary conditions (9.17) become

$$E_1 = E_2 = 0 \qquad \text{and} \qquad H_3 = 0 \qquad \text{at } z = 0 \,, \tag{9.19}$$

it proves simplest to consider separately the cases (a) of incident waves with $E_2 \equiv 0$ and (b) of incident waves with $H_2 \equiv 0$ (superposition then yields the general case). In either case, the incident wave has a phase factor $\exp\imath(k_1x + k_3z - \omega t) \equiv \exp(\imath\chi_I)$ (say), such that $k_1^2 + k_3^2 = \epsilon\mu\omega^2$. Hence, the conditions at $z = 0$ can be satisfied over a range of values of x and t only if the reflected wave has phase factor $\exp\imath(k_1x - k_3z - \omega t) \equiv \exp(\imath\chi_R)$ (say). Thus, just as in acoustics, the reflected wave must have wave vector $k_1\boldsymbol{e}_1 - k_3\boldsymbol{e}_3 = k(\boldsymbol{e}_1 \sin\psi_I + \boldsymbol{e}_3\cos\psi_I)$.

The building blocks for the solutions have non-zero components of the form: (a) waves with transverse magnetic (TM) excitation

$$E_1 = \cos\psi\, \mathrm{Re}\{\mathcal{E}e^{\imath\chi}\}, \quad H_2 = -(\epsilon/\mu)^{1/2}\mathrm{Re}\{\mathcal{E}e^{\imath\chi}\}, \quad E_3 = \sin\psi\, \mathrm{Re}\{\mathcal{E}e^{\imath\chi}\},$$

(b) waves with transverse electric (TE) field

$$H_1 = (\epsilon/\mu)^{1/2}\cos\psi\, \mathrm{Re}\{\mathcal{E}e^{\imath\chi}\}, \; E_2 = \mathrm{Re}\{\mathcal{E}e^{\imath\chi}\}, \; H_3 = (\epsilon/\mu)^{1/2}\sin\psi\, \mathrm{Re}\{\mathcal{E}e^{\imath\chi}\},$$

with $\psi = \psi_I$, $\chi = \chi_I \equiv k_1x + k_3z - \omega t$ for incident waves, and $\psi = \pi - \psi_R = \pi - \psi_I$, $\chi = \chi_R \equiv k_1x - k_3z - \omega t$ for reflected waves (remember that $k_3 < 0$).
(a) Superposing an *incident* TM wave with $\boldsymbol{H} = (\epsilon/\mu)^{1/2}\boldsymbol{e}_2\mathrm{Re}\{\mathcal{E}\mathrm{e}^{\imath\chi_I}\}$ and a *reflected* wave with $\boldsymbol{H} = (\epsilon/\mu)^{1/2}\boldsymbol{e}_2\mathrm{Re}\{\mathcal{E}_R\mathrm{e}^{\imath\chi_R}\}$ gives

$$E_1 = \cos\psi_I\mathrm{Re}\{\mathcal{E}\mathrm{e}^{\imath\chi_I} - \mathcal{E}_R\mathrm{e}^{\imath\chi_R}\}, \quad H_2 = -(\epsilon/\mu)^{1/2}\mathrm{Re}\{\mathcal{E}\mathrm{e}^{\imath\chi_I} + \mathcal{E}_R\mathrm{e}^{\imath\chi_R}\}$$
$$E_3 = \sin\psi_I\mathrm{Re}\{\mathcal{E}\mathrm{e}^{\imath\chi_I} + \mathcal{E}_R\mathrm{e}^{\imath\chi_R}\},$$

with all other field components zero. Since $\chi_I = \chi_R = k_1x - \omega t$ on $z = 0$, all boundary conditions (9.19) are satisfied when $\mathcal{E}_R = \mathcal{E}$. Thus, the reflected wave has the same amplitude as the incident wave, with total fields given by

$$\boldsymbol{E} = \mathrm{Re}\{\mathcal{E}[(\boldsymbol{e}_1\cos\psi_I + \boldsymbol{e}_3\sin\psi_I)\mathrm{e}^{\imath\chi_I} + (-\boldsymbol{e}_1\cos\psi_I + \boldsymbol{e}_3\sin\psi_I)\mathrm{e}^{\imath\chi_R}]\},$$
$$\boldsymbol{H} = -(\epsilon/\mu)^{1/2}\boldsymbol{e}_2\mathrm{Re}\{\mathcal{E}\,(\mathrm{e}^{\imath\chi_I} + \mathrm{e}^{\imath\chi_R})\}.$$

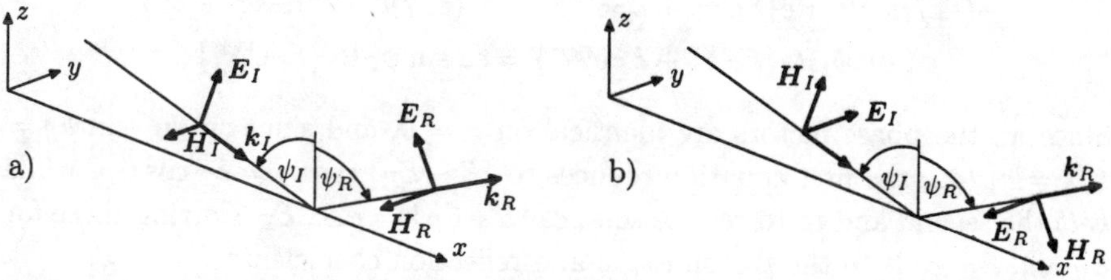

Figure 9.2 Reflection of a) a TM plane wave and b) a TE plane wave at a conducting surface $z = 0$.

(b) By superposing an *incident* TE wave with $\boldsymbol{E} = \boldsymbol{e}_2\mathrm{Re}\{\mathcal{E}\mathrm{e}^{\imath\chi_I}\}$ (and with magnetic field $\boldsymbol{H}$ parallel to $\boldsymbol{e}_1\cos\psi_I + \boldsymbol{e}_3\sin\psi_I$) and a *reflected* TE wave with $\boldsymbol{E} = \boldsymbol{e}_2\mathrm{Re}\{\mathcal{E}_R\mathrm{e}^{\imath\chi_R}\}$ (and with $\boldsymbol{H}$ parallel to $-\boldsymbol{e}_1\cos\psi_I + \boldsymbol{e}_3\sin\psi_I$), it is found that the choice $\mathcal{E}_R = -\mathcal{E}$ makes both $E_2 = 0$ and $H_3 = 0$ at $z = 0$. This shows that the combined incident and reflected fields describing the reflection of TE waves are

$$\boldsymbol{E} = \boldsymbol{e}_2\mathrm{Re}\{\mathcal{E}\,[\exp\imath(k_1x + k_3z - \omega t) - \exp\imath(k_1x - k_3z - \omega t)]\},$$
$$\boldsymbol{H} = (\epsilon/\mu)^{1/2}\mathrm{Re}\{\mathcal{E}\,[(\boldsymbol{e}_1\cos\psi_I + \boldsymbol{e}_3\sin\psi_I)\exp\imath(k_1x + k_3z - \omega t)$$
$$+(\boldsymbol{e}_1\cos\psi_I - \boldsymbol{e}_3\sin\psi_I)\exp\imath(k_1x - k_3z - \omega t)]\}.$$

Refraction at a material interface. Again, the cases (a) of TM waves and (b) of TE waves may be treated separately. In $z > 0$, *incident* and *reflected* plane waves are taken with respective wave vectors $k_1\boldsymbol{e}_1 \pm k_3\boldsymbol{e}_3$ and phase factors $\chi_I = k_1x + k_3z - \omega t$ and $\chi_R = k_1x - k_3z - \omega t$, respectively (with $k_3 < 0$).

The fields have the same form as above, but with $\epsilon = \epsilon_+$, $\mu = \mu_+$. In $z < 0$, any *refracted* plane wave has the correct phase factor on $z = 0$ only if it has wave vector $k_1\boldsymbol{e}_1 + \bar{k}_3\boldsymbol{e}_3$ and phase factor $\mathrm{e}^{\imath\chi_T} = \exp\imath(k_1x + \bar{k}_3z - \omega t)$, where $c_+^2(k_1^2 + k_3^2) = \omega^2 = c_-^2(k_1^2 + \bar{k}_3^2)$, with $\bar{k}_3 \le 0$ and $c_\pm = (\epsilon_\pm\mu_\pm)^{-1/2}$. The fields are taken so that (a) $\boldsymbol{H} = (\epsilon_-/\mu_-)^{1/2}\boldsymbol{e}_2\mathrm{Re}\{\mathcal{E}\mathrm{e}^{\imath\chi_T}\}$, or (b) $\boldsymbol{E} = \boldsymbol{e}_2\mathrm{Re}\{\mathcal{E}\mathrm{e}^{\imath\chi_T}\}$.

The conditions (9.18) at the interface $z = 0$ require that

(a)

$$E_{1+} = E_{1-}\,, \quad H_{2+} = H_{2-} \text{ and } \quad \epsilon_+E_{3+} = \epsilon_-E_{3-} \qquad \text{for TM waves,}$$

(b)

$$H_{1+} = H_{1-}\,, \quad E_{2+} = E_{2-} \text{ and } \quad \mu_+H_{3+} = \mu_-H_{3-} \qquad \text{for TE waves.}$$

(a) *Refraction of TM waves.* Applying these conditions yields

$$\begin{aligned} \cos\psi_I\mathrm{Re}\{\mathcal{E}\mathrm{e}^{\imath\chi_I} - \mathcal{E}_R\mathrm{e}^{\imath\chi_R}\} &= \cos\psi_T\mathrm{Re}\{\mathcal{E}_T\mathrm{e}^{\imath\chi_T}\}\,, \\ -(\epsilon_+/\mu_+)^{1/2}\mathrm{Re}\{\mathcal{E}\mathrm{e}^{\imath\chi_I} + \mathcal{E}_R\mathrm{e}^{\imath\chi_R}\} &= -(\epsilon_-/\mu_-)^{1/2}\mathrm{Re}\{\mathcal{E}_T\mathrm{e}^{\imath\chi_T}\}\,, \\ \epsilon_+\sin\psi_I\mathrm{Re}\{\mathcal{E}\mathrm{e}^{\imath\chi_I} + \mathcal{E}_R\mathrm{e}^{\imath\chi_R}\} &= \epsilon_-\sin\psi_T\mathrm{Re}\{\mathcal{E}_T\mathrm{e}^{\imath\chi_T}\}\,. \end{aligned}$$

Since all the phase factors are identical on $z = 0$, and since $\sin\psi_I/\sin\psi_T = \bar{k}/k = c_+/c_-$, the first equation reduces to $\{\mathcal{E} - \mathcal{E}_R\}\cos\psi_I = \mathcal{E}_T\cos\psi_T$, while *both* the second and third reduce to $\epsilon_+c_+\{\mathcal{E} + \mathcal{E}_R\} = \epsilon_-c_-\mathcal{E}_T$. Solving these for $\mathcal{E}_R$ and $\mathcal{E}_T$ leads to the transmission and reflection coefficients

$$\frac{\mathcal{E}_R}{\mathcal{E}} = \frac{\epsilon_-c_-\cos\psi_I - \epsilon_+c_+\cos\psi_T}{\epsilon_-c_-\cos\psi_I + \epsilon_+c_+\cos\psi_T}\,, \quad \frac{\mathcal{E}_T}{\mathcal{E}} = \frac{2\epsilon_+c_+\cos\psi_I}{\epsilon_-c_-\cos\psi_I + \epsilon_+c_+\cos\psi_T}\,. \tag{9.20}$$

In optics, it is conventional to define the *refractive index* of a material as $n \equiv c_0/c = (\epsilon\mu/\epsilon_0\mu_0)^{1/2}$, so that the relation $\sin\psi_T/c_- = \sin\psi_I/c_+$ between the inclinations ψ_T and ψ_I of the two wavenormals to $-\boldsymbol{e}_3$ (the normal to the interface) becomes $n_+\sin\psi_I = n_-\sin\psi_T$. This is *Snell's law of refraction* which states that, at the interface between any two materials, the product $n\sin\psi$ is equal in the two media. This applies independently of the polarization of the wave, since it results directly from the relation $c_-^2\bar{k}_3^2 = c_+^2k_3^2 + (c_+^2 - c_-^2)k_1^2$ connecting $\bar{k}_3$ to k_3 and k_1.

(b) *Refraction of TE waves.* By applying the interface conditions to the non-zero field components H_1, E_2 and H_3 and using the fact that $\chi_I = \chi_R = \chi_T$ on $z = 0$, it follows that

$$\begin{aligned} &(\epsilon_+/\mu_+)^{1/2}\cos\psi_I\{\mathcal{E} - \mathcal{E}_R\} = (\epsilon_-/\mu_-)^{1/2}(\cos\psi_T)\mathcal{E}_T\,, \quad \mathcal{E} + \mathcal{E}_R = \mathcal{E}_T\,, \\ &(\epsilon_+\mu_+)^{1/2}\sin\psi_I\{\mathcal{E} + \mathcal{E}_R\} = (\epsilon_-\mu_-)^{1/2}(\sin\psi_T)\mathcal{E}_T\,. \end{aligned}$$

Again, the last two equations are equivalent. Solving for the ratios $\mathcal{E}_R/\mathcal{E}$ and $\mathcal{E}_T/\mathcal{E}$ gives

$$\frac{\mathcal{E}_R}{\mathcal{E}} = \frac{\epsilon_+ c_+ \cos\psi_I - \epsilon_- c_- \cos\psi_T}{\epsilon_+ c_+ \cos\psi_I + \epsilon_- c_- \cos\psi_T}, \quad \frac{\mathcal{E}_T}{\mathcal{E}} = \frac{2\epsilon_+ c_+ \cos\psi_I}{\epsilon_+ c_+ \cos\psi_I + \epsilon_- c_- \cos\psi_T}. \tag{9.21}$$

Equations (9.20) and (9.21) give similar, but distinct, formulae for the reflection coefficients within the medium occupying $z > 0$. Also, by using Snell's law in the form $\sin\psi_T = (c_-/c_+)\sin\psi_I = (n_+/n_-)\sin\psi_I$, both coefficients may be expressed as functions of ψ_I. Moreover, in the widespread case of non-magnetic naterials for which $\mu_+ = \mu_- = \mu_0$ gives $\epsilon_+ c_+ = c_0\sqrt{\epsilon_0 \epsilon_+} = \epsilon_0 c_0 n_+$ and $\epsilon_- c_- = \epsilon_0 c_0 n_-$, Equations (9.21) simplify to

$$\frac{\mathcal{E}_R}{\mathcal{E}} = \frac{n_+ \cos\psi_I - n_- \cos\psi_T}{n_+ \cos\psi_I + n_- \cos\psi_T} = \frac{\sin(\psi_T - \psi_I)}{\sin(\psi_T + \psi_I)}, \quad \frac{\mathcal{E}_T}{\mathcal{E}} = \frac{\sin\psi_T \cos\psi_I}{\sin(\psi_T + \psi_I)}.$$

An important deduction for an air–water interface, where $c_+/c_- \approx 1.33$, but $\epsilon_-/\epsilon_+ \approx 81$ so that $\epsilon_- c_-/\epsilon_+ c_+ \gg 1$, is that there is a range of small glancing angles (i.e. $\psi_I \approx \pi/2$) for which $|\mathcal{E}_R/\mathcal{E}|$ predicted by (9.20) is much smaller than is predicted by (9.21). This implies that the majority of light reflected from such an interface at those angles will have TE polarization. Consequently, to reduce glare, sunglasses designed to filter out much of the horizontally polarized electric field can considerably reduce the perceived glare.

As in acoustics, the phenomenon of *total internal reflection* arises when $(n_+/n_-)\sin\psi_I > 1$. As in Section 8.1.5, the fields are determined by replacing $e^{\imath \bar{k}_3 z}$ by $e^{\gamma z}$.

9.2 Waveguides

Electromagnetic waves, just like acoustic and elastic waves, may be guided along layers, slabs or ducts. They are analysed by seeking solutions of Maxwell's equations (9.7)–(9.10) (with $\rho_c = 0$, $\boldsymbol{J} = \boldsymbol{0}$, $\boldsymbol{D} = \epsilon \boldsymbol{E}$, $\boldsymbol{B} = \mu \boldsymbol{H}$) in which *all* field components involve the phase factor $e^{\imath(\beta z - \omega t)}$ (cf. Section 8.2.2). Writing the components of $\boldsymbol{E}$ and $\boldsymbol{H}$ as $E_i = \mathcal{E}_i(x,y)e^{\imath(\beta z - \omega t)}$, $H_i = \mathcal{H}_i(x,y)e^{\imath(\beta z - \omega t)}$ then yields (from the $\boldsymbol{e}_1$ and $\boldsymbol{e}_2$ components of (9.9) and (9.10))

$$\begin{aligned} &\frac{\partial \mathcal{E}_3}{\partial y} - \imath\beta\mathcal{E}_2 - \imath\omega\mu\mathcal{H}_1 = 0\,, && \frac{\partial \mathcal{H}_3}{\partial y} - \imath\beta\mathcal{H}_2 + \imath\omega\epsilon\mathcal{E}_1 = 0\,, \\ &\imath\beta\mathcal{E}_1 - \frac{\partial \mathcal{E}_3}{\partial x} - \imath\omega\mu\mathcal{H}_2 = 0\,, && \imath\beta\mathcal{H}_1 - \frac{\partial \mathcal{H}_3}{\partial x} + \imath\omega\epsilon\mathcal{E}_2 = 0\,. \end{aligned} \tag{9.22}$$

Throughout this section, the symbol Re{ } is omitted, but is implied. As earlier, β denotes the *propagation constant.* Equations (9.22) may be used to express

$\mathcal{E}_1$, $\mathcal{E}_2$, $\mathcal{H}_1$ and $\mathcal{H}_2$ in terms of the axial field components as, for example, $(\omega^2 - \beta c)\mathcal{E}_2 = \imath(\beta c^2 \partial\mathcal{E}_3/\partial y - \epsilon^{-1}\omega\partial\mathcal{H}_3/\partial x)$. To simplify the representation, it is convenient to define the dimensionless propagation constant $P \equiv \beta c/\omega = (\beta/\omega)(\epsilon\mu)^{-1/2}$ (which equals 1 for plane waves). Then, writing

$$\mathcal{E}_3 = \imath\beta(P^{-1} - P)\,\psi(x,y)\,, \qquad \mathcal{H}_3 = \imath(\epsilon/\mu)^{1/2}\beta(P^{-1} - P)\,\phi(x,y)\,, \tag{9.23}$$

allows the transverse field components to be expressed using (9.22) as

$$\begin{aligned} \mathcal{E}_1 &= -\phi_y - P\psi_x\,, & \mathcal{E}_2 &= \phi_x - P\psi_y\,, \\ \mathcal{H}_1 &= -(\epsilon/\mu)^{1/2}(P\phi_x - \psi_y)\,, & \mathcal{H}_2 &= -(\epsilon/\mu)^{1/2}(P\phi_y + \psi_x)\,. \end{aligned} \tag{9.24}$$

Substitution into (9.7) and (9.8) then gives

$$\frac{\partial^2\phi}{\partial x^2} + \frac{\partial^2\phi}{\partial y^2} + (\omega^2\epsilon\mu - \beta^2)\phi = 0\,, \quad \frac{\partial^2\psi}{\partial x^2} + \frac{\partial^2\psi}{\partial y^2} + (\omega^2\epsilon\mu - \beta^2)\psi = 0 \tag{9.25}$$

(thereby also satisfying the $\mathbf{e}_3$ components of (9.10) and (9.9), respectively).

9.2.1 Rectangular Waveguides

The simplest type of waveguide to analyse is that in which microwaves are guided within a metallic duct of rectangular cross-section $0 \le x \le a$, $0 \le y \le b$, such that $E_2 = E_3 = 0$ on $x = 0, a$ and $E_1 = E_3 = 0$ on $y = 0, b$. For TE modes (with $\mathcal{E}_3 \equiv 0$ giving $\psi(x,y) \equiv 0$) it is necessary to choose $\phi(x,y)$ to satisfy (9.25) with boundary conditions

$$\frac{\partial\phi}{\partial x} = 0 \;\text{ on }\; x = 0, a\,, \qquad \frac{\partial\phi}{\partial y} = 0 \;\text{ on }\; y = 0, b\,.$$

The solutions are *exactly analogous* to those in Example 8.7. They give

$$\phi(x,y) = \cos\frac{m\pi x}{a}\,\cos\frac{n\pi y}{b} \equiv \phi_{mn}(x,y)\,,$$

so that the transverse fields are given by

$$\mathcal{E}_1 = \frac{n\pi}{b}\cos\frac{m\pi x}{a}\,\sin\frac{n\pi y}{b}\,, \qquad \mathcal{E}_2 = \frac{-m\pi}{a}\sin\frac{m\pi x}{a}\,\cos\frac{n\pi y}{b}\,, \tag{9.26}$$

with $\mathcal{H}_1 = -(\epsilon/\mu)^{1/2}P\mathcal{E}_2$, $\mathcal{H}_2 = (\epsilon/\mu)^{1/2}P\mathcal{E}_1$. For these the *dispersion relation* connecting the propagation constant β to the frequency ω and the modal numbers $\{m, n\}$ is

$$\omega^2\epsilon\mu = \omega^2/c^2 = \beta^2 + \pi^2\left(\frac{m^2}{a^2} + \frac{n^2}{b^2}\right) \tag{9.27}$$

just as in Example 8.7, with $m, n = 0, 1, 2, \ldots$.

A similar treatment yields the TM modes (with $\mathcal{H}_3 \equiv 0$, $\phi \equiv 0$), for which (9.23) requires that $\psi = 0$ on the conductor, so giving

$$\psi(x, y) = \sin\frac{m\pi x}{a}\sin\frac{n\pi y}{b} \equiv \psi_{mn}(x, y),$$

with the transverse fields given by

$$\mathcal{E}_1 = -P\frac{m\pi}{a}\cos\frac{m\pi x}{a}\sin\frac{n\pi y}{b}, \qquad \mathcal{E}_2 = -P\frac{n\pi}{b}\sin\frac{m\pi x}{a}\cos\frac{n\pi y}{b}, \quad (9.28)$$

with $\mathcal{H}_1 = -(\epsilon/\mu)^{1/2}P^{-1}\mathcal{E}_2$, $\mathcal{H}_2 = (\epsilon/\mu)^{1/2}P^{-1}\mathcal{E}_1$. The dispersion relation is identical to (9.27), but in the TM case neither m nor n may take the value zero.

The interpretation of the two cases is as follows:

- In a rectangular conducting waveguide, plane TE modes may exist for any $m, n = 0, 1, \ldots$ other than $\{m, n\} = \{0, 0\}$ (for which $P = 1$ and $\boldsymbol{E} = \boldsymbol{0} = \boldsymbol{H}$). In general, $P = \{1 + \pi^2\beta^{-2}(m^2/a^2 + n^2/b^2)\}^{-1/2}$ and the axial component of field is given by

$$\begin{aligned}\mathcal{H}_3 &= \imath(\epsilon/\mu)^{1/2}\beta(P^{-1} - P)\phi_{mn}(x, y) \\ &= \imath\frac{\pi^2}{\omega\mu}\left(\frac{m^2}{a^2} + \frac{n^2}{b^2}\right)\cos\frac{m\pi x}{a}\cos\frac{n\pi y}{b}.\end{aligned}$$

- Each $\{m, n\}$ mode has a cut-off frequency $\bar{\omega}_{mn} = \pi c(m^2/a^2 + n^2/b^2)^{1/2}$.
- If $a > b$, the lowest cut-off frequency ($\bar{\omega}_{10}$) arises for $m = 1$, $n = 0$. It gives

$$\mathcal{E}_1 = 0, \quad \mathcal{H}_2 = 0, \quad \mathcal{E}_2 = \frac{-\pi}{a}\sin\frac{\pi x}{a}, \quad \mathcal{H}_1 = -\sqrt{\frac{\epsilon}{\mu}\left(1 - \frac{\pi^2 c^2}{\omega^2 a^2}\right)}\,\mathcal{E}_2$$

  and so has electric field parallel to the *shorter side* of the rectangular cross-section, but of non-uniform strength, as in Figure 9.3b.
- There is an axial component $\mathcal{H}_3$ of magnetic field and also a transverse component $\mathcal{H}_1$ parallel to the *longer* side of the cross-section.
- Each $\{m, 0\}$ mode may be regarded as made up from TE plane waves reflected successively from *each* of the planes $x = 0$ and $x = a$ (with $k_1 = m\pi/a$, so that $x = 0, a$ are at nodes of a sine function).
- The TM mode with lowest cut-off frequency is the $\{1, 1\}$ mode. It has

$$\mathcal{H}_1 = (\epsilon/\mu)^{1/2}\frac{\pi}{b}\sin\frac{\pi x}{a}\cos\frac{\pi y}{b}, \quad \mathcal{H}_2 = -(\epsilon/\mu)^{1/2}\frac{\pi}{a}\cos\frac{\pi x}{a}\sin\frac{\pi y}{b}$$

  with $\mathcal{E}_1 = (\beta c/\omega)(\epsilon/\mu)^{-1/2}\mathcal{H}_2$, $\mathcal{E}_2 = -(\beta c/\omega)(\epsilon/\mu)^{-1/2}\mathcal{H}_1$ and with axial field $\mathcal{E}_3 = \imath(c/\omega)\sin(\pi x/a)\sin(\pi y/b)$.

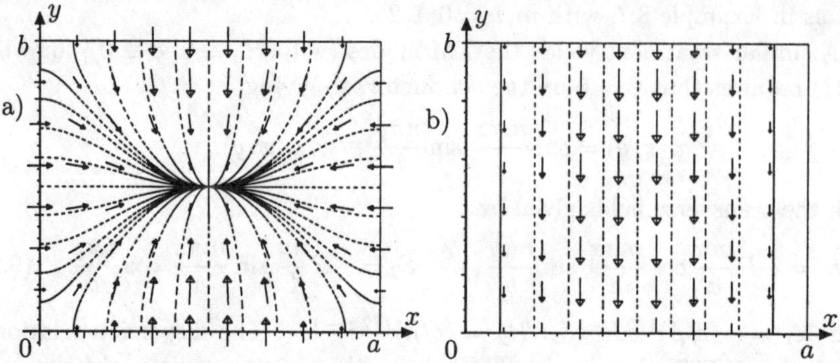

Figure 9.3 Transverse electric field lines for a) the lowest TM-polarized mode and b) the lowest TE-polarized mode in a guide having $a/b = 3/2$.

Example 9.3

For microwaves at frequency 5×10^{10} Hz (i.e. $\omega = 10^{11}\pi$ sec^{-1}), show that TE modes cannot propagate along a rectangular conducting duct unless one side of the rectangle has length exceeding 2.998 mm. If the duct is square with dimensions 4 mm, can the lowest TM mode propagate also?

Let the cross-section have dimensions a and b, with $a \geq b$. Then, no modes can propagate at frequencies below the lowest cut-off frequency $c\pi/a$. This gives the requirement $c\pi < \omega a$, or

$$a > c\pi/\omega = 10^{-11}(2.998 \times 10^8)\,\text{m} = 2.998\,\text{mm}.$$

For the $\{1,1\}$ TM mode, the cut-off frequency is given by $\bar{\omega}_{11} = c\pi\sqrt{a^{-2}+b^{-2}}$ $= \sqrt{2}c\pi/a = 2.998\pi \times 10^8/(2\sqrt{2} \times 10^{-3}) > 10^{11}\pi$ sec^{-1}. Hence, no TM modes can propagate at frequency ω.

9.2.2 Circular Cylindrical Waveguides

Many practical waveguides have circular cross-section. Their waveguide modes share many of the features seen in Section 9.2.1 for guides of rectangular cross-section. As in (9.24), the transverse field components may be expressed in terms of derivatives of the two axial components, but Equations (9.25) should be solved in terms of polar coordinates r, θ in the x, y plane. By standard transformations (see Exercise 9.5) of $\phi_{xx} + \phi_{yy}$, this requires that $\phi(r,\theta)$ and $\psi(r,\theta)$ which define $\boldsymbol{E}$ and $\boldsymbol{H}$ as in (9.23) and (9.24) should satisfy

$$\phi_{rr} + r^{-1}\phi_r + r^{-2}\phi_{\theta\theta} + K\phi = 0\,, \quad \psi_{rr} + r^{-1}\psi_r + r^{-2}\psi_{\theta\theta} + K\psi = 0\,, \tag{9.29}$$

where $K \equiv \omega^2\epsilon\mu - \beta^2 = \beta^2(P^{-2} - 1)$. For waves inside a metallic conductor of internal radius a, the boundary conditions $\boldsymbol{E} \times \boldsymbol{n} = \boldsymbol{0}$, $\boldsymbol{B} \cdot \boldsymbol{n} = 0$ require that at radius $r = a$, both $\psi = 0$ and $\phi_r = 0$. Thus, as for waves within the rectangular waveguide, TM and TE modes may be found by solving separately for ψ and for ϕ. In either case, solutions depending only upon r exist. They will be considered first.

For TM modes, choose $\psi = \Psi(r)$ and insert into (9.29) to give

$$r^2\Psi''(r) + r\Psi'(r) + Kr^2\Psi(r) = 0\,. \tag{9.30}$$

This linear, homogeneous ODE has solutions which are oscillatory in r when $K > 0$ just as, for spherically symmetric heat conduction, the solutions $R = r^{-1}\sin\kappa r$ to $r^2R'' + 2rR' + \kappa^2r^2R = 0$ are oscillatory and, in one dimension, the solutions $R = \sin\kappa r$ to $R'' + \kappa^2R = 0$ are oscillatory. Thus, writing $\kappa^2 = K > 0$ allows the general solution to (9.30) to be written in terms of the two fundamental solutions $J_0(s)$ and $Y_0(s)$ of *Bessel's equation of order zero*

$$s^2y''(s) + sy'(s) + s^2y(s) = 0\,,$$

as $R(r) \equiv AJ_0(\kappa r) + BY_0(\kappa r)$. Many details concerning the functions $J_0(s)$ and $Y_0(s)$ may be found in Chapter 1 of Evans, Blackledge and Yardley (1999), but the properties of particular relevance are that $Y_0(s)$ is singular as $s \to 0$, while $J_0(0) = 1, J_0'(0) = 0$. Thus, since $\Psi(r)$ must be finite at $r = 0$, the required solution must have the form $\psi = \Psi(r) = AJ_0(\kappa r)$. The parameter κ must be chosen so that $\Psi(a) = 0$, so giving $J_0(\kappa a) = 0$. Figure 9.4 shows how this condition selects a sequence of values $\kappa_1 a, \kappa_2 a, \ldots$ as the zeros $s_1, s_2, \ldots$ of the *Bessel function* $J_0(s)$. Corresponding to *each* of the choices $\kappa = \kappa_n = s_n/a$, there is an axisymmetric TM mode in which the transverse electric field is radial and the magnetic field is azimuthal. The field components in a mode are given, through (9.23) and (9.24), as

$$\begin{gathered}\boldsymbol{E}_\perp = -AP\boldsymbol{e}_r\kappa_n J_0'(\kappa_n r)\mathrm{e}^{\imath(\beta z - \omega t)}\,, \quad \boldsymbol{H} = -A(\epsilon/\mu)^{1/2}\boldsymbol{e}_\theta\kappa_n J_0'(\kappa_n r)\mathrm{e}^{\imath(\beta z - \omega t)}\,, \\ E_3 = \imath A\beta(P^{-1} - P)J_0(\kappa_n r)\mathrm{e}^{\imath(\beta z - \omega t)} = \imath AP\beta^{-1}\kappa_n^2 J_0(\kappa_n r)\mathrm{e}^{\imath(\beta z - \omega t)}\,, \end{gathered} \tag{9.31}$$

where $\boldsymbol{e}_r$ and $\boldsymbol{e}_\theta$ are the unit vectors in the directions of increasing r and of increasing θ respectively, while the propagation constant β and frequency ω are interrelated through the dispersion relation

$$\omega^2/c^2 - \beta^2 = \kappa_n^2 \quad \text{with} \quad c = (\epsilon\mu)^{-1/2}\,,$$

so that $P = \{1 + (\kappa_n/\beta)^2\}^{-1/2} = \{1 - (\kappa_n c/\omega)^2\}^{1/2}$. Not surprisingly, the radial electric field vanishes on the axis (where $\boldsymbol{e}_r$ is ill-defined), while the axial electric field vanishes at the conductor $r = a$.

For axisymmetric TE modes, the electromagnetic fields are written in terms of $\phi = \bar{A}J_0(\bar{\kappa}_n r)e^{\imath(\beta z-\omega t)}$ as

$$\begin{aligned} \boldsymbol{E} = \boldsymbol{E}_\perp &= \bar{A}\boldsymbol{e}_\theta \bar{\kappa}_n J_0'(\bar{\kappa}_n r)e^{\imath(\beta z-\omega t)}, \\ \boldsymbol{H}_\perp &= -\bar{A}\bar{P}(\epsilon/\mu)^{1/2}\boldsymbol{e}_r \bar{\kappa}_n J_0'(\bar{\kappa}_n r)e^{\imath(\beta z-\omega t)}, \\ H_3 &= \imath\bar{A}\beta(\epsilon/\mu)^{1/2}(\bar{P}^{-1}-\bar{P})J_0(\bar{\kappa}_n r)e^{\imath(\beta z-\omega t)} \\ &= \imath\bar{A}\bar{P}(\epsilon/\mu)^{1/2}\beta^{-1}\bar{\kappa}_n^2 J_0(\bar{\kappa}_n r)e^{\imath(\beta z-\omega t)} \end{aligned} \tag{9.32}$$

where the eigenvalues $\bar{\kappa}_n$ are, in this case, determined as solutions $\kappa = \bar{\kappa}_n$ of the equation $J_0'(\kappa a) = 0$. As is seen from Figure 9.4, there is an infinite set of possibilities, which may be labelled so that $\kappa_1 < \bar{\kappa}_1 < \kappa_2 < \bar{\kappa}_2 < \ldots$. Associated with each mode, the *dispersion relation* is

$$\omega^2/c^2 - \beta^2 = \bar{\kappa}_n^2,$$

while $\bar{P} \equiv \{1+(\bar{\kappa}_n/\beta)^2\}^{-1/2} = \{1-(\bar{\kappa}_n c/\omega)^2\}^{1/2}$. Since the *cut-off frequencies* for the various TM modes are $c\kappa_n$ and for the TE modes are $c\bar{\kappa}_n$, it is clear that amongst these modes the first TM mode is the one with the lowest cut-off frequency.

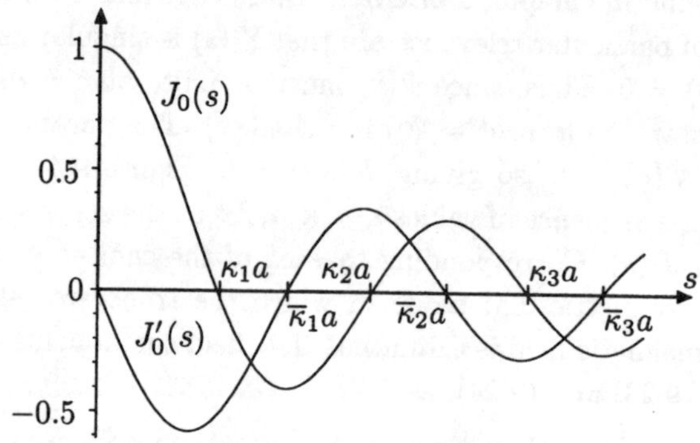

Figure 9.4 The Bessel function $J_0(s)$. Setting $s = \kappa a$ shows how the values $\kappa_1, \kappa_2, \ldots$ and $\bar{\kappa}_1, \bar{\kappa}_2, \ldots$ relevant for TM and for TE modes are identified.

There are, in addition, many other solutions of (9.29) which lend themselves to satisfying the boundary condition on $r = a$. This may be seen by realizing that, from any solutions $\phi(x, y)$ and $\psi(x, y)$ of (9.25), many others may be constructed by taking successive x– or y–derivatives (cf. Section 4.4). Indeed, since it is known that (9.25) has solutions $\psi = \Psi(r)$, it is readily seen that $\partial\psi/\partial x = \Psi'(r)x/r = \Psi'(r)\cos\theta$ and $\partial\psi/\partial y = \Psi'(r)\sin\theta$ also are solutions. For $K > 0$, it then follows that both $\psi = A_1 J_0'(\kappa r)\cos\theta$ and $\phi = \bar{A}_1 J_0'(\kappa r)\cos\theta$ are special solutions to Equations (9.29) which are bounded as $r \to \infty$. Since

$\nabla(J_0'(\kappa r)\cos\theta) = \boldsymbol{e}_r\kappa J_0''(\kappa r)\cos\theta - \boldsymbol{e}_\theta r^{-1}J_0'(\kappa r)\sin\theta$, it is found that within TM modes (i.e. $\phi \equiv 0$) the fields derived from the above choice of $\psi(r,\theta)$ are

$$\begin{aligned} \boldsymbol{E}_\perp &= -\mathcal{A}_1 P\{\boldsymbol{e}_r\kappa J_0''(\kappa r)\cos\theta - \boldsymbol{e}_\theta r^{-1}J_0'(\kappa r)\sin\theta\}\mathrm{e}^{\imath(\beta z-\omega t)}\,, \\ \boldsymbol{H} = \boldsymbol{H}_\perp &= -\mathcal{A}_1(\epsilon/\mu)^{1/2}\{\boldsymbol{e}_r r^{-1}J_0'(\kappa r)\sin\theta + \boldsymbol{e}_\theta\kappa J_0''(\kappa r)\cos\theta\}\mathrm{e}^{\imath(\beta z-\omega t)}\,, \\ E_3 &= \imath\mathcal{A}_1\beta(P^{-1}-P)J_0'(\kappa r)\cos\theta\mathrm{e}^{\imath(\beta z-\omega t)}\,, \end{aligned} \tag{9.33}$$

with $\kappa = \kappa_{n1}$ satisfying $J_0'(\kappa_{n1}a) = 0$ $(n = 1,2,\ldots)$ and with *dispersion relation* $\omega^2/c^2 - \beta^2 = \kappa_{n1}^2$ defining $P = \{1+(\kappa_{n1}/\beta)^2\}^{-1/2}$ (observe how these *first-order* TM modes have identically the same dispersion relation as do the *axisymmetric* TE modes given by (9.32)). The *first-order* TE modes are derived similarly from $\phi = \bar{A}_1 J_0'(\kappa r)\cos\theta$ as

$$\begin{aligned} \boldsymbol{E} = \boldsymbol{E}_\perp &= \bar{A}_1\{\boldsymbol{e}_r r^{-1}J_0'(\kappa r)\sin\theta + \boldsymbol{e}_\theta\kappa J_0''(\kappa r)\cos\theta\}\mathrm{e}^{\imath(\beta z-\omega t)}\,, \\ \boldsymbol{H}_\perp &= -\bar{A}_1(\epsilon/\mu)^{1/2}\bar{P}\{\boldsymbol{e}_r\kappa J_0''(\kappa r)\cos\theta - \boldsymbol{e}_\theta r^{-1}J_0'(\kappa r)\sin\theta\}\mathrm{e}^{\imath(\beta z-\omega t)}\,, \\ H_3 &= \imath\bar{A}_1(\epsilon/\mu)^{1/2}\beta(\bar{P}^{-1}-\bar{P})J_0'(\kappa r)\cos\theta\mathrm{e}^{\imath(\beta z-\omega t)}\,, \end{aligned} \tag{9.34}$$

with $\kappa = \bar{\kappa}_{n1}$ $(n = 1,2,\ldots)$ satisfying $J_0''(\bar{\kappa}_{n1}a) = 0$. Again the dispersion relation has the form $\omega^2/c^2 - \beta^2 = \bar{\kappa}_{n1}^2 = \text{const.}$ and defines $\bar{P} \equiv \{1+(\bar{\kappa}_{n1}/\beta)^2\}^{-1/2}$. Moreover, since for this family the lowest cut-off frequency arises (for $\kappa = \bar{\kappa}_{11}$) at the first inflection point of the graph of $J_0(s)$ in Figure 9.4 (because $J_0''(\kappa_1 a) = -a^{-1}J_0'(\kappa_1 a) - J_0(\kappa_1 a) = -a^{-1}J_0'(\kappa_1 a) > 0$, while $J_0''(0) < 0$), the first of these *first-order TE modes* can propagate at frequencies *below* the cut-off frequency for the first axisymmetric TM mode. Electromagnetic waves are essentially transverse phenomena, so it should not be surprising that modes having a preferred transverse orientation (here chosen as along Ox) propagate readily. In fibre optics, this transverse polarization is even more crucial.

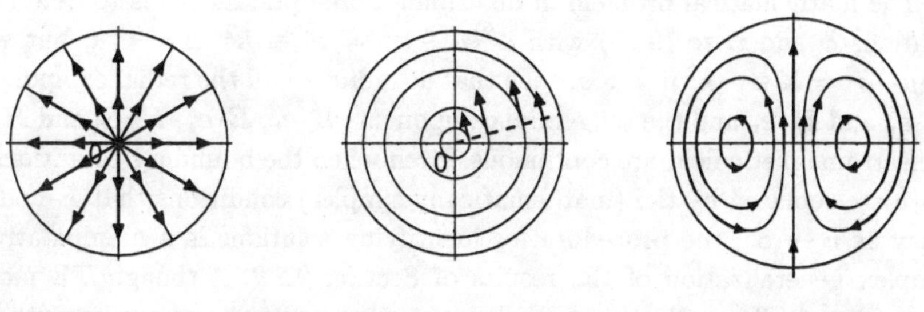

Figure 9.5 Modal fields for $\boldsymbol{E}_\perp$ for the fundamental TM and TE modes, and for the first-order TE mode.

9.2.3 An Introduction to Fibre Optics

The phenomenon of light guiding within optical fibres is closely analogous to underwater acoustic guiding by a 'slow' layer. The simplest case is for a cylindrically symmetric fibre in which the 'core' $0 \le r < a$ has permittivity $\tilde{\epsilon}$, while the 'cladding' $a < r < b$ has a slightly smaller permittivity $\hat{\epsilon} < \tilde{\epsilon}$ (so that $\tilde{c} \equiv (\tilde{\epsilon}\mu)^{-1/2} < \hat{c} \equiv (\hat{\epsilon}\mu)^{-1/2}$, since $\mu = \mu_0$ throughout the fibre). Then, a mode having frequency ω and field components $E_i = \mathcal{E}_i(x,y)\mathrm{e}^{\imath(\beta z - \omega t)}$, $H_i = \mathcal{H}_i(x,y)\mathrm{e}^{\imath(\beta z - \omega t)}$ may be represented as in (9.23)–(9.25), but with the parameter $K = \omega^2\epsilon\mu - \beta^2$ taking a larger value $K = \tilde{K}$ in the *core* $r < a$ than its value $K = \hat{K}$ in the *cladding.* Moreover, if a mode can be found such that $\omega^2\tilde{\epsilon}\mu_0 > \beta^2 > \omega^2\hat{\epsilon}\mu_0$, then $\tilde{K}$ is positive, but $\hat{K}$ is negative. This allows relevant solutions ϕ or ψ to be given in terms of the function $J_0(\kappa r)$ in $0 \le r < a$, while, in the cladding $a < r < b$, they may be chosen so as to decay as r increases. A typical radial scale for this decay is $\simeq |\hat{K}|^{-1/2}$. Optical frequencies are sufficiently high ($\omega \simeq 4 \times 10^{15}$ sec^{-1}) that, even for fibres having refractive index contrast as small as $(\tilde{n} - \hat{n})/\hat{n} \simeq 0.3\%$, so that $(\tilde{\epsilon} - \hat{\epsilon})/\hat{\epsilon} \simeq 0.006$, the difference $\tilde{K} - \hat{K}$ is given by

$$\tilde{K} - \hat{K} = \omega^2\mu_0(\tilde{\epsilon} - \hat{\epsilon}) \simeq 0.006\omega^2\mu_0\hat{\epsilon} \simeq 10^{-2} \times (4 \times 10^{15}/3 \times 10^8)^2 \simeq 10^{12}\,\mathrm{m}^{-2}.$$

Thus, a typical scale for the radial decay of field strengths within the cladding is comparable with $10^{-6}\,\mathrm{m} = 1\,\mu\mathrm{m}$ (= 1 micron), which is small compared to $b - a$ for typical fibre dimensions ($a \sim 5 \times 10^{-6}\,\mathrm{m}$, $b \sim 50 \times 10^{-6}\,\mathrm{m}$). Hence, light propagates as modes in which the electromagnetic field is concentrated within the core and its immediately surrounding region of cladding (i.e. light is guided by the 'slower' region). The field is negligible near the outer boundary $r = b$ of the cladding (which has the desirable effect that any imperfections near $r = b$ cause negligible deleterious effect on the light propagation).

The mathematical problem in describing fibre-optic modes is to determine solutions ϕ and ψ to (9.29) with $\omega^2\tilde{\epsilon}\mu - \beta^2 = \tilde{K} \equiv \tilde{\kappa}^2$ in $r < a$, but with $\omega^2\hat{\epsilon}\mu - \beta^2 = \hat{K} \equiv -\hat{\kappa}^2$ in $r > a$, such that at radius $r = a$ the radial components $\epsilon\boldsymbol{E}\cdot\boldsymbol{e}_r$ and $\boldsymbol{H}\cdot\boldsymbol{e}_r$ and the tangential components $\boldsymbol{E}\cdot\boldsymbol{e}_\theta$, $\boldsymbol{E}\cdot\boldsymbol{e}_z$, $\boldsymbol{H}\cdot\boldsymbol{e}_\theta$ and $\boldsymbol{H}\cdot\boldsymbol{e}_z$ of electromagnetic field are continuous. Even when the boundary conditions at $r = b$ are replaced by the (mathematically simpler) conditions that $\boldsymbol{E}$ and $\boldsymbol{H}$ decay as $r \to \infty$, the procedure for identifying solutions is a manipulatively complex generalization of the results of Section 9.2.2. Although TE modes ($\psi \equiv 0$) and TM modes ($\phi \equiv 0$) do exist, they are not the modes with the lowest cut-off frequency. For practical reasons, fibre communications systems are usually designed to allow optical transmission in a single mode only (so that the transmission speed associated with the *carrier frequency* ω is uniquely

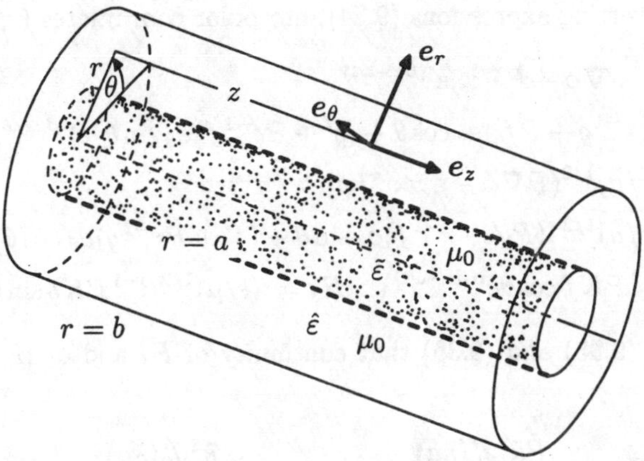

Figure 9.6 Core and cladding regions in a typical optical fibre, showing the basis vectors e_r, e_θ and e_z for cylindrical polars.

defined). In practice, this means that solutions involving both $\phi(r,\theta)$ and $\psi(r,\theta)$ are needed.

In order that the continuity equations at $r = a$ for $\epsilon \boldsymbol{E} \cdot \boldsymbol{e}_r$, E_3 and $\boldsymbol{H} \cdot \boldsymbol{e}_\theta$ contain the factor $\cos\theta$ while those for $\epsilon \boldsymbol{E} \cdot \boldsymbol{e}_\theta$, $\boldsymbol{H} \cdot \boldsymbol{e}_r$ and H_3 contain the factor $\sin\theta$ as in (9.33), the functions $\psi(r,\theta)$ and $\phi(r,\theta)$ are taken in the forms

$$\psi(r,\theta) = f(r)\cos\theta\,, \qquad \phi(r,\theta) = g(r)\sin\theta\,. \tag{9.35}$$

Then $f(r) = \Psi'(r)$ and $g(r) = \Phi'(r)$, where both $\Psi(r)$ and $\Phi(r)$ satisfy (9.30), with $K = \tilde{\kappa}^2 > 0$ in the core $r < a$ and with $K = -\hat{\kappa}^2$ in the cladding $a < r$. As in (9.33) and (9.34), the solutions which are bounded at $r = 0$ are multiples of $J_0(\tilde{\kappa}r)$, so that

$$f(r) = A_1 J_0'(\tilde{\kappa}r)\,, \qquad g(r) = B_1 J_0'(\tilde{\kappa}r)\,, \qquad r < a\,. \tag{9.36}$$

In $a < r$, where $K = -\hat{\kappa}^2 < 0$, solutions to Equation (9.30) may be written as $\Psi(r) = CI_0(\hat{\kappa}r) + DK_0(\hat{\kappa}r)$, where the *modified Bessel functions of order zero* $I_0(s)$ and $K_0(s)$ are two fundamental solutions of the ODE

$$s^2 y''(s) + sy'(s) - s^2 y(s) = 0\,.$$

The function $I_0(s)$ is finite at $s = 0$ (actually, $I_0(0) = 1$) but grows exponentially as $s \to \infty$. On the contrary, $K_0(s)$ decays as $s \to \infty$, but is (logarithmically) singular as $s \to 0$. Hence, solutions bounded for $r > a$ must be multiples of $K_0(s)$ alone, so showing that fields in the cladding are determined by using[1]

$$f(r) = C_1 K_0'(\hat{\kappa}r)\,, \qquad g(r) = D_1 K_0'(\hat{\kappa}r)\,, \qquad a < r\,. \tag{9.37}$$

[1] Just as $J_0'(s)$ and $Y_0'(s)$ are solutions to *Bessel's equation of order one*

$$s^2 y''(s) + sy'(s) + (s^2 - 1)y(s) = 0\,,$$

After converting expressions (9.24) into polar coordinates (r,θ) so that

$$\begin{aligned}
\boldsymbol{E}_\perp &= (\boldsymbol{e}_z \times \nabla\phi - P\nabla\psi)\mathrm{e}^{\imath(\beta z-\omega t)} \\
&= \{-(r^{-1}g + Pf')\boldsymbol{e}_r\cos\theta + (g' + Pr^{-1}f)\boldsymbol{e}_\theta\sin\theta\}\mathrm{e}^{\imath(\beta z-\omega t)}\,, \\
\boldsymbol{H}_\perp &= -(\epsilon/\mu)^{1/2}(P\nabla\phi + \boldsymbol{e}_z\times\nabla\psi)\mathrm{e}^{\imath(\beta z-\omega t)} \\
&= -(\epsilon/\mu)^{1/2}\{(Pg' + r^{-1}f)\boldsymbol{e}_r\sin\theta + (f' + Pr^{-1}g)\boldsymbol{e}_\theta\cos\theta\}\mathrm{e}^{\imath(\beta z-\omega t)}\,, \\
E_3 &= \imath\beta^{-1}PKf\cos\theta\mathrm{e}^{\imath(\beta z-\omega t)}\,, \quad H_3 = \imath(\epsilon/\mu)^{1/2}\beta^{-1}PKg\sin\theta\mathrm{e}^{\imath(\beta z-\omega t)}\,,
\end{aligned}$$

we find from (9.35) and (9.36) that continuity of E_3 and of H_3 across $r=a$ gives

$$C_1 = -\frac{\tilde{P}\tilde{\kappa}^2 J_0'(\tilde{\kappa}a)}{\hat{P}\hat{\kappa}^2 K_0'(\hat{\kappa}a)}A_1\,, \quad D_1 = -\frac{\tilde{\kappa}^2 J_0'(\tilde{\kappa}a)}{\hat{\kappa}^2 K_0'(\hat{\kappa}a)}B_1\,.$$

Using these in the conditions arising from the continuity of $\boldsymbol{E}\cdot\boldsymbol{e}_\theta$ and $\boldsymbol{H}\cdot\boldsymbol{e}_\theta$ leads to a pair of equations which may be reduced to

$$\begin{aligned}
&\left\{\frac{\tilde{\kappa}aJ_0''(\tilde{\kappa}a)}{J_0'(\tilde{\kappa}a)} + \frac{\hat{\epsilon}}{\tilde{\epsilon}}\frac{(\tilde{\kappa}a)^2K_0''(\hat{\kappa}a)}{\hat{\kappa}aK_0'(\hat{\kappa}a)}\right\}A_1 + \left(\frac{\hat{\kappa}^2+\tilde{\kappa}^2}{\hat{\kappa}^2}\right)\tilde{P}B_1 = 0\,, \\
&\left(\frac{\hat{\kappa}^2+\tilde{\kappa}^2}{\hat{\kappa}^2}\right)\tilde{P}A_1 + \left\{\frac{\tilde{\kappa}aJ_0''(\tilde{\kappa}a)}{J_0'(\tilde{\kappa}a)} + \frac{(\tilde{\kappa}a)^2K_0''(\hat{\kappa}a)}{\hat{\kappa}aK_0'(\hat{\kappa}a)}\right\}B_1 = 0\,.
\end{aligned} \tag{9.38}$$

In these expressions, it is convenient to introduce the *contrast* parameter Δ defined by $\tilde{\epsilon}/\hat{\epsilon} = 1 + 2\Delta = \tilde{n}^2/\hat{n}^2$, so that in terms of $\hat{P} \equiv (1-\hat{\kappa}^2/\beta^2)^{-1/2}$ it is possible to write

$$\frac{\hat{\kappa}^2+\tilde{\kappa}^2}{\hat{\kappa}^2} = \frac{2\Delta}{\hat{P}^2-1}\quad , \qquad \tilde{P} = \frac{\hat{P}}{\sqrt{1+2\Delta}}$$

and so to write the compatibility condition for solution of the system (9.38) as

$$\begin{aligned}
(\tilde{\kappa}a)^2\left\{\frac{J_0''(\tilde{\kappa}a)}{J_0'(\tilde{\kappa}a)} + \frac{\tilde{\kappa}}{\hat{\kappa}}\frac{K_0''(\hat{\kappa}a)}{K_0'(\hat{\kappa}a)}\right\}\left\{\frac{J_0''(\tilde{\kappa}a)}{J_0'(\tilde{\kappa}a)} + \frac{\tilde{\kappa}}{(1+2\Delta)\hat{\kappa}}\frac{K_0''(\hat{\kappa}a)}{K_0'(\hat{\kappa}a)}\right\}& \\
= \frac{4\Delta^2\hat{P}^2}{(1+2\Delta)(\hat{P}^2-1)}&\,.
\end{aligned} \tag{9.39}$$

This complicated equation requires considerable numerical investigation; however, certain features are readily deduced:

---

so $I_0'(s)$ and $K_0'(s)$ are solutions to the *modified Bessel equation of order one*

$$s^2y''(s) + sy'(s) - (s^2+1)y(s) = 0\,.$$

These Bessel equations of order one are also readily obtained from (9.29) by seeking separated solutions of the form (9.35).

- To every solution of (9.39), there corresponds a solution for the ratios $A_1 : B_1 : C_1 : D_1$. One of these parameters (say, A_1) can be chosen as the *complex amplitude* of the corresponding modal field.
- The continuity conditions for the radial components of $\boldsymbol{E}$ and of $\boldsymbol{H}$ are satisfied when Equations (9.38) are satisfied.
- For each choice of Δ and $\hat{P}$ (> 1), the ratio $\tilde{\kappa}/\hat{\kappa}$ is known.
- In (9.39), the right-hand side is then fixed. The left-hand side is a function of $\tilde{\kappa}a$ with asymptotes at all roots of $J_0'(\tilde{\kappa}a) = 0$. Note that these are the values $\tilde{\kappa} = \kappa_{n1}$ corresponding to the first-order TM modes for a cylinder composed of the 'core' material, with a conducting boundary at $r = a$. [The function $K_0''(\hat{\kappa}a)/K_0'(\hat{\kappa}a)$ is bounded, non-zero and virtually constant over a wide range of values of $\hat{\kappa}a$.]
- In consequence, Equation (9.39) relates $\tilde{\kappa}$ (and $\hat{\kappa}$) to $\tilde{P}$ for each choice of the contrast parameter Δ. Through $\tilde{\kappa}^2 = \omega^2\tilde{\epsilon}\mu - \beta^2$ and $\tilde{P}^2 = \beta^2/\omega^2\tilde{\epsilon}\mu$, it encodes the *dispersion relation* between ω and β.
- It can be shown that one of the modes has no cut-off frequency. It can propagate at all frequencies, for all core radii a. It is the mode exploited in *single-mode fibres.*

The above theory yields the modal fields and corresponding dispersion relation (dependence of wave speed on frequency), for the idealized case of a *step-index fibre* (i.e. a fibre with an abrupt change of refractive index at the core/cladding interface). For more realistic fibres, having refractive index which varies smoothly with radius, qualitatively similar predictions are possible. Moreover, these predictions are an important ingredient of the theory for pulsed light signals, which are essential in modern digital communications. The details are, of course, much more intricate to calculate than those outlined here, but the existence of many modes, each with their individual dispersion relation and cut-off frequency, is generic. The reader interested in further theoretical or practical details may consult Snyder and Love (1983) or Midwinter and Guo (1992).

EXERCISES

9.1. Using the vector identity $\operatorname{div}(\operatorname{curl}\boldsymbol{F}) = \boldsymbol{\nabla}\cdot(\boldsymbol{\nabla}\times\boldsymbol{F}) = 0$, show that Gauss's law (9.7) and Ampère's law (9.10) imply that the local form of the charge conservation law (9.2) is satisfied.

9.2. Given the vector identity $\boldsymbol{\nabla}\times(\boldsymbol{\nabla}\times\boldsymbol{F}) = \boldsymbol{\nabla}(\boldsymbol{\nabla}\cdot\boldsymbol{F}) - \nabla^2\boldsymbol{F}$ for any suitably differentiable vector field $\boldsymbol{F}$, show that, when $\rho = 0$, $\boldsymbol{J} = \boldsymbol{0}$, $\boldsymbol{D} = \epsilon\boldsymbol{E}$ and $\boldsymbol{B} = \mu\boldsymbol{H}$, Equations (9.7)–(9.10) imply that

$$\text{(a)} \quad \frac{\partial^2 \boldsymbol{E}}{\partial t^2} = \frac{1}{\epsilon\mu}\nabla^2\boldsymbol{E}\,, \qquad \text{(b)} \quad \frac{\partial^2 \boldsymbol{H}}{\partial t^2} = \frac{1}{\epsilon\mu}\nabla^2\boldsymbol{H}\,.$$

[Hence, each cartesian component of either $\boldsymbol{E}$ or $\boldsymbol{H}$ satisfies the wave equation. As illustrated by Section 9.1.2, they may not be chosen independently.]

9.3. If $\boldsymbol{a}$ is any real, constant vector satisfying $\boldsymbol{a} \cdot \boldsymbol{e}_1 = 0$, confirm that $\boldsymbol{E} = \mathrm{Re}\{\boldsymbol{a}\,\mathrm{e}^{\imath(kx-\omega t)}\}$ is consistent with Gauss's law (9.7) and with the wave equation $\epsilon\mu\partial^2\boldsymbol{E}/\partial t^2 = \nabla^2\boldsymbol{E}$, when $\boldsymbol{D} = \epsilon\boldsymbol{E}$, $\boldsymbol{B} = \mu\boldsymbol{H}$, $\boldsymbol{J} = \boldsymbol{0}$ and $\rho_e = 0$. Interpret $\boldsymbol{E}$ as a linearly polarized wave. Find the corresponding magnetic field $\boldsymbol{H}$.

By writing $\boldsymbol{a} = \frac{1}{2}(\boldsymbol{a} + \imath\boldsymbol{a} \times \boldsymbol{e}_1) + \frac{1}{2}(\boldsymbol{a} - \imath\boldsymbol{a} \times \boldsymbol{e}_1)$, show that each linearly polarized electromagnetic wave is the sum of two circularly polarized waves.

9.4. A plane electromagnetic wave with wavevector $\boldsymbol{k} = k(\boldsymbol{e}_1 \sin\beta - \boldsymbol{e}_3 \cos\beta)$ and electric field $\boldsymbol{E}_I = E_0[(\boldsymbol{e}_1 \cos\beta + \boldsymbol{e}_3 \sin\beta)\cos\alpha + \boldsymbol{e}_2 \sin\alpha]\cos(\boldsymbol{k}\cdot\boldsymbol{x} - \omega t)$ is incident from $z > 0$ upon the conducting surface $z = 0$. By regarding this wave as a superposition of a TM and a TE wave, determine the reflected wave.

Is it linearly polarized? If so, describe how its polarization vector is related to that of the incident wave.

9.5. Using the identities $r_x = x/r$, $r_y = y/r$, $\theta_x = -yr^{-2}$ and $\theta_y = xr^{-2}$, verify the results $\partial\phi/\partial x = xr^{-1}\partial\phi/\partial r - yr^{-2}\partial\phi/\partial\theta$, $\partial\phi/\partial y = yr^{-1}\partial\phi/\partial r + xr^{-2}\partial\phi/\partial\theta$ and $\phi_{xx} + \phi_{yy} = \phi_{rr} + r^{-1}\phi_r + r^{-2}\phi_{\theta\theta}$.

9.6. Verify that the equation $\phi_{rr} + r^{-1}\phi_r + r^{-2}\phi_{\theta\theta} + K\phi = 0$ has solutions $\phi = F(r)\cos\theta$, determine the resulting ODE governing $F(r)$ and show that, when $K = \kappa^2 > 0$, any function $F(r) = CJ_0'(\kappa r) + DY_0'(\kappa r)$ is a solution (actually, it is the *general solution*, since it involves two arbitrary constants C and D).

10
Chemical and Biological Models

Many ideas and techniques developed in earlier chapters prove useful for understanding the diffusion and reaction of chemical species and for describing the mixing and spreading of biological populations and organisms. As with all *continuum models*, the description applies when the numbers of individual chemical molecules or biological cells, insects or animals in a region being modelled is sufficiently large that movements, births or deaths of individuals need not be traced, but changes in spatial averages give sufficient information.

10.1 Diffusion of Chemical Species

In everyday activities, such as cooking, washing and eating, and in many manufacturing processes, the spreading of chemical constituents and their reaction with others are crucial. In many situations, molecules of the various chemical species move through a bulk material by a diffusion process very similar to that of heat conduction in a solid. Let C_1, $C_2, \ldots, C_n$ label the chemicals active in a process. Then, the number of molecules of species C_i per unit of volume is usually expressed as the *concentration* $c_i(\boldsymbol{x}, t)$ (measured by chemists in 'moles' per unit of volume).[1] It can change due to essentially two processes – the *flow* of C_i through the region and the conversion from and to other species through

[1] A *mole* is the amount of the species having as many elementary entities (atoms or molecules, as appropriate) as there are atoms in 0.012 kg of C_{12} carbon.

reaction. If, for species C_i, $\boldsymbol{q}_i$ is the flux vector and R_i is the rate of formation per unit of volume, the *balance law* for species C_i is

$$\frac{\mathrm{d}}{\mathrm{d}t}\iiint_{\mathcal{R}} c_i(\boldsymbol{x},t)\mathrm{d}V + \iint_{\partial\mathcal{R}} \boldsymbol{q}_i \cdot \boldsymbol{n}\mathrm{d}S = \iiint_{\mathcal{R}} R_i \mathrm{d}V$$

or, in point form (cf. (4.4) and (4.5)),

$$\frac{\partial c_i}{\partial t} + \boldsymbol{\nabla}\cdot\boldsymbol{q}_i = R_i\,, \qquad i = 1, 2, \ldots, n. \tag{10.1}$$

To complete the model, rules relating R_i to the various concentrations $c_i(\boldsymbol{x}, t)$ (and probably also to temperature $T(\boldsymbol{x}, t)$) and relating $\boldsymbol{q}_i$ to spatial variations in the concentrations are necessary.

10.1.1 Fick's Law of Diffusion

In many situations, the flux vector for species C_i is well modelled by

$$\boldsymbol{q}_i = -D_i \boldsymbol{\nabla} c_i \qquad \text{(no summation over } i\text{)}. \tag{10.2}$$

This is known as *Fick's law of diffusion.* It is similar to Fourier's law of heat conduction, since it states that flow is parallel to and proportional to the negative of the concentration gradient. The coefficient D_i is the *diffusivity* of the species C_i.

When reaction (conversion of C_i from or to other species) may be neglected, substitution of Fick's law into (10.1) gives

$$\frac{\partial c_i}{\partial t} = \operatorname{div}(D_i \operatorname{grad}\, c_i) = D_i \nabla^2 c_i\,. \tag{10.3}$$

Thus, the spreading of constituent C_i is mathematically analogous to the flow of heat without sources (see Chapter 3). Important applications are the diffusion (percolation) of moisture through soil and of gases through membranes. Methods given in Chapters 3 and 4 for solution of both steady and unsteady problems are easily adapted (in appropriately shaped regions).

Example 10.1

A porous, spherical solid initially in air is placed at time $t = 0$ in an atmosphere rich in CO_2 (carbon dioxide). Assuming that the concentration c of CO_2 within the solid obeys Fick's law of diffusion $\partial c/\partial t = D\nabla^2 c$ in $0 \le r < a$, with $c(\boldsymbol{x}, 0) = c_0$ for $|\boldsymbol{x}| < a$ and with $c = c_1$ at $|\boldsymbol{x}| = a$ for $t > 0$, adapt the solution of Example 2.8 to find the spherically symmetric solution $c(r, t)$ for $t > 0$.

In view of the spherical symmetry, it follows that $c = c(r,t)$, with $r \equiv |\boldsymbol{x}|$ and $(rc)_t = D(rc)_{rr}$. Write $u = r(c(r,t) - c_1)$, so that $(rc)_t = u_t$, $(rc)_{rr} = u_{rr}$ and $u_t = Du_{rr}$. As in Example 2.8, $u(r,t)$ satisfies the one-dimensional heat equation with coefficient D and with $u(0,t) = 0$ and $u(r,0) = (c_0 - c_1)r$. Also, for $t > 0$, $u(a,t) = 0$. In the superposition of separable solutions given in Example 2.8, replace ν by D, T_1 by $c_1 - c_0$, T_0 by c_0 and T by c to give

$$c = c(r,t) = c_1 + \frac{2(c_1 - c_0)a}{\pi} \sum_{n=1}^{\infty} \frac{(-1)^n}{nr} \exp\left(\frac{-Dn^2\pi^2 t}{a^2}\right) \sin\frac{n\pi r}{a}\,.$$

10.1.2 Self-similar Solutions

It was observed in Chapter 2 that typical heating times for similarly shaped objects increase as the square of the linear dimensions. An important consequence is that, when a flat surface $x = 0$ of a solid is suddenly increased in temperature, a specified fraction ($\frac{1}{2}$, say) of that change does not reach depth x until after an elapsed time proportional to x^2. This suggests that there may be special solutions of $u_t = \nu u_{xx}$ (or, equivalently, of $c_t = Dc_{xx}$) which depend only upon the ratio x^2/t. If this were so, the solution would have the form $u = F(x^2/t)$, or equivalently $u = f(xt^{-1/2})$. This surmise is in fact correct, as will now be demonstrated. Moreover, the relevant function f has applications which are important both physically and mathematically.

Self-similar solutions $u = f(xt^{-1/2})$ are sought to the one-dimensional heat equation $u_t = \nu u_{xx}$ (solutions $c = f(xt^{-1/2})$ describing one-dimensional diffusion may be sought similarly) by writing

$$\begin{aligned}
&xt^{-1/2} \equiv \eta \qquad \text{so that} \quad u = f(\eta)\,, \quad \eta_x = t^{-1/2}\,, \quad \eta_t = -\tfrac{1}{2}xt^{-3/2}\,,\\
&u_x = t^{-1/2}f'(\eta)\,, \qquad u_{xx} = (t^{-1/2})^2 f''(\eta) = t^{-1}f''(\eta)\,,\\
&u_t = -\tfrac{1}{2}xt^{-3/2}f'(\eta) = -\tfrac{1}{2}t^{-1}\eta f'(\eta)\,.
\end{aligned}$$

Substituting these into $u_t = \nu u_{xx}$, it is seen that the factor t^{-1} cancels, so leaving just

$$\nu f''(\eta) = -\tfrac{1}{2}\eta f'(\eta)\,,$$

which may be recognized as a first-order ODE for $f'(\eta) \equiv g(\eta)$. Moreover, this equation may be written in separated form as $g^{-1}\mathrm{d}g/\mathrm{d}\eta = -(2\nu)^{-1}\eta$, so that its general solution is readily found to be

$$f'(\eta) = g(\eta) = A\exp\frac{-\eta^2}{4\nu} = A\exp\frac{-x^2}{4\nu t}\,, \tag{10.4}$$

with A an arbitrary constant. The function $p(x) = (\pi\sigma)^{-1/2}\exp(-x^2/\sigma^2)$ is very important in probability theory, as the *probability density function* for the so-called *normal distribution.* The parameter σ, known as the *standard deviation*, is a measure of how much the distribution is *spread* about its mean value zero. The factor $(\pi\sigma)^{-1/2}$ is included to scale $p(x)$ so that $\int_{-\infty}^{\infty} p(x)\,\mathrm{d}x = 1$ (a standard exercise in multiple integration is to show that $\int_{-\infty}^{\infty} \mathrm{e}^{-s^2}\mathrm{d}s = 2\int_0^{\infty} \mathrm{e}^{-s^2}\mathrm{d}s = \sqrt{\pi}$, see e.g. McCallum *et al.*, 1997, p.271). The definite integral of $p(x)$

$$\operatorname{erf} x \equiv \frac{2}{\sqrt{\pi}}\int_0^x \mathrm{e}^{-s^2}\,\mathrm{d}s \tag{10.5}$$

is known as the *error function.* It has the properties $\operatorname{erf} 0 = 0$, $\operatorname{erf}(-x) = -\operatorname{erf} x$ and $\operatorname{erf} x \to 1$ as $x \to +\infty$. Using expression (10.4) in the self-similar solution $u = f(\eta)$ gives

$$\int_{-\infty}^{\infty} u_x\,\mathrm{d}x = At^{-1/2}\int_{-\infty}^{\infty} \exp\frac{-x^2}{4\nu t}\mathrm{d}x = \sqrt{2\nu}A\int_{-\infty}^{\infty} \mathrm{e}^{-s^2}\mathrm{d}s = \sqrt{2\nu\pi}A\,.$$

The choice $A = \sqrt{2/(\nu\pi)}$ gives $u(x,t) = \operatorname{erf}(x/\sqrt{4\nu t})$, which has the properties $u = 0$ at $x = 0$ for all $t > 0$, with $u \to 1$ as $t \to 0+$ for all $x > 0$ and with $u \to 1$ as $x \to +\infty$ for all $t > 0$. The functions $\exp(-x^2/4\nu t)$ and $\operatorname{erf}(x/\sqrt{4\nu t})$ are shown in Figure 10.1 for various values of νt. They illustrate how the *temperature profiles* have similar shapes at different instants t, but with scaling proportional to $t^{1/2}$, so that the profiles broaden as t increases.

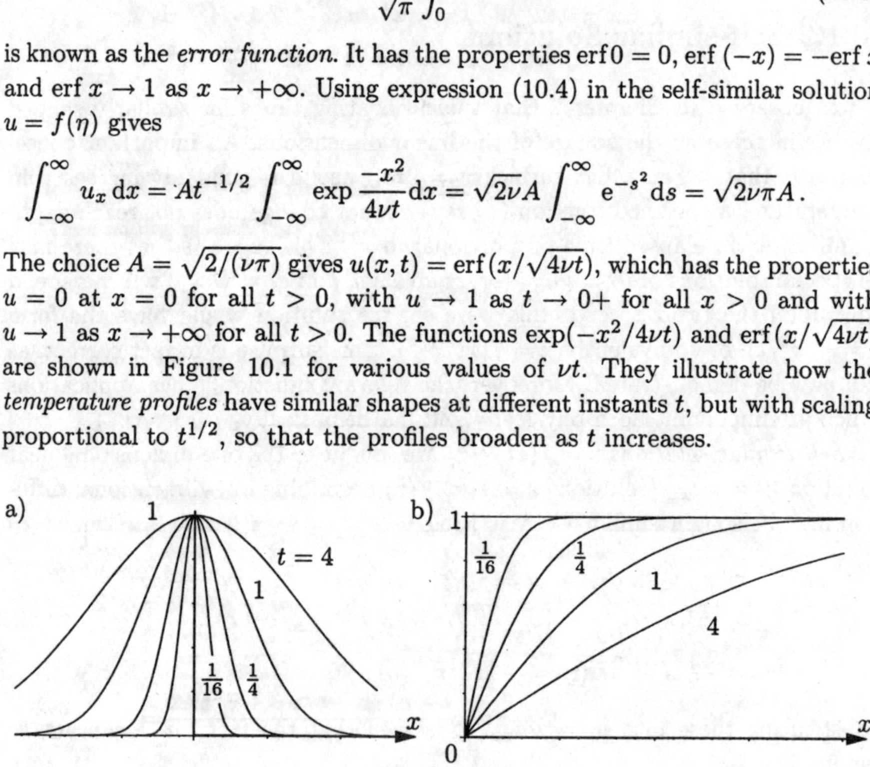

Figure 10.1 The functions a) $\exp(-x^2/4\nu t)$ and b) $\operatorname{erf}(x/\sqrt{4\nu t})$ for $t = \frac{1}{16}, \frac{1}{4}, 1, 4$.

Example 10.2

In the semiconductor industry, diffusion is widely used for *doping* a material. Assuming that, at fabrication temperature, diffusion is governed by $c_t = Dc_{xx}$

in $x > 0$, with $c(x, 0) = 0$ and $c(0, t) = C$ for $t > 0$, for how long is it necessary to run the fabrication process in order that $c > \frac{1}{2}C$ throughout the layer $0 \leq x < d$? If the diffusion coefficient at fabrication temperature is $D = 0.5\,(\mu\text{m})^2/\text{sec}$, find this time when $d = 8\,\mu\text{m}\quad (1\,\mu\text{m} = 10^{-6}\,\text{m})$.

We require a self-similar solution $c(x, t)$, such that $c \to 0$ as $x \to +\infty$. Hence, choose $c \propto 1 - \text{erf}\,(x/\sqrt{4Dt})$. The constant of proportionality must be C, so that $c(0, t) = C$. Hence, the function $c = C\{1 - \text{erf}\,(x/\sqrt{4Dt})\}$ is the required solution to $c_t = Dc_{xx}$, with $c \to C$ as $t \to 0+$ throughout $x > 0$. The region in which $c > \frac{1}{2}C$ is determined by $\text{erf}\,(x/\sqrt{4Dt}) < \frac{1}{2}$, i.e. $x < \sqrt{4Dt}\,\text{erf}^{-1}0.5$. Thus, this region includes the whole of $x < d$ when $4Dt \geq (d/\text{erf}^{-1}0.5)^2$. Since $\text{erf}\,0.477 = 0.5$, the required fabrication time is $t_f = (d/0.477)^2/(4D)$.

For the figures given, $t_f = (8 \times 10^{-6}/0.477)^2/(4 \times 0.5 \times 10^{-12}) = 17.6$ sec.

10.1.3 Travelling Wavefronts

When both reaction and diffusion must be taken into account, equations such as (10.1) rarely possess self-similar solutions describing a spreading process. However, there is another important class of solutions described by functions depending only upon a single combination of the coordinates. These are *travelling waves*, which often describe a fairly sharp transition between regions of two distinct states of chemical equilibrium.

For simplicity, consider just two species, with concentrations c_1, c_2 and with reaction rates R_1 and R_2 which depend only on the local values of both c_1 and c_2. A typical case is $R_1 = \gamma c_1 c_2^m = -R_2$ (m = integer), describing an autocatalytic reaction in which creation of species C_1 necessarily causes equivalent decrease in the quantity of C_2 present. This leads to the pair of coupled equations

$$\begin{aligned} \frac{\partial c_1}{\partial t} &= D_1 \nabla^2 c_1 + \gamma c_1 c_2^m\,, \\ \frac{\partial c_2}{\partial t} &= D_2 \nabla^2 c_2 - \gamma c_1 c_2^m\,. \end{aligned} \tag{10.6}$$

This example (like many others of the type (10.1)) allows solutions of the form

$$c_1 = X(\theta)\,, \qquad c_2 = Y(\theta)\,, \qquad \text{with} \quad \theta \equiv \boldsymbol{x} \cdot \boldsymbol{n} - Vt \quad \text{and} \quad |\boldsymbol{n}| = 1\,,$$

where θ is a variable which is constant on each member of a family of planes advancing in the direction of the unit vector $\boldsymbol{n}$ at speed V (real). To check this, note that $\partial c_1/\partial t = -VX'(\theta)$, $\boldsymbol{\nabla} c_1 = \boldsymbol{n}X'(\theta)$, $\nabla^2 c_1 = X''(\theta)$ (since $\boldsymbol{n} \cdot \boldsymbol{n} = 1$), with similar expressions involving c_2. Insertion into (10.6) then yields the coupled pair of ODEs

$$D_1X''(\theta) + VX'(\theta) + \gamma XY^m = 0\,,$$
$$D_2Y''(\theta) + VY'(\theta) - \gamma XY^m = 0\,, \qquad (10.7)$$

so that any solution $X(\theta), Y(\theta)$ to (10.7) describes special solutions to (10.6) having both c_1 and c_2 constant on parallel planes each moving at speed V (real).

The system (10.7) of ODEs is nonlinear and of fourth order and so must in general be solved using numerical methods. However, a few important features of its solutions can be deduced independently of any computation. Firstly, the system possesses solutions $(X, Y) = (0, Y_0)$ which describe the chemical state $c_1 = 0$ with $c_2 = Y_0$ being arbitrary. Similarly, there are states with $(X, Y) = (X_0, 0)$ in which the concentration of C_2 is zero but the concentration of C_1 is arbitrary. A travelling wavefront is described by a solution $c_1 = X(\theta)$, $c_2 = Y(\theta)$ making a transition between any pair of these states. Secondly, addition of the two equations in (10.7) eliminates the reaction terms, so yielding an equation which may be integrated to give

$$D_1X'(\theta) + D_2Y'(\theta) + VX + VY = \text{ constant}.$$

In the special case of equal diffusivities ($D_2 = D_1 \equiv D$), this equation allows the solution $Y + X = k$. Making this choice allows $Y(\theta)$ to be eliminated from (10.7), so giving the single equation

$$DX''(\theta) + VX'(\theta) + \gamma X(k - X)^m = 0\,. \qquad (10.8)$$

This equation is *autonomous* and may be analysed using *phase plane methods* (Simmons, 1972; Arrowsmith and Place, 1992) by writing $Z \equiv X'(\theta)$ and seeking curves in the (X, Z) plane which satisfy the first-order ODE

$$\frac{\mathrm{d}Z}{\mathrm{d}X} = \frac{Z'(\theta)}{X'(\theta)} = \frac{-\{\gamma X(k-X)^m + VZ\}}{DZ}$$

and which join the *stationary points* $(X, Z) = (0, 0)$ and $(X, Z) = (k, 0)$. For wide ranges of the parameters D, γ, k and m, there exist speeds V for which such curves exist. The corresponding function $X(\theta)$ describes the *waveform* for the concentration of C_1. A typical prediction of the waveform is shown in Figure 10.2. For more general examples of the type (10.1), investigation of possible transition fronts is a currently active research topic. Mathematically similar *reaction–diffusion equations* arise widely in models of biological processes, which will be treated next.

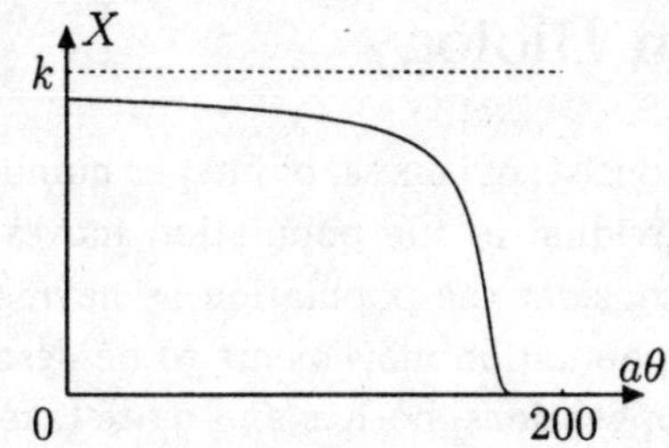

Figure 10.2 A travelling wavefront $X(\theta)$ computed from (10.8) for $m = 3$, with horizontal scale $a\theta$ where $a = (\gamma k^3/D)^{1/2}$.

EXERCISES

10.1. Apply to the function $c \equiv \operatorname{erf}(x/\sqrt{4Dt})$ the result that, if $c(x,t)$ satisfies $c_t = Dc_{xx}$, then so does $c_x(x,t)$. Thereby deduce that the function $c = (\pi Dt)^{-1/2}\exp(-x^2/4Dt)$ is a solution to $c_t = Dc_{xx}$.

At instant t, (a) what is the maximum value of $c(x,t)$, (b) where does that maximum occur and (c) what is the value of $\int_{-\infty}^{\infty} c(x,t)\,\mathrm{d}x$? Sketch the graph of c versus x at some typical values of t. Interpret the solution in terms of one-dimensional dispersal of an initially concentrated species.

[Notice that $c(x,t)$ has the form $t^{-1/2}g(\eta)$. In fact, $c_t = Dc_{xx}$ possesses solutions of the form $c = t^{-n/2}F(\eta)$ for all integers n.]

10.2. Some diffusive processes, for example diffusion of moisture through a powder, have a diffusion coefficient which depends upon the concentration c. When $D(c) = Bc^2$, the relevant *nonlinear diffusion equation* is $c_t = B(c^2c_x)_x$. Investigate the possibility that this equation allows solutions of the form $c = t^{-1/4}F(xt^{-1/4})$.
First write $xt^{-1/4} = \zeta$ and confirm that $c_t = \frac{-1}{4}t^{-5/4}\mathrm{d}(\zeta F)/\mathrm{d}\zeta$ and that $(c^2c_x)_x = t^{-5/4}(F^2F')'$. Next substitute these into $c_t = B(c^2c_x)_x$ and so deduce that $\zeta F + 4BF^2F'$ is a constant. Choose this constant so as to allow the solution to have $F = 0$ for some value of ζ and so show that this leads to a solution

$$c(x,t) = t^{-1/4}\sqrt{a^2 - \frac{x^2}{4Bt^{1/2}}}\,.$$

[Note that c is non-zero only for $|x| < 2aB^{1/2}t^{1/4}$. Concentration is non-zero only on a finite interval, which expands with time.]

10.2 Population Biology

Biological populations consist, of course, of integer numbers of insects, animals or fish. While each individual in the population moves separately, for many purposes it suffices to consider the population as having a *population density* $u(\boldsymbol{x}, t)$ and to consider population movements to be described by a *population flux* $\boldsymbol{q}(\boldsymbol{x}, t)$. In many applications, both $\boldsymbol{x}$ and $\boldsymbol{q}$ are taken as two-dimensional, measuring only horizontal position and flux, though when flying insects and micro-organisms are being considered, both $\boldsymbol{x}$ and $\boldsymbol{q}$ are sometimes taken as three-dimensional.

A resulting *continuum model* cannot attempt to predict idiosyncracies of individual movements, but it can well predict collective properties over regions containing many hundreds or thousands of individuals and over timescales allowing considerable numbers of individuals to cross the boundary of such a region. Furthermore, biological individuals undergo both birth and death. These events, localized in time, are modelled by including suitable laws for birth rates and death rates. The resulting balance law for the species then takes the form

$$\frac{\partial u}{\partial t} + \boldsymbol{\nabla} \cdot \boldsymbol{q} = S\,, \tag{10.9}$$

where S is the difference between the birth and death rates, measured per unit of area (or volume). Laws (constitutive relations) relating S to densities and $\boldsymbol{q}$ to densities and density gradients of one (or more) populations are determined less precisely than in the physical sciences, by judicious matching of prediction to experimental observation.

10.2.1 Growth and Dispersal

A simple (mathematically idealized) model will be considered first. Let $u(\boldsymbol{x}, t)$ be the population of a single species in an environment with food supply sufficient that, in the absence of migrations, the natural growth rate is α, independently of population density u. In a spatially homogeneous region, the population then grows exponentially according to $\mathrm{d}u/\mathrm{d}t = \alpha u$, but if u depends upon position, it is natural to expect there to be a tendency for smearing out of spatial non-uniformities. Physical analogies suggest the rule $\boldsymbol{q} = -D\boldsymbol{\nabla} u$, for some constant D, known as the *dispersivity*. In fact, it can be shown rigorously that a *random walk* incorporating a large number of individual steps is described asymptotically by a diffusion process. This motivates consideration of the equation

$$u_t = D\nabla^2 u + \alpha u\,, \tag{10.10}$$

combining the effects of both growth and dispersion. This is a linear PDE, so that solutions may be constructed by separation of variables. In fact, (10.10) is closely related to the heat equation or the diffusion equation, since writing $u = c(\boldsymbol{x}, t)\mathrm{e}^{\alpha t}$ gives $u_t = (c_t + \alpha c)\mathrm{e}^{\alpha t}$ and $\nabla^2 u = \mathrm{e}^{\alpha t}\nabla^2 c$. Thus,

$$(c_t + \alpha c)\mathrm{e}^{\alpha t} = D\mathrm{e}^{\alpha t}\nabla^2 c + \alpha c\mathrm{e}^{\alpha t}\,,$$

which simplifies to $c_t = D\nabla^2 c$. Hence, $u(\boldsymbol{x}, t)$ equals some solution $c(\boldsymbol{x}, t)$ of the diffusion equation, multiplied by an exponentially growing term.

Equation (10.10) may be taken to model growth of algae or plankton colonies on a lake. Suppose that it applies within the strip $0 < x < L$ and that, at both $x = 0$ and $x = L$ no algae survive (for example, being carried away by swift currents). Then, analysing one-dimensional solutions $u = u(x, t)$ to (10.10) with $u(0, t) = 0 = u(L, t)$ reveals a criterion determining whether or not the algae will *bloom* (i.e. grow in abundance).

Separable solutions $u = X(x)T(t)$ to (10.10) must satisfy

$$\frac{T'}{T} = D\frac{X''}{X} + \alpha\,,$$

in which the solutions to $X''(x) \propto X(x)$ must also satisfy $X(0) = 0 = X(L)$. Thus, the relevant possibilities are $X(x) = \sin n\pi x/L$, so giving

$$T'(t)/T = \alpha - D\pi^2 n^2/L^2 \equiv \alpha_n \qquad \text{(say)}.$$

The general solution $u(x, t)$ is thus

$$\begin{aligned} u(x,t) &= \sum_{n=1}^{\infty} B_n \mathrm{e}^{\alpha_n t} \sin\frac{n\pi x}{L} \\ &= \mathrm{e}^{\alpha t}\sum_{n=1}^{\infty} B_n \mathrm{e}^{-D\pi^2 n^2 t/L^2} \sin\frac{n\pi x}{L}\,. \end{aligned} \tag{10.11}$$

As expected, this is $\mathrm{e}^{\alpha t}$ multiplying the general solution $c(x, t)$ to $c_t = Dc_{xx}$, $c(0, t) = 0 = c(L, t)$. If $\alpha_n < 0$ for all $n = 1, 2, \ldots$, it is seen that any clump of algae will die away, but if $\alpha_1 > 0$ there is at least one growing mode. Hence, algae can be expected to bloom in strips, if the width exceeds $\pi\sqrt{D/\alpha}$. Thus, $L > \pi\sqrt{D/\alpha}$ is the *survival criterion* for this model. In a strip of specified width L, increase in dispersivity D and decrease in α will each inhibit growth. For $L < \pi\sqrt{D/\alpha}$, any algae introduced will eventually disappear.

In an unbounded region, it is expected that any algae introduced locally will cause a colony to both grow and spread. An illustrative example is provided by using the solution $c = M(4\pi Dt)^{-1/2}\exp(-x^2/4Dt)$ (see Exercise 10.3) to the diffusion equation $c_t = Dc_{xx}$, so that

$$u = \frac{M}{\sqrt{4\pi Dt}}\exp\left(\alpha t - \frac{x^2}{4Dt}\right) \tag{10.12}$$

corresponds to spreading and growth from an initial mass M of algae concentrated at $x = 0$ at $t = 0$. At later times, the maximum concentration $M(4\pi D)^{-1/2}t^{-1/2}e^{\alpha t}$ first subsides to a minimum $M(\alpha e/2\pi D)^{1/2}$ at $t = (2\alpha)^{-1}$, then grows. The spatial extent of the colony is estimated by considering the evolution of a *front of isoconcentration* $u(x,t) = \tilde{u}$, say. Rearrangement of (10.12) gives

$$\frac{x^2}{t^2} - 4\alpha D = -4Dt^{-1}\ln\left(\sqrt{4\pi Dt}\,\frac{\tilde{u}}{M}\right)$$

or

$$\frac{x}{t} = \pm\left\{4D\alpha - 2Dt^{-1}\left[\ln t + \ln(4\pi D\tilde{u}^2/M^2)\right]\right\}^{1/2},$$

which shows that, as $t \to \infty$, each locus of constant $\tilde{u}$ moves at speed given asymptotically by $\pm(4D\alpha)^{1/2}$.

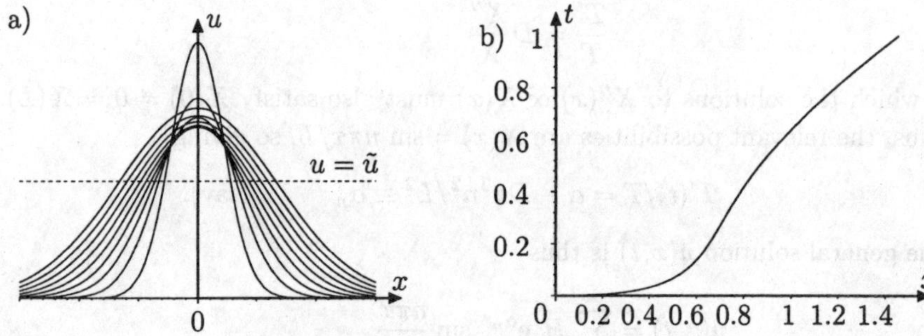

Figure 10.3 a) Spreading of an initially concentrated, growing colony described by (10.12), with $\alpha = 1$, portrayed for $Dt = 0.05, 0.1, \ldots, 0.45$, showing emergence of *fronts* at speed $\pm(4D\alpha)^{1/2}$. b) Location $\tilde{x}(t)$ of a typical concentration value $\tilde{u}$.

10.2.2 Fisher's Equation and Self-limitation

Equation (10.10) is unrealistic as a model in many situations, since (10.12) predicts unlimited growth of population density u. In practice, growth of many biological populations is limited by the availability of food (or nutrients). Population limitation is described, for spatially uniform distributions $u = u(t)$, not by solutions to $\mathrm{d}u/\mathrm{d}t = \alpha u$ but by solutions to the ODE

$$\frac{\mathrm{d}u}{\mathrm{d}t} = f(u),$$

where $f(u)$ is a function for which $f > 0$ in $0 < u < u_*$ and $f(0) = 0 = f(u_*)$. The simplest consistent choice is $f(u) = \alpha u - \beta u^2$ with $\alpha > 0, \beta > 0$, for which $u_* = \alpha/\beta$. This gives $du/dt = \alpha u - \beta u^2$, which has general solution

$$\alpha t = \int \frac{u_*}{u(u_* - u)} du = \int \frac{du}{u} - \int \frac{du}{u - u_*} = \ln\left|\frac{u}{u - u_*}\right| + \alpha t_0 .$$

When $u(0) = a u_*$, with $0 < a < 1$, this solution may be written as

$$u = \frac{a u_*}{a + (1 - a)e^{-\alpha t}}$$

which is known as the *logistic growth* law, showing exponential growth $\propto e^{\alpha t}$ for small values of u, but also showing how u cannot increase above a limiting value u_*. Moreover, $u_* - u \propto e^{-\alpha t}$ as u approaches u_*. The time $t = t_0$, corresponding to $u = \frac{1}{2}u_*$, is related to a through $1 + e^{\alpha t_0} = a^{-1}$.

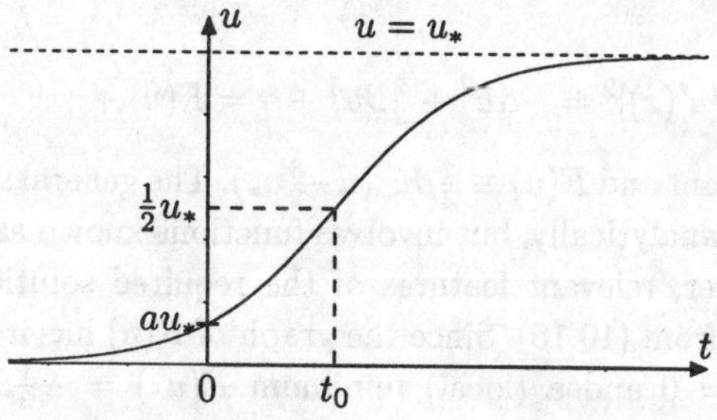

Figure 10.4 Logistic growth, showing the significance of u_*, a and t_0.

Using this growth law $f(u)$ to replace the linear term αu in (10.10) gives the nonlinear diffusion equation

$$u_t = D u_{xx} + \alpha u - \beta u^2 , \tag{10.13}$$

known as *Fisher's equation*, after R.A. Fisher, one of the pioneers of biometrics and biological statistics. If Fisher's equation is applied to the problem of algal bloom in the strip $0 \leq x \leq L$, it is found that there exist *steady solutions* satisfying *both* the ODE

$$D u''(x) = -\alpha u + \beta u^2 = \beta u(u - u_*) \tag{10.14}$$

and the boundary conditions $u(0) = 0 = u(L)$ for all values $L > L_0 \equiv \pi\sqrt{D/\alpha}$. This is unlike the situation for $\beta = 0$ ($u_* \to \infty$), for which (10.10) predicts that solutions exist only if $L = nL_0$, with n being an integer. Moreover, even for $L > L_0$ so that (10.10) predicts that small initial distributions $u(x, 0)$ can lead to unbounded growth, it can be shown that unsteady solutions to (10.13)

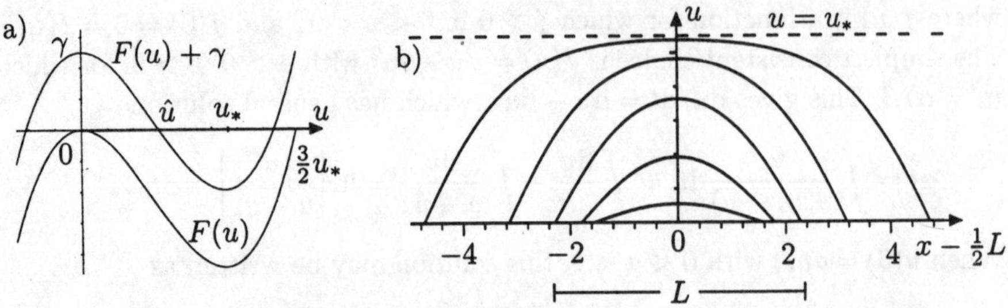

Figure 10.5 a) The functions $F(u)$ and $F(u) + \gamma$; b) steady solutions for various values of γ (with centreline $x = \frac{1}{2}L$ shifted to the origin).

remain everywhere bounded by u_*. They tend towards the steady solution, which is stable.

Solutions to (10.14) are found by first multiplying by $2u'(x)$ and then integrating to yield

$$D[u'(x)]^2 = -\alpha u^2 + \tfrac{2}{3}\beta u^3 + \gamma = F(u) + \gamma \,, \qquad (10.15)$$

where γ is a constant and $F(u) \equiv \frac{2}{3}\beta u^2(u - \frac{3}{2}u_*)$. The general solution to (10.15) may be obtained analytically, but involves functions known as *Jacobian elliptic functions.* However, relevant features of the required solutions $u(x)$ may be deduced directly from (10.15). Since the graph of $F(u)$ has a (local) maximum $F(0) = 0$ at $u = 0$ and a (local) minimum $F(u_*) = -\frac{1}{3}\beta u_*^3 = -\frac{1}{3}\alpha u_*^2$ at $u = u_*$, *any* choice of γ in $0 < \gamma < \frac{1}{3}\alpha u_*^2$ ensures that there is *just one* value $\hat{u}$ satisfying $0 < \hat{u} < u_*$, such that $[u']^2 > 0$ for $0 < u < \hat{u}$ with $u' = 0$ where $u = \hat{u}$. Thus, for each $u \in [0, \hat{u})$, exactly two values $\pm\sqrt{[F(u)+\gamma]/D}$ exist for $u'(x)$. This shows that if a solution $u(x)$ to (10.15) has $u = \hat{u}$ and $u' = 0$ at location $x = \hat{x}$, then $u(x)$ is symmetric about $x = \hat{x}$. The required solutions with $u(0) = 0 = u(L)$ thus must have $\hat{x} = \frac{1}{2}L$, so that the population density is symmetric about the centreline of the strip. By varying γ (and hence varying the maximum density $\hat{u}$) it is found that the half-width given by

$$\tfrac{1}{2}L = \int_0^{\hat{u}} \frac{\mathrm{d}u}{u'(x)} = D^{1/2}\int_0^{\hat{u}} [F(u)+\gamma]^{-1/2}\mathrm{d}u$$

exceeds $\frac{1}{2}L_0 = \frac{1}{2}\pi\sqrt{D/\alpha}$. Moreover, L increases without bound as $\gamma \to \frac{1}{3}\alpha u_*^2$, so that for *all* values $L > L_0$ Equation (10.14) has solutions satisfying $u(0) = 0 = u(L)$. In these, the maximum population density u_* depends upon γ, which itself depends upon L. In particular, as $L \to L_0$ the density maximum u_* tends to zero. The value $L = L_0$ is the *threshold* below which a steady strip of bloom cannot form.[2]

[2] The analysis involves comparison with the limiting case $\beta \to 0$, for which

10.2.3 Population-dependent Dispersivity

Just as some physical diffusion processes have a concentration-dependent diffusivity (see Exercise 10.2), in population dynamics and ecology where pressures to disperse are enhanced at high population densities, it is plausible to modify the flux term $-D\nabla u$ to a term proportional to $-u^m\nabla u$, for some constant m. By appropriate scaling of the density u, this replaces (10.10) by

$$u_t = \nabla\cdot(u^m\nabla u) + \alpha u\,, \tag{10.16}$$

an equation incorporating the growth term αu together with the population-dependent dispersivity.

To analyse (10.16), first write $u(x,t) = \mathrm{e}^{\alpha t}\rho$, so yielding

$$\mathrm{e}^{-m\alpha t}\frac{\partial\rho}{\partial t} = \nabla\cdot(\rho^m\nabla\rho)\,.$$

Then define a new time-like variable $\tau(t)$ through $\mathrm{d}\tau/\mathrm{d}t = \mathrm{e}^{m\alpha t}$, so that

$$\tau = \frac{1}{m\alpha}\left[\mathrm{e}^{m\alpha t} - 1\right] \quad (\text{with } \tau(0) = 0\,). \tag{10.17}$$

The equation governing $\rho \equiv \rho(x,\tau)$ then becomes

$$\frac{\partial\rho}{\partial\tau} = \nabla\cdot(\rho^m\nabla\rho)\,. \tag{10.18}$$

In one space dimension, this gives $\rho_\tau = (\rho^m\rho_x)_x$. Comparing with Exercise 10.2 (for which $m = 2$) suggests how self-similar solutions may be constructed.

Seek solutions $\rho(x,\tau) = \tau^{-k}F(\zeta)$ with $\zeta \equiv x\tau^{-q}$ (with constants k, q yet to be chosen). Then

$$\frac{\partial\zeta}{\partial x} = \tau^{-q}\,, \qquad \frac{\partial\zeta}{\partial\tau} = -qx\tau^{-(q+1)} = -q\zeta/\tau\,,$$

so that

$$\frac{\partial\rho}{\partial\tau} = -k\tau^{-(k+1)}F(\zeta) + \tau^{-k}\frac{\partial\zeta}{\partial\tau}F'(\zeta) = -\tau^{-(k+1)}\left\{kF(\zeta) + q\zeta\frac{\mathrm{d}F}{\mathrm{d}\zeta}\right\}\,,$$

$$\frac{\partial\rho}{\partial x} = \tau^{-k}\tau^{-q}F'(\zeta)\,, \quad \text{so giving} \quad \rho^m\rho_x = \tau^{-(k+q+mk)}F^m\frac{\mathrm{d}F}{\mathrm{d}\zeta}\,.$$

$\hat{u} = (\gamma/\alpha)^{1/2}$, so that the substitution $u = \hat{u}\sin\phi$ allows evaluation as $\frac{1}{2}L = \sqrt{D/\alpha}\int_0^{\hat{u}}(\hat{u}^2 - u^2)^{-1/2}\mathrm{d}u = \sqrt{D/\alpha}\int_0^{\pi/2}\mathrm{d}\phi = \frac{1}{2}\pi\sqrt{D/\alpha}$. Using the same substitution $u = \hat{u}\sin\phi$ in the general case, for which $\gamma = \alpha\hat{u}^2 - \frac{2}{3}\beta\hat{u}^3$, gives $F(u) + \gamma = \alpha(\hat{u}^2 - u^2) + (2\alpha/3u_*)(u^3 - \hat{u}^3) = \alpha\hat{u}^2[\cos^2\phi + \frac{2}{3}(\hat{u}/u_*)(\sin^3\phi - 1)] < \alpha\hat{u}^2\cos^2\phi$ throughout $0 \le \phi < \frac{1}{2}\pi$. Hence $\frac{1}{2}L > \frac{1}{2}L_0$. Moreover, as $\gamma \to \frac{1}{3}\alpha u_*^2$ so that $\hat{u} \to u_*$ and $F(u) + \gamma \to (2\alpha/3u_*)(u - u_*)^2(u + \frac{1}{2}u_*)$, the limiting value is found by using the substitution $u + \frac{1}{2}u_* = \frac{3}{2}u_*y^2$ as $\frac{1}{2}L \to \sqrt{D/\alpha}\int_{1/\sqrt{3}}^{1} 2\mathrm{d}y/(1 - y^2)$, which is infinite.

Then, inserting

$$\frac{\partial}{\partial x}(\rho^m \rho_x) = \tau^{-(k+q+mk)}\tau^{-q}\frac{\mathrm{d}}{\mathrm{d}\zeta}\left(F^m \frac{\mathrm{d}F}{\mathrm{d}\zeta}\right)$$

and the expression for $\partial\rho/\partial\tau$ into the equation $\rho_\tau = (\rho^m \rho_x)_x$ gives

$$\tau^{-(k+1)}\left\{kF(\zeta) + q\zeta\frac{\mathrm{d}F}{\mathrm{d}\zeta}\right\} + \tau^{-(k+mk+2q)}\frac{\mathrm{d}}{\mathrm{d}\zeta}\left(F^m\frac{\mathrm{d}F}{\mathrm{d}\zeta}\right) = 0.$$

This equation is self-consistent only if powers of τ in each term are equal, so giving $k+1 = k+mk+2q$, or, equivalently,

$$mk + 2q = 1\,. \tag{10.19}$$

When (10.19) is satisfied, the governing equation reduces to the ODE

$$\frac{\mathrm{d}}{\mathrm{d}\zeta}\left(F^m\frac{\mathrm{d}F}{\mathrm{d}\zeta}\right) + q\zeta\frac{\mathrm{d}F}{\mathrm{d}\zeta} + kF = 0\,.$$

While this possesses solutions for all pairs k, q satisfying (10.19), an important simplification arises when $q = k$, since the ODE for $F(\zeta)$ then becomes

$$\frac{\mathrm{d}}{\mathrm{d}\zeta}\left(F^m\frac{\mathrm{d}F}{\mathrm{d}\zeta} + k\zeta F\right) = 0\,.$$

Hence, for $q = k = 1/(m+2)$, the function $F(\zeta)$ must satisfy

$$F^m\frac{\mathrm{d}F}{\mathrm{d}\zeta} + k\zeta F = A\,,$$

for some constant A. Notice, however, that if $F \to 0$ for any finite value of ζ, the integration constant A must vanish. This suggests analysis of

$$(m+2)F^{m-1}\frac{\mathrm{d}F}{\mathrm{d}\zeta} + \zeta = 0\,,$$

which itself is readily integrated to give

$$\frac{m+2}{m}F^m + \tfrac{1}{2}\zeta^2 = \text{constant} = \tfrac{1}{2}\zeta_0^2 \quad \text{(say)}.$$

Finally, this yields for $F(\zeta)$ the expression

$$F(\zeta) = \sqrt[m]{\frac{m}{2(m+2)}(\zeta_0^2 - \zeta^2)} \qquad \text{for} \quad |\zeta| \le \zeta_0$$

with $F(\zeta) \equiv 0$ for $|\zeta| \le \zeta_0$, where ζ_0 is adjustable.

The population density then is

$$u(x,t) = e^{\alpha t}\rho(x,\tau)$$

$$= e^{\alpha t}\tau^{-1/(m+2)} \begin{cases} \left\{\frac{m}{2(m+2)}\left(\zeta_0^2 - x^2/\tau^{2k}\right)\right\}^{1/m} & \text{for } |x| < \zeta_0\tau^k \\ 0 & \text{otherwise,} \end{cases} \tag{10.20}$$

with τ given by (10.17). In contrast to the solution (10.12) to (10.10) (with constant dispersivity), the self-similar solution (10.20) (for $m > 0$) describes population dispersal for which the population is confined to a finite, but expanding, interval $|x| < \zeta_0\tau^k = \zeta_0^{1/(m+2)}$. There is a sharp boundary to the colonized region (and, moreover, $|\partial u/\partial x|$ is infinite at that boundary, for $m > 1$). The population spreads rather like a puddle on a smooth, non-absorbent surface.

10.2.4 Competing Species

The classical mathematical description of competition between two species is due to V. Volterra (1931), who considered interactions between a predator having total population $Y(t)$ and its prey having total population $X(t)$, without considering any spatial variations in population density. The corresponding *predator–prey* equations are frequently taken as

$$\dot{X} \equiv \frac{dX}{dt} = aX - bXY \quad , \quad \dot{Y} \equiv \frac{dY}{dt} = -cY + dXY \tag{10.21}$$

with a, b, c and d each positive. The product terms (involving XY) model the frequency of encounters between species, which lead to deaths of prey, thereby providing food and sustenance for the predator. In the absence of encounters, the prey population $X(t)$ would grow exponentially, while the predator population $Y(t)$ would decay to extinction.

A natural generalization of (10.21), when prey and predator species have population densities $u(\boldsymbol{x},t)$ and $v(\boldsymbol{x},t)$ depending on position $\boldsymbol{x}$ as well as upon time, is to set

$$u_t = D_1\nabla^2 u + au - buv \quad , \quad v_t = D_2\nabla^2 v - cv + duv\,. \tag{10.22}$$

The coefficients D_1 and D_2 measure the tendency of each population to spread or disperse, thereby reducing the dependence upon position. In particular, observe that every solution to (10.21) also describes spatially–uniform solutions $u(\boldsymbol{x},t) = X(t)$, $v(\boldsymbol{x},t) = Y(t)$ to (10.22).[3] Hence, knowledge of the behaviour

[3] The eagle-eyed will note that the change from total populations in (10.21) to poulation densities in (10.22) implies a rescaling of b and d.

of solutions to (10.21) is useful for understanding the dynamics of species competition.

System (10.21) may be regarded as governing the motion of a point in the X, Y plane. The coordinates (X, Y) describe the *state* of the system. It has *stationary points* where $X(a - bY) = 0$ and $(c - dX)Y = 0$. This yields two possibilities $(X, Y) = (0, 0)$ and $(X, Y) = (c/d, a/b)$. The first describes an *unstable equilibrium state* since, with $Y \equiv 0$, $X(0) = X_0 > 0$, the solution $X(t) = X_0 e^{at}$ to (10.21) grows without bound, however small is X_0. On the contrary, the solution $(X, Y) = (c/d, a/b)$ is found to be *stable*, since any small perturbation at time $t = 0$ leads only to bounded perturbations at later times. This is found by writing $X = c/d + x$, $Y = a/b + y$ so that

$$\dot{x} = \frac{dx}{dt} = -by\left(\frac{c}{d} + x\right), \qquad \dot{y} = \frac{dy}{dt} = dx\left(\frac{a}{b} + y\right).$$

Then, neglecting the product terms xy, yields the *linearized system*

$$\dot{x} = -(bc/d)y, \qquad \dot{y} = (ad/b)x,$$

which leads to the SHM equation $\ddot{x}(t) + acx = 0$. Since this has general solution $x = A\cos\sqrt{ac}t + B\sin\sqrt{ac}t$, with $y = (d/b)\sqrt{a/c}[A\sin\sqrt{ac}t - B\cos\sqrt{ac}t]$ $(= -(d/bc)\dot{x})$, the solution with arbitrary initial conditions $x(0) = A$, $y(0) = -(d/b)\sqrt{a/c}B$ remains bounded.

In fact, the system (10.21) leads readily to

$$\frac{dY}{dX} = \frac{dY/dt}{dX/dt} = \frac{Y(dX - c)}{X(a - bY)}, \tag{10.23}$$

which is a separable ODE relating the populations X and Y (compare this with the *phase plane* analysis in Section 10.1.3). Its solution may be found (see also Exercise 10.4) as $F(X, Y) \equiv a\ln Y - bY + c\ln X - dX = K$, $K =$ constant. Since the functions $a\ln Y - bY$ and $c\ln X - dX$ each have single stationary values, which are minima at $Y = a/b$ and at $X = c/d$, respectively, the function $F(X, Y)$ has only one stationary value in the relevant quarter-plane $X > 0$, $Y > 0$, namely a minimum at $(c/d, a/b)$. The solutions to (10.23) are thus the contours $F(X, Y) = K$, which must be nested curves surrounding the stationary point, as in Figure 10.6. Along each curve, the predator population Y grows whenever $X > c/d$, but decays when $X < c/d$. As the state (X, Y) traverses any loop, the two populations evolve periodically. For example, with $X > c/d$ the predator population grows until, for $Y > a/b$, the population of prey starts to reduce, so causing X to fall below the value c/d necessary to maintain the predator population Y. Since Y then decreases along with X, there comes a stage when Y falls below a/b, so allowing the prey population to recover (while Y is small). The process then repeats cyclically.

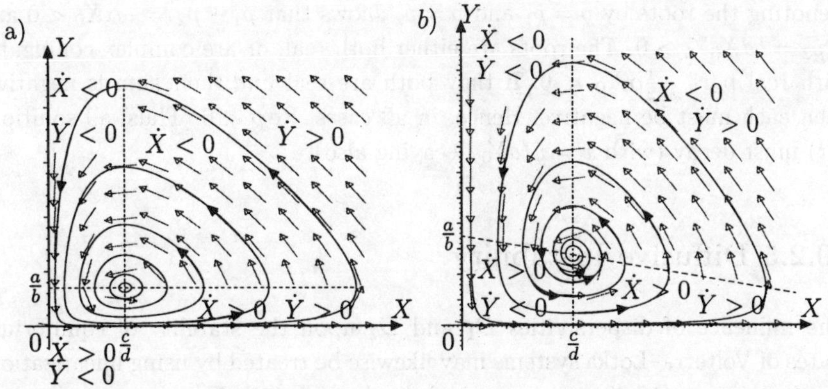

Figure 10.6 a) Typical solution curves for (10.23). b) Equilibrium point (X_0, Y_0) and a typical solution curve for Example 10.3.

Example 10.3

A modification to Equations (10.21), including self-limitation of the prey population, is (with $\alpha > 0$)

$$\dot{X} = aX - \alpha X^2 - bXY \ , \quad \dot{Y} = -cY + dXY$$

(like (10.21), a *Volterra–Lotka model*). Identify the equilibrium point (X_0, Y_0) in $X > 0$, $Y > 0$. Find also the four regions $\dot{X} > 0$, $\dot{Y} > 0$; $\dot{X} > 0$, $\dot{Y} < 0$; $\dot{X} < 0$, $\dot{Y} > 0$ and $\dot{X} < 0$, $\dot{Y} < 0$. Near the equilibrium point, write $X = X_0 + x(t)$, $Y = Y_0 + y(t)$, derive a linear second-order ODE for $y(t)$ and show that its solutions of the form $y \propto e^{pt}$ must have $\operatorname{Re} p < 0$. Deduce that the stationary solution $X = X_0$, $Y = Y_0$ is stable.

Since $Y_0(dX_0 - c) = 0$ and $X_0(a - \alpha X_0 - bY_0) = 0$, the equilibrium point within $X > 0$, $Y > 0$ has $X_0 = c/d$, $Y_0 = a/b - \alpha X_0/b = (ad - \alpha c)bd$ (this lies within $Y > 0$ only if $\alpha c < ad$). The regions for $\dot{X} < 0$, $\dot{X} > 0$, $\dot{Y} < 0$ and $\dot{Y} > 0$ are shown in Figure 10.6b.

From

$$\dot{x} = \dot{X} = (X_0 + x)(a - \alpha X_0 - bY_0 - \alpha x - by) = -(X_0 - x)(\alpha x + by)\,,$$
$$\dot{y} = \dot{Y} = (Y_0 + y)(-c + dX_0 + dx) = d(Y_0 + y)x\,,$$

linearization (i.e. neglecting all terms except those linear in x and y) gives

$$\dot{x} = -X_0(\alpha x + by)\,, \quad \dot{y} = dY_0 x \qquad \text{so that } \ddot{y} = dY_0\dot{x} = -\alpha X_0 \dot{y} - bdX_0Y_0 y\,.$$

Inserting $y = ke^{pt}$ into $\ddot{y} + \alpha X_0 \dot{y} + bdX_0Y_0 y = 0$ gives

$$p^2 + \alpha X_0 p + bdX_0Y_0 = 0\,.$$

Denoting the roots by $p = p_1$ and $p = p_2$ shows that $p_1 + p_2 = -\alpha X_0 < 0$ and $p_1 p_2 = bdX_0Y_0 > 0$. The roots are either both real, or are complex conjugates with real part $-\frac{1}{2}\alpha X_0 < 0$. If they both are real *and* their sum is negative, then each must be negative. Hence, in all cases, $\operatorname{Re} p < 0$. Thus, all solutions $y(t)$ must decay (with $x = \dot{y}/dY_0$ decaying also).

10.2.5 Diffusive Instability

The influence of dispersivities D_1 and D_2 upon the stability of equilibrium states of Volterra–Lotka systems may likewise be treated by using linearization. For the system (10.22) in two spatial dimensions with $\nabla^2 u = u_{xx} + u_{yy}$ and with (x, y) as cartesian coordinates, write $u(\boldsymbol{x}, t) = X_0 + \hat{u}(x, y, t)$, $v(\boldsymbol{x}, t) = Y_0 + \hat{v}(x, y, t)$ where $(X_0, Y_0) = (c/d, a/b)$. Then, retaining only the terms linear in $\hat{u}$, $\hat{v}$ and their partial derivatives yields

$$\begin{aligned} \hat{u}_t &= D_1(\hat{u}_{xx} + \hat{u}_{yy}) - (bc/d)\hat{v}\,, \\ \hat{v}_t &= D_2(\hat{v}_{xx} + \hat{v}_{yy}) - (ad/b)\hat{u}\,. \end{aligned} \tag{10.24}$$

Although this is a system of two *coupled* partial differential equations, separable solutions (cf. Sections 4.2, 8.4 and 9.2) are easily sought. They reveal important information about the time evolution of fluctuations around the *equilibrium state* (X_0, Y_0).

The substitution $\hat{u} = \bar{X}(x)\bar{Y}(y)U(t)$, $\hat{v} = \bar{X}(x)\bar{Y}(y)V(t)$ is readily shown to be consistent with (10.24) and to have $\bar{X}(x)$ and $\bar{Y}(y)$ bounded if $\bar{X}''(x) + l^2\bar{X} = 0$, $\bar{Y}''(y) + m^2\bar{Y} = 0$ with

$$\begin{aligned} \dot{U}(t) &= -k^2 D_1 U \; - \; (bc/d)V \\ \dot{V}(t) &= (ad/b)U \; - \; k^2 D_2 V \end{aligned} \tag{10.25}$$

and $k^2 \equiv l^2 + m^2$. For ODE systems such as (10.25) which may be written in matrix form as $\dot{\mathsf{U}}(t) = \mathsf{AU}$, with $\mathsf{U} \equiv (U \;\; V)^T$ and with A a constant square matrix, the general solution may be constructed by seeking special solutions $\mathsf{U} = \mathsf{U}_0 e^{pt}$, so giving the matrix eigenvalue problem $\mathsf{AU}_0 = p\mathsf{U}_0$. Hence, possible parameters p must satisfy $\det(\mathsf{A} - p\mathsf{I}) = 0$, where I is the identity matrix, so that, for (10.25), the equation determining p is

$$0 = \begin{vmatrix} -k^2 D_1 - p & -bc/d \\ ad/b & -k^2 D_2 - p \end{vmatrix} = p^2 + k^2(D_1 + D_2)p + k^4 D_1 D_2 + ac\,.$$

Since this gives

$$\{p + \tfrac{1}{2}(D_1 + D_2)k^2\}^2 = \tfrac{1}{4}(D_1 - D_2)^2 k^4 - ac\,,$$

the two roots p will be complex conjugates for $k^2 < 2\sqrt{ac}/|D_1 - D_2|$, but will be real and negative for $k^2 > 2\sqrt{ac}/|D_1 - D_2|$. In either case, all roots p have negative real part (cf. Example 10.3), so that fluctuations die out. (The one exception is $k = 0$, for which $p^2 = -ac$ so that small amplitude spatially uniform perturbations oscillate at frequency $\sqrt{ac}$ with neither growth nor decay.) The system (10.22) does *not* suffer diffusive instability.

Example 10.4 (Diffusive instability)

For any Volterra–Lotka model of the form

$$u_t = D_1 \nabla^2 u + f(u,v)\,, \qquad v_t = D_2 \nabla^2 v + g(u,v)$$

with an equilibrium state $(u,v) = (u_0, v_0)$ satisfying $f(u_0,v_0) = 0$, $g(u_0,v_0) = 0$, small fluctuations $\hat{u} \equiv u - u_0$, $\hat{v} \equiv v - v_0$ satisfy a linearized system of the form

$$\hat{u}_t = D_1 \nabla^2 \hat{u} + a_{11}\hat{u} + a_{12}\hat{v}\,, \qquad \hat{v}_t = D_2 \nabla^2 \hat{v} + a_{21}\hat{u} + a_{22}\hat{v}\,,$$

where $a_{11} = f_u(u_0,v_0)$, $a_{12} = f_v(u_0,v_0)$, $a_{21} = g_u(u_0,v_0)$ and $a_{22} = g_v(u_0,v_0)$.

Investigate conditions on the dispersivities D_1, D_2 and the coefficients a_{ij} such that some separable solutions $\hat{u}$, $\hat{v}$ for which $\nabla^2\hat{u} = -k^2\hat{u}$, $\nabla^2\hat{v} = -k^2\hat{v}$ are unstable, even when the state (u_0, v_0) is stable to spatially uniform perturbations.

For separable solutions (e.g. $\hat{u} = \bar{X}(x)\bar{Y}(y)U(t)$, as above), the system generalizing (10.25) is

$$\begin{aligned} \dot{U}(t) &= (a_{11} - k^2 D_1)U + a_{12}V\,, \\ \dot{V}(t) &= a_{21}U + (a_{22} - k^2 D_2)V\,. \end{aligned} \tag{10.26}$$

Seeking solutions $(U \;\; V) = (U_0 \;\; V_0)e^{pt}$ gives the determinantal condition

$$\begin{aligned} 0 &= \begin{vmatrix} a_{11} - k^2 D_1 - p & a_{12} \\ a_{21} & a_{22} - k^2 D_2 - p \end{vmatrix} \\ &= p^2 - \{a_{11} + a_{22} - k^2(D_1 + D_2)\}p + (a_{11} - k^2 D_1)(a_{22} - k^2 D_2) - a_{12}a_{21}\,. \end{aligned}$$

Spatially uniform solutions ($k = 0$) are stable only if $a_{11} + a_{22} \le 0$ and $\Delta \equiv a_{11}a_{22} - a_{12}a_{21} > 0$ (so ensuring that $\operatorname{Re} p \le 0$, for $k = 0$). Thus, for *all* k, the sum of roots $p_1 + p_2 = a_{11} + a_{22} - k^2(D_1 + D_2)$ is non-positive. Instability can arise only for the case in which both roots p_1 and p_2 are real, but $p_1 p_2 < 0$. Since $p_1 p_2 = \Delta - (D_1 a_{22} + D_2 a_{11})k^2 + D_1 D_2 k^4$, if $D_1 a_{22} + D_2 a_{11} > 0$ the

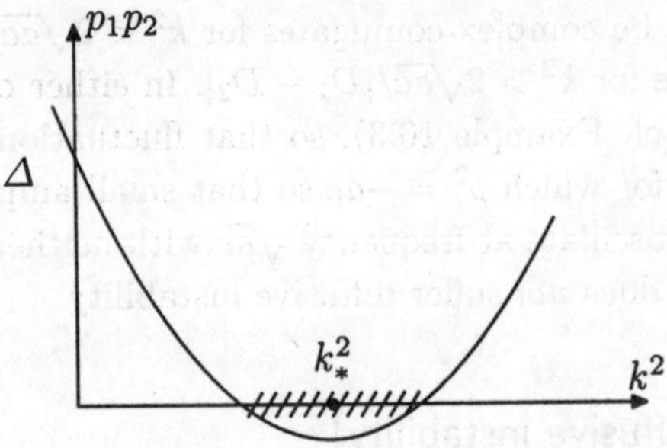

Figure 10.7 Dependence of p_1p_2 on k^2, showing (shaded) the instability window.

product p_1p_2 has a minimum at $k^2 = (D_1a_{22} + D_2a_{11})/(2D_1D_2) \equiv k_*^2$ (say) and this minimum is negative if

$$D_1a_{22} + D_2a_{11} > \sqrt{4D_1D_2\Delta}\,. \tag{10.27}$$

This inequality is the condition determining situations in which the system suffers *diffusive instability.*

If the condition (10.27) is satisfied, there is a range of values of k^2 surrounding k_*^2 for which separable solutions, of the type considered in Example 10.4 with $l^2 + m^2 = k^2$, are *unstable* (i.e. grow with time). This phenomenon arises only for $D_1 \neq D_2$, since $a_{11} + a_{22} < 0$. Thus, differences in the dispersivity can give rise to diffusive instability. Similar deductions arise widely in biological and chemical systems (and indeed in oceanography, meteorology, materials science, laser science and other physical systems). A spatially uniform state may be in equilibrium and be deemed to be stable if only spatially uniform perturbations are considered. However, fluctuations having spatial scale characterized by $2\pi/k_*$ emerge. According to the linearized analysis they grow without bound. However, a fuller more intricate, theoretical and numerical investigation of the nonlinear equations frequently reveals that growth is held in check, so that patterns of certain familiar types – stripes, hexagons (e.g. honeycombs) and spirals – emerge.

Note the similarity with the self-limitation of algae concentration provided by the nonlinear term βu^2 in Equation (10.14), for steady solutions having $u = 0$ at $x = 0, L$ with $L \approx \pi\sqrt{D/\beta u_*}$ ($k \approx \sqrt{\beta u_*/D}$). Just as those solutions are closely approximated by part of a sinusoid, one-dimensional patterns (*stripes*) may be found in many systems possessing diffusive instability, when the *instability window* is small, by seeking $\hat{u} \approx \mathrm{Re}\{U_0\mathrm{e}^{ikx}\}$, $\hat{v} \approx \mathrm{Re}\{V_0\mathrm{e}^{ikx}\}$ with $k \approx k_*$ and with small (complex) U_0 and V_0. Detailed analysis (which is beyond the scope of this text) relates the amplitudes and $k - k_*$ to small corrections to the system for $\hat{u}$ and $\hat{v}$ provided by the fully nonlinear system. For example, a superposition of three one-dimensional waves with the complex factors $\exp ikx$, $\exp i\frac{1}{2}k(-x \pm \sqrt{3}y)$ gives a function which is symmetric under rotation through angles $\pm 2\pi/3$ about the origin. It is the basic approximation to a hexagonal pattern.

Example 10.5 (Hexagonal patterns)

Show that the function $u = \cos kx + \cos \frac{1}{2}k(-x + \sqrt{3}y) + \cos \frac{1}{2}k(-x - \sqrt{3}y)$ has stationary values where $\sin kx = \sin \frac{1}{2}k(x - \sqrt{3}y) = \sin \frac{1}{2}k(x + \sqrt{3}y)$. From this deduce that the possibilities are $\cos \frac{1}{2}kx = 0$ with $\cos \frac{\sqrt{3}}{2}ky = 0$ and $\sin \frac{\sqrt{3}}{2}ky = 0$ with either $\sin \frac{1}{2}kx = 0$ or $\cos \frac{1}{2}kx = -\frac{1}{2}\cos \frac{\sqrt{3}}{2}ky$. Determine the locations of these various stationary points and evaluate u at each of them. Confirm that u is periodic in both x and y with maximum value 3, minimum value $\frac{-3}{2}$ at points distant $2\pi/k$ from the origin at angles $p\pi/3$ to the x-axis (with p integer).

Since $u = \cos kx + \cos \frac{1}{2}k(x - \sqrt{3}y) + \cos \frac{1}{2}k(x + \sqrt{3}y)$, stationary values arise where

$$0 = u_x = -k \sin kx - \tfrac{1}{2}k\{\sin \tfrac{1}{2}k(x - \sqrt{3}y) + \sin \tfrac{1}{2}k(x + \sqrt{3}y)\}$$

$$\text{and} \qquad 0 = u_y = \tfrac{\sqrt{3}}{2}k\{\sin \tfrac{1}{2}k(x - \sqrt{3}y) - \sin \tfrac{1}{2}k(x + \sqrt{3}y)\}\,.$$

Thus, $\sin \frac{1}{2}k(x - \sqrt{3}y) = \sin \frac{1}{2}k(x + \sqrt{3}y) = -\sin kx$.
The equation $u_x = 0$ then gives $\cos \frac{1}{2}kx \sin \frac{\sqrt{3}}{2}ky = 0$; while $u_y = 0$ gives $\sin \frac{1}{2}kx \cos \frac{\sqrt{3}}{2}ky + 2\sin \frac{1}{2}kx \cos \frac{1}{2}kx = 0$.

The possibilities then are $\cos \frac{1}{2}kx = 0$ which gives (since $\sin \frac{1}{2}kx \neq 0$) $\cos \frac{\sqrt{3}}{2}ky = 0$. Otherwise, $\sin \frac{\sqrt{3}}{2}ky = 0$ with *either* of $\sin \frac{1}{2}kx = 0$ *or* $\cos \frac{1}{2}kx = -\frac{1}{2}\cos \frac{\sqrt{3}}{2}ky$. The stationary points and values then are:

a) $kx = (2m + 1)\pi$ with $ky = \frac{2}{\sqrt{3}}(2n + 1)\pi$, m, n integers. At these points $\cos kx = -1$, and $\cos \frac{1}{2}k(x \mp \sqrt{3}y) = 0$ so that $u = -1$.

b) $kx = 2m\pi$ with $ky = \frac{2}{\sqrt{3}}n\pi$ so that $\cos \frac{1}{2}k(x \mp \sqrt{3}y) = (-1)^{m \mp n}$. Hence, at points with $(kx, ky) = (2m\pi, \frac{2}{\sqrt{3}}n\pi)$ the values are $u = 3$ if $m + n$ is even, $u = -1$ if $m + n$ is odd.

c) Finally, if $ky = \frac{2}{\sqrt{3}}n\pi$ so that $\cos \frac{\sqrt{3}}{2}ky = (-1)^n$ there are stationary points with $\cos \frac{1}{2}kx = \frac{-1}{2}(-1)^n$, so that $kx = 2(2m\pi \pm \pi/3)$ for n odd, but $kx = 2(2m\pi \pm 2\pi/3)$ for n even. At all these points, it is found that $u = \frac{-3}{2}$.

All terms in the definition of u have period $4\pi/k$ in x and period $\frac{4}{\sqrt{3}}\pi/k$ in y, so u is periodic in both x and y. Contours of the function $u(x, y)$ are shown in Figure 10.8, revealing the hexagonal pattern. An elementary hexagonal cell is centred at the origin, where u takes its maximum value 3. Six local minima ($u = \frac{-3}{2}$) surround the origin at distance $2\pi/k$, at angles $p\pi/3$ to the x-axis. The midpoints of the sides of the hexagon are saddles, where u has the stationary value -1.

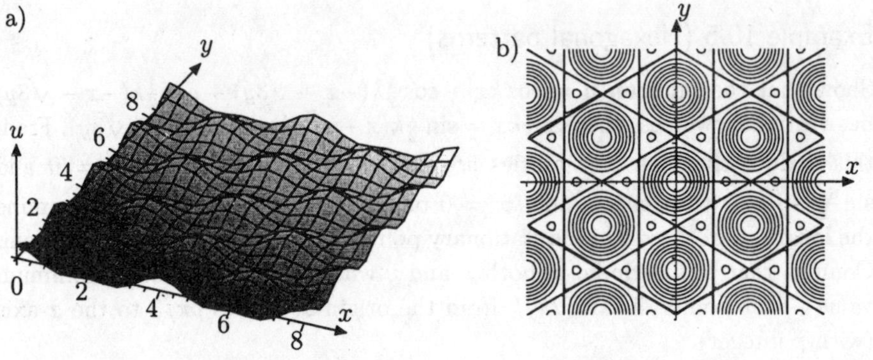

Figure 10.8 a) The function $u(x, y)$ of Example 10.5 with $k = \pi$ and b) the contours $u = -1.45, -1, -0.5, 0, 0.5, 1, 1.5, 2$ and 2.5, showing how superposition of three sinusoidal perturbations oriented at $\pm 60^o$ to each other creates a pattern of hexagonal cells.

The hexagonal patterns found in many physical and biological systems are closely approximated by those shown in Figure 10.8. Their scale is determined by $k \approx k_*$, but their location and orientation may be altered without violating the governing partial differential equations. In practice, the pattern is only approximately hexagonal, having distortion near lateral boundaries and possibly containing flaws, or dislocations, internally. However, judicious use of a complex exponential formulation (cf. treatments of waves in Chapter 8) and invoking small nonlinear corrections to the linearized analysis of unstable modes (near the threshold $k = k_*$ of diffusive instability) has proved in recent years to be very informative in describing biological, chemical and physical pattern formation – for example, in explaining why adjustment of some external (*control*) parameter may cause stripes to form in preference to hexagonal patterns of spots.

10.3 Biological Waves

10.3.1 The Logistic Wavefront

In biology, just as in chemical systems, travelling disturbances and waves are widespread. Important examples which have been treated mathematically are nerve impulse transmission (the *Hodgkin–Huxley equations* and *Fitzhugh–Nagumo equations*), the activation of heart muscle, the spread of epidemics and invasion by fungi, bacteria or intruding species. A simple, but illustrative, example uses the generalization of Fisher's equation (10.13)

$$u_t = Du_{xx} + u(u-a)(1-u) \qquad \text{with} \quad 0 < a < 1\,. \tag{10.28}$$

The final term can be taken to model a competition between dissipative effects and energy sources (e.g. chemical), with dissipation dominant for $0 < u < a$, but energy input dominant for $a < u < 1$. Observe that although Equation (10.28) defines three equilibrium states $u = 0, a, 1$, only the states $u = 0, 1$ are stable. Study of the equation $\mathrm{d}u/\mathrm{d}t = u(a-u)(1-u)$ governing spatially uniform solutions shows that $\mathrm{d}u/\mathrm{d}t < 0$ for $0 < u < a$ and $\mathrm{d}u/\mathrm{d}t > 0$ for $a < u < 1$, so that the state $u = a$ is unstable.

Equation (10.28), like (10.13), possesses *travelling wave* solutions of the form $u = U(\theta)$, with $\theta = x - ct$ (cf. Section 10.1.3) so that c is the (as yet unknown) propagation speed. These satisfy the ODE

$$DU''(\theta) + cU'(\theta) + U(U-a)(1-U) = 0$$

which, like (10.8), is usefully analysed in the *phase plane*, as follows. Define $V \equiv U'(\theta)$ and investigate the properties of curves in the (U, V) plane defined as solutions to the first-order ODE

$$\left(\frac{U''(\theta)}{U'(\theta)} =\right) \quad \frac{\mathrm{d}V}{\mathrm{d}U} = \frac{-cV - U(U-a)(1-U)}{DV}\,. \tag{10.29}$$

This equation gives unique values for $\mathrm{d}U/\mathrm{d}V$ at all points except the *critical points* identified by $V = 0$, with $cV + U(U-a)(1-U) = 0$. Thus, there are just three critical points $(0,0)$, $(a,0)$ and $(1,0)$ corresponding to the three equilibrium states. Linearization near each point $(U_0, 0)$ in turn, using $U = U_0 + \hat{u}$, gives $D\hat{u}''(\theta) + c\hat{u}'(\theta) + B\hat{u} = 0$, where $B = -a$ for $U_0 = 0$, $B = a(1-a)$ for $U_0 = a$ and $B = -(1-a)$ for $U_0 = 1$. For either of the stable equilibria $U_0 = 0, 1$, such that $B < 0$, the two exponents p in assumed solutions $\hat{u} \propto \mathrm{e}^{p\theta}$ are real and of opposite sign. Denote these by $p_0^{\pm}$ and $p_1^{\pm}$, respectively. Then, a travelling front connecting a state with $U \to 0$ as $\theta \to +\infty$ to a state $U \to 1$ as $\theta \to -\infty$ will have $(U, V) \propto (1, p_0^-)\mathrm{e}^{p_0^- \theta}$ as $\theta \to +\infty$ and $(U-1, V) \propto (1, p_1^+)\mathrm{e}^{p_1^+ \theta}$ as $\theta \to -\infty$.

The required solution $U(\theta)$ corresponds to a solution curve to (10.29) connecting the points $(0,0)$ and $(1,0)$. This exists only when the curve which leaves $(0,0)$ with $\mathrm{d}V/\mathrm{d}U = p_0^-$ reaches the point $(1,0)$ (with slope $\mathrm{d}V/\mathrm{d}U = p_1^+$). This can happen only for specially chosen values of c. Fortuitously, for equation (10.29) these values may be obtained explicitly. Since $V = 0$ at $U = 0, 1$, make the *guess* $V = kU(1-U)$, observing that the right-hand side of (10.29) then simplifies to $-\{c + k^{-1}(U-a)\}/D$. This agrees with the linear function $\mathrm{d}V/\mathrm{d}U = k(1-2U)$ when $-(Dk)^{-1} = -2k$ and $(ak^{-1} - c)/D = k$. The choice $k < 0$ (so that $\mathrm{d}U/\mathrm{d}\theta < 0$) gives $k = -(2D)^{-1/2}$ with $c = (a - \frac{1}{2})/k = -(a-\frac{1}{2})\sqrt{2D}$, so that the required solution to (10.29) is

$$V = kU(1-U) \qquad \left(= \frac{\mathrm{d}U}{\mathrm{d}\theta}\right).$$

This first-order ODE for $U(\theta)$ is also readily solved, so showing that the profile of the travelling wave is the logistic curve (cf. Section 10.2.2)

$$u(x,t) = U(\theta) = \frac{1}{1 + \mathrm{e}^{\sqrt{2D}(\theta - \theta_0)}} \tag{10.30}$$

with $\theta - \theta_0 = x - \theta_0 - ct$, in which the speed is $c = -(a - \frac{1}{2})\sqrt{2D}$ and θ_0 is arbitrary. Notice that $c > 0$ for $a < \frac{1}{2}$, but $c < 0$ for $a > \frac{1}{2}$. Thus, the wavefront advances into the region $u \approx 0$ when the interval $a < u < 1$ of energy input dominates the interval $0 < u < a$ of dissipation, while, for $a > \frac{1}{2}$ so that the interval of dissipation dominates, the front recedes into the region where $u \approx 1$. (Observe that the choice $k = (2D)^{-1/2}$ corresponds to $U \to 0$ as $\theta \to -\infty$ and $U \to 1$ as $\theta \to +\infty$. Although the sign of c is also reversed, the front still advances *into* the region $u \approx 0$, for $a < \frac{1}{2}$.)

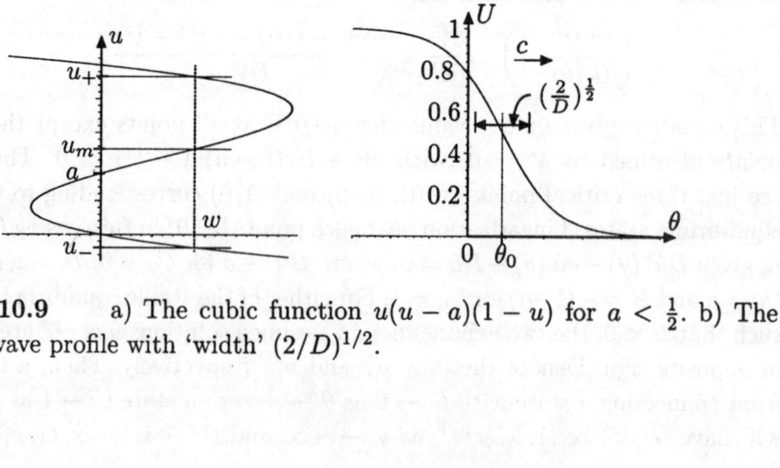

Figure 10.9 a) The cubic function $u(u-a)(1-u)$ for $a < \frac{1}{2}$. b) The logistic wave profile with 'width' $(2/D)^{1/2}$.

10.3.2 Travelling Pulses and Spiral Waves

The travelling solution (10.30) to (10.28) reveals how an irreversible change from state $u = 0$ to $u = 1$ (or vice versa) might advance. It shows, however, that Equation (10.28) is too simple to describe phenomena such as nerve impulses, in which a succession of spikes or pulses propagate. Physiologically, a nerve permits transmission of pulses of electrical potential associated with transport of ions (electrically charged molecules) of potassium and sodium through a cylindrical membrane, or sheath. The membrane permeabilities for the two species are sensitive to the electric potential, so allowing the chemical balance to recover, on a timescale significantly longer than the duration of the pulse. A system including this phenomenon is

$$u_t = Du_{xx} + u(u-a)(1-u) - w\,, \tag{10.31}$$

$$w_t = \gamma(bu - w)\,, \tag{10.32}$$

known as the Fitzhugh–Nagumo equations.[4]

Observe that, when w is treated as constant, Equation (10.31) is very similar to (10.28), since $u(u-a)(1-u) - w$ is still a cubic function of u. For small values of w, the cubic has zeros at $u = u_-$, u_{m} and u_+ (see Figure 10.9), with $u_\pm$ being stable equilibria for spatially uniform solutions. Moreover, for $u_{\mathrm{m}} > \frac{1}{2}(u_- + u_+)$ the wave will travel *into* the region where $u \approx u_+$. In the system (10.31) and (10.32), the state $u = 0$ is consistent with $w = 0$ (and, moreover, for $b > \frac{1}{4}(1-a)^2$ the only equilibrium is $(u, w) = (0, 0)$). Assume that $a < \frac{1}{2}$ and that a front through which u increases towards $u = 1$ is advancing in the direction of increasing x. This front disturbs w from its value $w = 0$, driving w towards the value $w = bu$ on a timescale typified by γ^{-1}. As w increases, the values $u_\pm$ will decrease, but u_{m} will increase (see Figure 10.9a). This both perturbs the profile from the logistic curve of Figure 10.9b by reducing the upper value $u_+(w)$ and also brings the system into a situation in which a wave of decrease from u_+ to u_- may advance. Thus, the zone of elevated values of u will have limited duration. Moreover, as u reduces, (10.32) causes w to reduce also. This reduction, though initially rapid, eventually (with $u \approx 0$) becomes exponential, with decay rate γ. In physiological cases, this *recovery* process is relatively slow when compared with the time over which most of a transition from $u = 0$ to $u \approx 1$ takes place (cf. $2|c\sqrt{2D}|^{-1} = D^{-1}|a - \frac{1}{2}|^{-1}$, for (10.30)). Thus, a nerve impulse can propagate with a well-defined profile, but followed by a recovery length within which no following pulse can approach. This allows

[4] These equations were proposed by Fitzhugh in 1961 and then investigated by Nagumo and others. They are a substantial mathematical simplification of those devised by Hodgkin and Huxley in 1952, following careful experimentation in a physiological laboratory.

the nerve to carry a patterned sequence of spiked pulses, which are prevented from colliding with one another.

An individual pulse is a travelling wave, which may be analysed much as in Section 10.1.3. Seek solutions $u = U(\theta)$, $w = W(\theta)$ with $u_x = U'(\theta) \equiv V(\theta)$ and $\theta \equiv x - ct$, so that the system (10.31) and (10.32) yields the ODE system

$$\begin{aligned} U'(\theta) &= V\,, \\ V'(\theta) &= D^{-1}\{U(U-a)(U-1) - cV + W\}\,, \\ W'(\theta) &= (\gamma/c)(-bU + W)\,. \end{aligned} \tag{10.33}$$

The desired pulse corresponds to a solution in which $(U, V, W) \to (0, 0, 0)$ for both $\theta \to -\infty$ and $\theta \to \infty$ (in the language of dynamical systems, a *homoclinic orbit*, Arrowsmith and Place, 1992). Just as the solution (10.30) (describing curves $(U(\theta), V(\theta))$ joining two distinct critical points $(0, 0)$ and $(1, 0)$ in the phase plane – a *heteroclinic orbit*) exists only for certain values of c $(= \pm(a - \frac{1}{2})\sqrt{2D})$, pulselike solutions to (10.33) exist only for selected values of c. These values and corresponding solutions cannot be determined analytically. They are found by careful numerical integration of the system (10.33), using the fact that, as $\theta \to +\infty$, the behaviour is $(U, V, W) \sim (1, p_-, \gamma b/(\gamma - cp_-))\,e^{p_- \theta}$, where $p = p_-$ is the single negative root of the equation

$$(Dp^2 + cp - a)(cp - \gamma) + \gamma b = 0$$

arising for solutions $\propto e^{p\theta}$ to the linearization $DU'' + cU' - aU + W = 0$, $cW' = \gamma(W - bU)$ of the system (10.33). A solution for typical parameter values is shown in Figure 10.10.

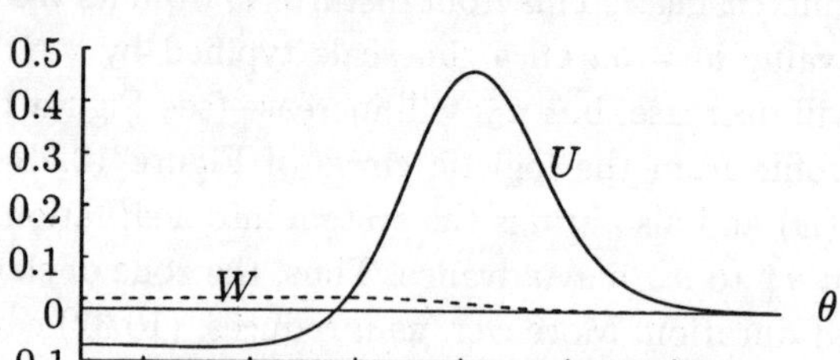

Figure 10.10 The variables $U(\theta)$ and $W(\theta)$ for a nerve pulse governed by the Fitzhugh–Nagumo equations (10.31) and (10.32).

The phenomenon of gradual recovery following passage of a pulse or 'front' is quite widespread in biological, ecological and chemical systems. In two dimensions, it gives rise to the beautiful phenomenon of spiral waves. If, in (10.31) and (10.32), the term u_{xx} is generalized to $\nabla^2 u$, the new system governing $u(\boldsymbol{x}, t)$ and $w(\boldsymbol{x}, t)$ possesses solutions depending on only $\phi \equiv \theta - \omega t$

and r, where (r, θ) are plane polar coordinates. Any such solution describes patterns $u = u(r, \phi)$, $w = w(r, \phi)$ which rotate about the origin at (radian) frequency ω (the functions u and w must be 2π-periodic in ϕ). If, at large radii r, the variables u, w are well approximated by functions of the single variable $kr + (\theta - \omega t) = kr + \phi$, then, at each angle θ, they become periodic also in r. Hardly surprisingly, the behaviour at large r is like a succession of nerve impulses. This, however, is just the behaviour as $r \to \infty$. At finite r, the solution is essentially two-dimensional and must be determined as a solution to the coupled system of PDEs

$$D(u_{rr} + r^{-1}u_r + r^{-2}u_{\phi\phi}) + \omega u_\phi + u(u-a)(1-u) - w = 0\,,$$
$$\omega w_\phi + \gamma b u - \gamma w = 0$$

(with bounded behaviour as $r \to 0$).

More generally, systems of Volterra–Lotka type allow solutions depending only upon r and $\phi \equiv \theta - \omega t$ (with the origin $r = 0$ arbitrarily chosen in the plane). The system given in Example 10.4 then becomes

$$D_1(u_{rr} + r^{-1}u_r + r^{-2}u_{\phi\phi}) + \omega u_\phi + f(u, v) = 0\,,$$
$$D_2(v_{rr} + r^{-1}v_r + r^{-2}v_{\phi\phi}) + \omega v_\phi + g(u, v) = 0\,.$$

For wavenumbers $k \approx k_*$ in a window of diffusive instability, these may allow spiral waves which, as $r \to \infty$, are approximately periodic in $kr \pm \phi$ (the curves $kr \pm \theta =$ constant are Archimedean spirals, clockwise or anticlockwise). As these spiral waves rotate, they appear to spread. The instability mechanism is the means by which chemical energy is 'supplied' to the expanding pattern. The spirals maintain their shape remarkably well up to the zones in which they encounter other spirals. Such patterns of interfering spirals are now regarded as a paradigm in biological and chemical systems, so that their continuum modelling and their numerical computation and display has recently been investigated widely. This, like the fibre optics outlined in Chapter 9, is an instance where fields and waves have much relevance to modern science.

EXERCISES

10.3. Confirm that, if $u(x, t)$ satisfies $u_t = Du_{xx} + \alpha u$, then $c(x, t) \equiv e^{-\alpha t}u(x, t)$ satisfies $c_t = Dc_{xx}$. Use the properties (see exercise 10.1) of the function $c(x, t) = (\pi Dt)^{-1/2}\exp(-x^2/4Dt)$ to show that, for all $t > 0$, the function $u(x, t)$ given in Equation (10.12) describes one-dimensional blooming and that the total population grows exponentially according to $\int_{-\infty}^{\infty} u(x, t)\,dx = Me^{\alpha t}$.

10.4. For a modified predator–prey model

$$\dot{X} = aX - \alpha X^2 - bXY \ , \qquad \dot{Y} = -cY + dXY + \beta Y^2$$

with each of a, b, c, d, α and β positive, determine the regions within $X > 0, Y > 0$ where (a) $\dot{X} > 0$, (b) $\dot{X} < 0$, (c) $\dot{Y} > 0$ and (d) $\dot{Y} < 0$. Show that, if $\alpha < ad/c$ and $\beta < cb/a$ exactly one equilibrium state (X_0, Y_0) exists in $X > 0$, $Y > 0$. Find it.

10.5. Using the fact that, in polar coordinates (r, θ), the Bessel function $\phi = J_0(kr)$ satisfies $\nabla^2\phi \equiv \phi_{rr} + r^{-1}\phi_r = -k^2\phi$ (c.f. (9.30)), deduce that (10.24) possesses axisymmetric solutions $\hat{u} = J_0(kr)U(t)$, $\hat{v} = J_0(kr)V(t)$ whenever $U(t)$ and $V(t)$ satisfy the system (10.25).

10.6. Using Taylor series expansions of $f(u, v)$ and $g(u, v)$ about the values (u_0, v_0), derive the expressions quoted in Example 10.4 for each a_{ij}.

Verify the values $a_{11} = -\alpha u_0$, $a_{12} = -bu_0$, $a_{21} = dv_0$, $a_{22} = 0$, $u_0 = c/d$ and $v_0 = (ad - \alpha c)/bd$ for the system

$$u_t = D_1\nabla^2 u + au - \alpha u^2 - buv \ , \quad v_t = D_2\nabla^2 v - cv + duv$$

(cf. Example 10.3). Confirm that in this case the quadratic equation defining p in Example 10.4 has no roots with positive real part.

10.7. Show that if the coefficients at (u_0, v_0) arising from the Volterra-Lotka model in Example 10.4 are $D_1 = 1$, $D_2 = 8$; $a_{11} = 1$, $a_{12} = -1$, $a_{21} = 3$ and $a_{22} = -2$, verify that $k_*^2 = \frac{3}{8}$ and that diffusive instability arises only for $\frac{1}{4} < k^2 < \frac{1}{2}$. At $k = k_*$, show that the growth rate of the instability is $p = 0.02839$.

10.8. Show that, if a system such as (10.24) has solutions $\hat{u} = U \cos kx\, e^{\imath\omega t}$, $\hat{v} = V \cos kx\, e^{\imath\omega t}$ for a certain pair of values (ω, k), then also it possesses solutions $\hat{u} = U\cos(k_1x + k_2y)e^{\imath\omega t}$, $\hat{v} = V\cos(k_1x + k_2y)e^{\imath\omega t}$, whenever $k_1^2 + k_2^2 = k^2$.

By superposing two such solutions to give $\hat{u} = U(\cos kx + \cos ky)e^{\imath\omega t}$, show that an oscillating chequerboard pattern is obtained (identify the locations where $\cos kx + \cos ky$ is maximum, minimum and zero). What is the equivalent pattern for $\hat{v}(x, y, t)$?

10.9. Show that, if U remains finite, the solution of $(10.33)_3$ with $W \to 0$ as $\theta \to \infty$ may be written as

$$W(\theta) = \frac{b\gamma}{c} e^{\gamma\theta/c} \int_0^\infty \exp\frac{-\gamma s}{c} U(s)\, ds \ . \qquad (10.34)$$

By approximating the $U(\theta)$ pulse by the ramp function $U(\theta) = -\sqrt{2D}\theta$ for $-(2D)^{-1/2} \le \theta \le 0$ with $U = 0$ for $\theta > 0$ and $U = 1$ for $\theta < -(2D)^{-1/2}$, obtain the equivalent approximation for $W(\theta)$ (either from (10.34), or by direct integration of $(10.33)_3$). Confirm that it has $W(\theta) = 0$ in $\theta \ge 0$, that $W(-(2D)^{-1/2}) = b(\mathrm{e}^{-\Gamma} - 1 + \Gamma)/\Gamma \equiv W_1$, where $\Gamma \equiv (\gamma/c)(2D)^{-1/2}$. Confirm also that, even though $W \to b$ exponentially as $\theta \to -\infty$, if $\Gamma \ll 1$ then $W/b \ll 1$ at the end $\theta = -(2D)^{-1/2}$ of the ramp.

Solutions

Chapter 1

1.1 Mass $= ab\int_0^h \rho\, dz = abAh - abB\alpha^{-1}\left(1 - e^{-\alpha h}\right)$;
Solve $\rho_0 = A + B$, $\rho_1 = A + Be^{-\alpha h}$ for ρ_0 and ρ_1.

1.2 (a)Temperature *drops* ΔT_1 and ΔT_2 satisfy $q = K_1\Delta T_1/h_1 = K_2\Delta T_2/h_2$. Hence $\Delta T_1 = h_1 q/K_1$, $\Delta T_2 = h_2 q/K_2$. (b) Thus, $\Delta T \equiv \Delta T_1 + \Delta T_2$, so that $\Delta T = q(h_1K_2 + h_2K_1)/K_1K_2$ gives $q = K_1K_2\Delta T/(h_1K_2 + h_2K_1)$.

1.3 (a) In each slab, where $K = K_1$, temperature drop is $\Delta T_1 = qh/K_1$.
(b) Across the air layer, $\Delta T_2 = qH/K_2$, so $\Delta T = q(2hK_1^{-1} + HK_2^{-1})$.

1.4 Inserting the data gives $\Delta T = q(2h+10H)/K_1 = 12hq/K_1$, so $q = K_1\Delta T/(12h)$. For a single pane of thickness $2h$, the flux is $q_0 = K_1\Delta T/(2h)$.
Then $q/q_0 = \frac{1}{6} < 20\%$.

1.5 From $-KT'(r)2\pi r = Q$ for $a < r < b$, it follows that

$$T_0 - T_1 = T(a) - T(b) = \frac{-Q}{2\pi K}(\ln a - \ln b) = \frac{Q}{2\pi K}\ln(b/a)\ .$$

For $b = a + h$, use $\ln(b/a) = \ln(1 + h/a) \approx h/a$ to give $Q \approx 2\pi aK(T_0 - T_1)/h$.

1.6 Writing $K(T_0 - T_1)/h \equiv \hat{Q}$ (the heat flow rate through unit area of slab) gives $Q = 2\pi a\hat{Q}$, where $2\pi a$ is the area of unit length of cylinder of radius a.
Using $\ln(1 + h/a) \approx h/a - \frac{1}{2}h^2/a^2$ gives $Q \approx 2\pi Kah^{-1}(T_0 - T_1)(1 - h/2a)^{-1}$ $\approx 2\pi Kah^{-1}(T_0 - T_1)(1 + h/2a) = \pi Kh^{-1}(T_0 - T_1)(a + b)$. Since this equals $\pi(a + b)\hat{Q}$, it is the flux through the stated area of slab.

1.7 For pipe wall, $T_0 - T(b) = Q_2(2\pi K)^{-1}\ln(b/a)$;
for cladding $T(b) - T_1 = Q_2(2\pi K_2)^{-1}\ln(c/b)$.
Thus, $2\pi(T_0 - T_1) = Q_2\{K^{-1}\ln(b/a) + K_2^{-1}\ln(c/b)\}$,
giving $Q_2/Q = \{1 + (K/K_2)\ln(c/b)/[\ln(b/a)]\}^{-1}$.

1.8 From $Q = 4\pi r^2 q(r) = -4\pi r^2 KT'(r)$, obtain $T(r) - T(a) = Q(r^{-1} - a^{-1})/4\pi K$. Then put $r = b$, giving $Q = 4\pi K(T_1 - T_2)/(a^{-1} - b^{-1}) = 4\pi Kab(T_1 - T_2)/(b - a)$. Inserting this value for $Q/4\pi K$ gives $T(r) = \{ab(T_1 - T_2)r^{-1} + bT_2 - aT_1\}/(b - a)$.

1.9 Rate of supply of heat to sphere of radius a is $\frac{4}{3}\pi a^3 S_0 = 4\pi a^2 q(a) = Q$. Hence, $T_1 - T_2 = (b - a)Q/(4\pi Kab) = (b - a)a^2 S_0/(3bK)$.

1.10 Radial heat flux is $-K_cT'(r) = 2\alpha K_cT_cre^{-\alpha r^2}$.
Then $S(r) = r^{-2}2\alpha K_cT_c(r^3e^{-\alpha r^2})' = 2\alpha K_cT_c(3-2\alpha r^2)e^{-\alpha r^2}$. Let Q be total heat outflow rate from core, then $Q = 4\pi a^2[-K_cT'(a)] = 8\pi a^3\alpha K_cT_ce^{-\alpha a^2}$. Use this in Exercise 1.8 for $a < r < b$, giving
$T(b) = T(a) - Q(2\pi K_c)^{-1}(a^{-1} - b^{-1}) = \{1 - 4\alpha a^2(b-a)/b\}T_ce^{-\alpha a^2}$.

Chapter 2

2.1 (a) Heat equation becomes $Ax+B = \nu(6Cx+2D)$, so giving $A = 6\nu C$, $B = 2\nu D$ with C, D, E and F arbitrary.
(b) From $Ae^{At+Bx+C} = \nu B^2e^{At+Bx+C}$, it follows that $A = \nu B^2$ with B and C arbitrary.
(c) From $Ae^{At}\cos(Bx+C) = -\nu B^2e^{At}\cos(Bx+C)$, it follows that $A = -\nu B^2$, again with B and C arbitrary.
(d) $ABe^{Bt}\cos Cx + DEe^{Et}\sin Fx = -\nu AC^2e^{Bt}\cos Cx - \nu DF^2e^{Et}\sin Fx$. Comparing coefficients gives $B = -\nu C^2$ (for $A \neq 0$) and $E = -\nu F^2$ (for $D \neq 0$). C, F, A and D are arbitrary.
(e) $\partial T/\partial t = A(Bt^{B-1} - x^2t^{B-2}/C)\exp(x^2/Ct)$ and
$\partial^2T/\partial x^2 = A(2C^{-1}t^{B-1} + 4x^2t^{B-2}/C^2)\exp(x^2/Ct)$. Comparing coefficients in $\partial T/\partial t = \nu\partial^2T/\partial x^2$ gives $B = \nu 2C^{-1}$, $-C^{-1} = 4\nu C^{-2}$ (assuming $A \neq 0$). Hence $C = -4\nu$, $B = -\frac{1}{2}$, giving solutions $T(x,t) = At^{-1/2}\exp(-x^2/4\nu t)$.

2.2 Differentiate term-by-term, giving e.g.

$$\partial T/\partial t = \sum_{n=1}^{\infty}\frac{-\nu n^2\pi^2}{h^2}b_ne^{-\nu n^2\pi^2t/h^2}\sin\frac{n\pi x}{h}.$$

(a) $T(h/2,0) = 32\pi^{-3}(1-\frac{1}{27}) = 0.9938$. (The graph is symmetric about $x = \frac{1}{2}h$ and is close to the parabola $f(x)$.)
(b) When $\nu t = h^2\pi^{-2}$, $T = 32\pi^{-3}\{e^{-1}\sin(\pi x/h) + \frac{1}{27}e^{-9}\sin(3\pi x/h)\}$. The term involving b_3 has (between $t = 0$ and $t = h^2\nu^{-1}\pi^{-2}$) reduced in relative importance by a factor $e^{-8} \approx 0.000335$.
(c) Generally, $T(x,t)$ eventually reduces approximately to the decaying, sinusoidal distribution $b_1e^{-\nu\pi^2t/h^2}\sin(\pi x/h)$.

2.3 For all p, $T(0,t) = 0$. Since $\partial T/\partial x = pe^{-\nu p^2t}\cos px$, setting $x = h$ gives $pe^{-\nu p^2t}\cos ph = 0$, for all t. Thus, for $p \neq 0$, this gives $\cos ph = 0$; hence $ph = \frac{1}{2}\pi + n\pi$, or $p = (n+\frac{1}{2})\pi/h$, for n integer. Hence, superposing gives

$$T(x,t) = \sum_{n=0}^{\infty}b_n\exp\left(\frac{-\nu(2n+1)^2\pi^2t}{4h^2}\right)\sin(2n+1)\frac{\pi x}{2h}.$$

2.4 $\partial T/\partial t = \omega v(x)\cos\omega t - \omega u(x)\sin\omega t = \nu\partial^2T/\partial x^2 = \nu u''(x)\cos\omega t + \nu v''(x)\sin\omega t$. Comparing coefficients: $u''(x) = \omega\nu^{-1}v$, $v''(x) = -\omega\nu^{-1}u$; $\Rightarrow u^{(4)}(x) = -\omega^2\nu^{-2}u$. Seeking $u \propto e^{kx}$ gives $k^4 = -\omega^2/\nu^2$. Roots are $k = k_0\exp i(\frac{\pi}{4} + n\frac{\pi}{2})$, where $k_0 \equiv \sqrt{\omega/\nu}$. Write $k_0 = \sqrt{2}\lambda$, so the four distinct roots are $k = \lambda(1\pm i)$ (for which $|e^{kx}|$ *decays* as $x \to -\infty$) and $k = \lambda(-1\pm i)$ (for which $|e^{kx}|$ *grows* as $x \to -\infty$). Thus, bounded solutions in $x < 0$ are found using
$u = e^{\lambda x}(Ae^{i\lambda x} + Be^{-i\lambda x})$, $v = (\nu/\omega)u''(x) = e^{\lambda x}(iAe^{i\lambda x} - iBe^{-i\lambda x})$.
Setting $T(0,t) = T_0\cos\omega t$ gives $u(0) = T_0$, $v(0) = 0$, so that $B = A = \frac{1}{2}T_0$. Hence, $u(x) = T_0e^{\lambda x}\cos\omega x$, $v(x) = -T_0e^{\lambda x}\sin\omega x$, giving the required solution

$$T(x,t) = T_0e^{\lambda x}\cos(\omega t + \lambda x) \qquad \text{where} \quad \lambda = \sqrt{\omega/2\nu}.$$

At $x = 0$, inward flux $= KT_x(0,t) = K\lambda T_0 \cos(\omega t + \frac{1}{4}\pi)$, which *precedes* the temperature cycle by $\frac{1}{8}\times$period (i.e. by 3 hours). (Alternatively, the temperature cycle *lags* 3 hours behind the cycle of flux.)

2.5 Rate of heat loss $= -Mc_wT'(t) = \beta(T-20)$. Solving, with $T(0) = 98$ gives $T - 20 = 78\exp(-\beta t/Mc_w)$. If $t = t_1$ when $T = 36^\circ$C, then $16 = 78\exp(-\beta t_1/Mc_w)$ so that cooling to 36°C requires time $t_1 = Mc_w\beta^{-1}\ln 4.875$.

2.6 Solve $-KT''(x) = S_0$ in $0 < x < a$ with $T'(0) = 0$ and solve $-KT''(x) = 0$ in $a < x < 4a$ with $T(4a) = T_1$, keeping T and T' continuous at $x = a$. In $0 < x < a$, $KT'(x) = -S_0x$ so $T(x) = T_0 - \frac{1}{2}S_0x^2/K$. At $x = a$; $T(a) = T_0 - \frac{1}{2}S_0a^2/K$, $KT'(a) = -S_0a$, so that $T'(x) = -S_0a/K$ in $(a, 4a)$. Thus $T(x) = T_1 + aS_0K^{-1}(4a - x)$ in $[a, 4a]$. Solving $T_0 - \frac{1}{2}a^2S_0K^{-1} = T(a) = T_1 + 3a^2S_0K^{-1}$ for T_0 gives $T(x) = T_1 + \frac{1}{2}(7a^2 - x^2)S_0K^{-1}$ in $[0, a]$.

2.7 (a) Solving $K\hat{T}''(x) = -S_0$, $\hat{T}(0) = T_0 = \hat{T}(h)$ gives $\hat{T} = T_0 + \frac{1}{2}S_0K^{-1}x(h-x)$.

(b) Since $\partial u/\partial t - \nu\partial^2u/\partial x^2 = \partial T/\partial t - \nu\partial^2T/\partial x^2 + \nu\hat{T}''(x) = S_0(\rho c)^{-1} - \nu S_0K^{-1} = 0$ (using $\nu = K/(\rho c)$) with $u(0,t) = 0 = u(h,t)$, the solution for $u(x,t)$ has the representation given in (2.11).

(c) The initial condition for u is $u(x,0) = T_0 - \hat{T}(x) = \frac{1}{2}S_0K^{-1}(x^2 - hx)$. Hence the coefficients are given by

$$\frac{2K}{S_0}b_n = \frac{2}{h}\int_0^h (x^2 - hx)\sin\frac{n\pi x}{h}\,dx = 0 + \frac{2}{n\pi}\int_0^h (2x-h)\cos\frac{n\pi x}{h}\,dx$$

$$= 0 - \frac{2h}{(n\pi)^2}\int_0^h 2\sin\frac{n\pi x}{h}\,dx = \frac{4h^2}{(n\pi)^3}\left[\cos\frac{n\pi x}{h}\right]_0^h$$

$$= \begin{cases} \dfrac{-8h^2}{(n\pi)^3}, & n \text{ odd},\\ 0, & n \text{ even}.\end{cases}$$

$$T(x,t) = T_0 + \frac{S_0}{2K}x(h-x) - \frac{4h^2S_0}{K\pi^3}\left\{e^{-\nu\pi^2t/h^2}\sin\frac{\pi x}{h} + \frac{1}{27}e^{-9\nu\pi^2t/h^2}\sin\frac{3\pi x}{h} + \dots\right\}.$$

2.8 Write $rT = u(r,t)$, so that $\partial u/\partial t = \nu\partial^2u/\partial r^2$ (the familiar one-dimensional heat equation).

(a) For $u = e^{(Ar+\nu A^2t)}$, $\partial u/\partial t = \nu A^2e^{(Ar+\nu A^2t)} = \nu\partial^2u/\partial r^2$; hence $T = r^{-1}e^{(Ar+\nu A^2t)}$ is possible.

(b) Since $u = e^{-\nu p^2t}\sin pr$ satisfies $\partial u/\partial t = \nu\partial^2u/\partial r^2$ for all p, the function $r^{-1}u = r^{-1}e^{-\nu p^2t}\sin pr$ is a spherically symmetric solution for $T(r,t)$.

(c) Consider $u = r\{A(r + 2\nu t/r) + Br^{-1} + C\} = Ar^2 + 2\nu At + B + Cr$. Since $\partial u/\partial t = 2\nu A = \nu\partial^2u/\partial r^2$, the given function is a possible spherically symmetric form for $T(r,t)$.

2.9 Solve $-K(r\hat{T})'' = rS_0$ in $0 < r < a$; with $r\hat{T} = 0$ at $r = 0$ and $\hat{T}'(a) = BK^{-1}(T_0 - \hat{T}(a))$.
Then $-Kr\hat{T}(r) = \frac{1}{6}r^3S_0 + KAr$ (say) (using the condition at $r = 0$). From $\hat{T}'(r) = \frac{1}{6}S_0K^{-1}r^2 - A$ and $\hat{T}'(a) = \frac{1}{3}S_0K^{-1}a$, then $T_0 + A = \frac{1}{3}aS_0(B^{-1} + a/2K)$. Hence the steady temperature is $\hat{T}(r) = T_0 - \frac{1}{3}aS_0B^{-1} + \frac{1}{6}S_0K^{-1}(r^2 - a^2)$.
From $-K\partial\bar{T}/\partial r - K\hat{T}' = q = B(\bar{T} + \hat{T} - T_0)$ at $r = a$, it follows that

$\partial\bar{T}/\partial r = -BK^{-1}\bar{T}$ at $r = a$. Applying this to the separable solutions (see Exercise 2.8(b)) $\bar{T} = r^{-1}\mathrm{e}^{-\nu p^2 t}\sin pr$ gives the equation defining pa as

$$K(\sin pa + pa\cos pa) = aB\sin pa\,, \qquad \text{or} \qquad \tan pa = pa/(aBK^{-1}-1)\,.$$

(N.B. Graphical construction soon convinces that, for all values of $(aBK^{-1}-1)$, there is an infinite set of solutions for pa (and hence, for p).)

Chapter 3

3.1 $\nabla T = \boldsymbol{i}2Ax + \boldsymbol{j}2Ay + \boldsymbol{k}B$. Thus, $\boldsymbol{q} = -2KA(x\boldsymbol{i}+y\boldsymbol{j}) - KB\boldsymbol{k}$. The flux $-KB\boldsymbol{k}$ is uniform and parallel to Oz. The flux $-2KA(x\boldsymbol{i}+y\boldsymbol{j})$ at $\boldsymbol{x} = x\boldsymbol{i}+y\boldsymbol{j}+z\boldsymbol{k}$ points radially towards the Oz axis, with magnitude $2AK\sqrt{x^2+y^2}$.

3.2 Since $\Phi(\boldsymbol{x}-\boldsymbol{x}_n) = -G|\boldsymbol{x}-\boldsymbol{x}_n|^{-1}$, the expression for ϕ is $\sum -Gm_n|\boldsymbol{x}-\boldsymbol{x}_n|^{-1}$, and so describes the gravitational potential of a system of masses m_n at $\boldsymbol{x} = \boldsymbol{x}_n$.

(a) $\boldsymbol{F} = -\nabla\phi = -MG\boldsymbol{x}/|\boldsymbol{x}|^3$; the field due to a single mass M at $\boldsymbol{x} = \boldsymbol{0}$.

(b) This is the potential of M_1 at $\boldsymbol{x} = \boldsymbol{x}_1$ and M_2 at $\boldsymbol{x} = \boldsymbol{x}_2$, with gravitational field

$$\boldsymbol{F} = -M_1G\frac{\boldsymbol{x}-\boldsymbol{x}_1}{|\boldsymbol{x}-\boldsymbol{x}_1|^3} - M_2G\frac{\boldsymbol{x}-\boldsymbol{x}_2}{|\boldsymbol{x}-\boldsymbol{x}_2|^3}\,.$$

(c) The gravitational field $\boldsymbol{F} = -\boldsymbol{\nabla}\phi$ is

$$\boldsymbol{F} = -\frac{mG(\boldsymbol{x}-a\boldsymbol{i})}{|\boldsymbol{x}-a\boldsymbol{i}|^3} - \frac{mG(\boldsymbol{x}+a\boldsymbol{i})}{|\boldsymbol{x}+a\boldsymbol{i}|^3} - \frac{MG(\boldsymbol{x}-b\boldsymbol{j})}{|\boldsymbol{x}-b\boldsymbol{j}|^3} - \frac{MG(\boldsymbol{x}+b\boldsymbol{j})}{|\boldsymbol{x}+b\boldsymbol{j}|^3}\,.$$

Two masses m are at $\boldsymbol{x} = \pm a\boldsymbol{i}$, while two masses M are at $\boldsymbol{x} = \pm b\boldsymbol{j}$. At $\boldsymbol{x} = \boldsymbol{0}$, $\boldsymbol{F} = \boldsymbol{0}$, the masses m produce zero force at $\boldsymbol{0}$ (by symmetry) and likewise the forces due to M at $\boldsymbol{x} = \pm b\boldsymbol{j}$ cancel.

3.3 Potential is the sum $-m_1G|\boldsymbol{x}-a\boldsymbol{i}|^{-1} - m_2G|\boldsymbol{x}-(-a)\boldsymbol{i}|^{-1}$, so writing $x^2+y^2+z^2 = r^2$ in $|\boldsymbol{x}-a\boldsymbol{i}| = \{(x-a)^2+y^2+z^2\}^{1/2}$ gives

$$|\boldsymbol{x}-a\boldsymbol{i}| = (r^2-2ax+a^2)^{1/2}\,,\quad |\boldsymbol{x}+a\boldsymbol{i}| = (r^2+2ax+a^2)^{1/2}\,.$$

Hence

$$\begin{aligned}\phi &= -G\left\{m_1(r^2-2ax+a^2)^{-1/2} + m_2(r^2+2ax+a^2)^{-1/2}\right\}\\ &= -r^{-1}G\left\{m_1\left(1-2a\frac{x}{r^2}+\frac{a^2}{r^2}\right)^{-1/2} + m_2\left(1+2a\frac{x}{r^2}+\frac{a^2}{r^2}\right)^{-1/2}\right\}\\ &= -r^{-1}G\left\{m_1\left(1+a\frac{x}{r^2}+\ldots\right) + m_2\left(1-a\frac{x}{r^2}+\ldots\right)\right\}\end{aligned}$$

(binomial theorem, N.B. $|x/r| \leq 1$).

$$\text{Thus}\quad \phi \approx -\frac{m_1G}{r}\left(1+\frac{ax}{r^2}\right) - \frac{m_2G}{r}\left(1-\frac{ax}{r^2}\right) = -\frac{(m_1+m_2)G}{r} - \frac{(m_1-m_2)Gax}{r^3}\,.$$

(The first term is equivalent to a mass m_1+m_2 at $\boldsymbol{x} = \boldsymbol{0}$; the second describes a *dipole*, of strength $(m_1-m_2)a$, located at $\boldsymbol{x} = \boldsymbol{0}$ and oriented along Ox.)

3.4 Let $\mathcal{S}_\pm$ denote the discs $z = \pm b$, $R \equiv \sqrt{x^2+y^2} < a$, with respective unit normals $\pm\boldsymbol{k}$. Since $\boldsymbol{\nabla}\phi = \boldsymbol{x}/r^3$, then on $\mathcal{S}_\pm$ the integrand $\boldsymbol{\nabla}\phi\cdot\boldsymbol{n}$ is $\{R^2+b^2\}^{-3/2}\boldsymbol{x}\cdot(\pm\boldsymbol{k}) = b\{R^2+b^2\}^{-3/2}$.
The corresponding contributions to the total outward flux are

$$I_\pm = \iint_{\mathcal{S}_\pm} \frac{b}{(R^2+b^2)^{-3/2}}\mathrm{d}S = 2\pi b\int_0^a \frac{R}{(R^2+b^2)^{-3/2}}\mathrm{d}R$$

$$= -2\pi b\left[(R^2+b^2)^{-1/2}\right]_0^a = 2\pi\left(1 - b/\sqrt{a^2+b^2}\right).$$

On the curved surface $\mathcal{S}_3$ (i.e. $x^2+y^2=a^2$, $-b<z<b$), the vector $2x\boldsymbol{i}+2y\boldsymbol{j}$ is a normal. Hence, the outward unit normal is $\boldsymbol{n} = (x\boldsymbol{i}+y\boldsymbol{j})/a$, so that $\boldsymbol{\nabla}\phi\cdot\boldsymbol{n} = a^{-1}(x^2+y^2)/r^3 = a(a^2+z^2)^{-3/2}$. The total flux through $\mathcal{S}_3$ is then

$$I_3 = 2\pi a\int_{-b}^{b} a(a^2+z^2)^{-3/2}\mathrm{d}z\,.$$

By using the substitution $z = a\tan\psi$ this is evaluated as

$$I_3 = 4\pi\int_0^{\tan^{-1}b/a} \cos^3\psi\sec^2\psi\mathrm{d}\psi = 4\pi\sin\left(\tan^{-1}b/a\right) = \frac{4\pi b}{\sqrt{a^2+b^2}}.$$

Summing gives the total flux through $\mathcal{S}$ as $I_+ + I_- + I_3 = 4\pi$.

3.5 From $r^2 = x^2+y^2+z^2$, it follows that $\partial r/\partial x = x/r$, etc., so that for $\phi = \phi(r)$

$$\frac{\partial\phi}{\partial x} = \phi'(r)\frac{\partial r}{\partial x} = \phi'(r)\frac{x}{r}\,, \qquad \frac{\partial\phi}{\partial y} = \phi'(r)\frac{y}{r}\,, \qquad \frac{\partial\phi}{\partial z} = \phi'(r)\frac{z}{r}\,.$$

Thus $\dfrac{\partial^2\phi}{\partial x^2} = \left[\phi''(r)\dfrac{1}{r} + \phi'(r)\dfrac{-1}{r^2}\right]\dfrac{x}{r}x + \phi'(r)\dfrac{1}{r} = \dfrac{x^2}{r^2}\phi''(r) + \dfrac{r^2-x^2}{r^3}\phi'(r),$

with similar expressions for $\partial^2\phi/\partial y^2$ and $\partial^2\phi/\partial z^2$. Adding these gives $\nabla^2\phi = (x^2+y^2+z^2)r^{-2}\phi''(r) + (3r^2-x^2-y^2-z^2)r^{-3}\phi'(r) = \phi''(r) + 2r^{-1}\phi'(r)$.

3.6 Mass M within a sphere of radius r (where the gravitational field is radial with strength $-\mathrm{d}\phi/\mathrm{d}r$) satisfies

$$4\pi GM = 4\pi G\int_0^r 4\pi\hat{r}^2\rho(\hat{r})\,\mathrm{d}\hat{r} = 4\pi r^2\,\mathrm{d}\phi/\mathrm{d}r\,.$$

Differentiating gives $4\pi r^2 G\rho(r) = \mathrm{d}(r^2\mathrm{d}\phi/\mathrm{d}r)/\mathrm{d}r$. Thus, $4\pi Gr^2\rho(r) = r^2\phi''(r) + 2r\phi'(r)$. Rearranging, using Exercise 3.5, gives $4\pi G\rho(r) = \nabla^2(\phi(r))$.

3.7 A spherical shell of radius r and thickness δr has mass $\delta m \approx 4\pi r^2\rho(r)\delta r$, so that

$$m(r) = 4\pi\int_0^r \bar{r}^2\rho(\bar{r})\,\mathrm{d}\bar{r}\,.$$

For $\rho = \rho_0R^{-3}\{R^3-2Rr^2+r^3\}$, $m(r) = 4\pi\rho_0R^{-3}[R^3\frac{1}{3}r^3 - \frac{2}{5}Rr^5 + \frac{1}{6}r^6]$. Then, total mass is $M = m(R) = 4\pi\rho_0R^3(\frac{1}{3}-\frac{2}{5}+\frac{1}{6}) = 0.4\pi\rho_0R^3$. At $r = R$, gravitational field strength is $|\mathrm{d}\phi/\mathrm{d}r| = R^{-2}G\int_0^R 4\pi r^2\rho(r)\,\mathrm{d}r = R^{-2}Gm(R) = MGR^{-2}$.

3.8 (a) $\boldsymbol{E} = -(2x+4y)\boldsymbol{i} - (4x+4y)\boldsymbol{j} + 6z\boldsymbol{k}$, with
$\nabla^2\phi = \partial(2x+4y)/\partial x + \partial(4x+4y)/\partial y + \partial(-6z)/\partial z = 2+4-6 = 0.$
(b) $\partial\phi/\partial x = \frac{-1}{2}2(x-a)\{(x-a)^2+y^2+z^2\}^{-3/2}$, etc.
so that $\boldsymbol{E} = \{(x-a)\boldsymbol{i} + y\boldsymbol{j} + z\boldsymbol{k}\}/\{(x-a)^2+y^2+z^2\}^{3/2}$.
Using $\partial\left((x-a)\{(x-a)^2+y^2+z^2\}^{-3/2}\right)/\partial x = \{(x-a)^2+y^2+z^2\}^{-3/2}$ $-3(x-a)^2\{(x-a)^2+y^2+z^2\}^{-5/2}$ and $\partial\left(y\{(x-a)^2+y^2+z^2\}^{-3/2}\right)/\partial y$ $= \{(x-a)^2+y^2+z^2\}^{-3/2} - 3y^2\{(x-a)^2+y^2+z^2\}^{-5/2}$, etc., then adding

gives $-\nabla^2\phi = 0$ (as in Example 3.5, except at $\boldsymbol{x} = a\boldsymbol{i}$).
(c) Here $\partial\phi/\partial x = y(-3r^{-4})x/r = -3xyr^{-5}$ and $\partial\phi/\partial z = -3xzr^{-5}$, while $\partial\phi/\partial y = r^{-3} - 3y^2r^{-5}$.
Then $\partial^2\phi/\partial x^2 = -3yr^{-5} + 15x^2yr^{-7}$, $\partial^2\phi/\partial z^2 = -3yr^{-5} + 15z^2yr^{-7}$, while $\partial^2\phi/\partial y^2 = -3yr^{-5} - 6yr^{-5} + 15y^3r^{-7}$.
Adding gives $\nabla^2\phi = -15yr^{-5} + 15y(x^2 + y^2 + r^2)r^{-7} = 0$.

3.9 Everywhere in $r_1 < r < r_2$, the field is $\boldsymbol{E} = -\phi'(r)\boldsymbol{x}/r$. For any sphere $\mathcal{S}$: $r =$ constant, $\boldsymbol{n} = \boldsymbol{x}/r$ so that $\boldsymbol{E}\cdot\boldsymbol{n} = -\phi'(r)$. Then, within $r_1 < r < r_2$, $Q = 4\pi r^2\epsilon_0[-\phi'(r)]$ so that $\phi'(r) = -Q/(4\pi\epsilon_0)r^{-2}$. Integrating gives $\phi = Ar^{-1} + B$, where $A = Q/(4\pi\epsilon_0)$. Electrostatic field is $\boldsymbol{E} = (4\pi\epsilon_0)^{-1}Q\boldsymbol{x}/r^3$; potential difference is $V = A(r_1^{-1} - r_2^{-1})$, so capacitance is $C = Q/V = 4\pi\epsilon_0 r_1 r_2/(r_2 - r_1)$.

3.10 Write $\boldsymbol{x} - a\boldsymbol{i} \equiv \boldsymbol{x}_1$ (i.e. the position vector relative to the point $\boldsymbol{x} = a\boldsymbol{i}$). Then $A/(|\boldsymbol{x} - a\boldsymbol{i}|) = A|\boldsymbol{x}_1|^{-1}$ satisfies Laplace's equation except at $\boldsymbol{x}_1 = \boldsymbol{0}$ (i.e. at $\boldsymbol{x} = a\boldsymbol{i}$). Similarly, $A|\boldsymbol{x} + a\boldsymbol{i}|^{-1}$ satisfies Laplace's equation except at $\boldsymbol{x} = -a\boldsymbol{i}$. Hence $\nabla^2\phi = 0$ except at $\boldsymbol{x} = \pm a\boldsymbol{i}$.
For $\phi = 0$, the distances $|\boldsymbol{x} - a\boldsymbol{i}|$ and $|\boldsymbol{x} + a\boldsymbol{i}|$ must be equal.
Thus, $(x-a)^2 + y^2 + z^2 = (x+a)^2 + y^2 + z^2$. Hence, $x = 0$, with y and z arbitrary.
For *all* surfaces $\mathcal{S}$ surrounding just $\boldsymbol{x} = a\boldsymbol{i}$ the term $-A|\boldsymbol{x} + a\boldsymbol{i}|^{-1}$ gives no contribution to the total flux. The term $A|\boldsymbol{x} - a\boldsymbol{i}|^{-1}$ always gives flux $-4\pi A$.
Choose the flux $\iint_{\mathcal{S}} -\epsilon_0\boldsymbol{\nabla}\phi\cdot\boldsymbol{n}\,\mathrm{d}S = q$.
Thus, $4\pi A\epsilon_0 = q$, so that in $x \geq 0$ the potential

$$\phi = \frac{q}{4\pi\epsilon_0}\left(\frac{1}{|\boldsymbol{x} - a\boldsymbol{i}|} - \frac{1}{|\boldsymbol{x} + a\boldsymbol{i}|}\right)$$

satisfies $\nabla^2\phi = 0$ except at $\boldsymbol{x} = a\boldsymbol{i}$, has $\phi = 0$ on $x = 0$ and gives flux outward from $\boldsymbol{x} = a\boldsymbol{i}$ appropriate for a point charge q.
In $x > 0$,

$$\begin{aligned}\frac{\partial\phi}{\partial x} &= \frac{q}{4\pi\epsilon_0}\left(\frac{-(x-a)}{\{(x-a)^2 + y^2 + z^2\}^{3/2}} + \frac{x+a}{\{(x+a)^2 + y^2 + z^2\}^{3/2}}\right)\\ &\to \frac{q}{4\pi\epsilon_0}\frac{2a}{(a^2 + y^2 + z^2)^{3/2}} \qquad \text{as } x \to 0\,.\end{aligned}$$

Since $\partial\phi/\partial x = 0$ in $x < 0$, the induced surface charge density at $x = 0$ is $-\epsilon_0{}^{-1}\partial\phi/\partial x = -aq/\{2\pi(a^2 + y^2 + z^2)^{3/2}\}$. [N.B. The *total* charge on the surface is $\int_0^\infty -\frac{1}{2}aq\pi^{-1}(a^2 + R^2)^{-3/2}2\pi R\,\mathrm{d}R = aq[(a^2 + R^2)^{-1/2}]_0^\infty = -q$, as expected.]

Chapter 4

4.1 $0 = \nabla^2\phi = X''(x)Y(y) + X(x)Y''(y)$ gives $Y''/Y = -X''/X = \alpha$ (say). BCs $\partial\phi/\partial y = 0$ at $y = 0, b$ yield $Y'(0) = 0 = Y'(b)$. Solving $Y'' = \alpha Y$ for $\alpha = p^2$ (> 0) gives $Y = Ae^{py} + Be^{-py}$. Hence $Y'(0) = p(A - B) = 0 \Rightarrow B = A$. Also, $0 = Y'(b) = pA\left(e^{pb} - e^{-pb}\right)$, so that $e^{pb} = \pm 1$ or $A = 0$. This gives no viable possibilities.
For $\alpha = 0$, $Y = A + By$; while $0 = Y'(0) = Y'(b) \Rightarrow B = 0$. Hence, $Y =$ const. is *allowable*. Solve $X''(x) = 0$ to give $X = A_0 + B_0x$, so (w.l.o.g.) $\phi = A_0 + B_0x$.
For $\alpha = -p^2$ (< 0), so that $Y''(y) + p^2Y = 0$, possibilities are $Y = A\cos py + B\sin py$. Thus $Y'(y) = p(B\cos py - A\sin py)$. Setting $y = 0$ gives $B = 0$, while

$y = b$ gives $-pA\sin pb = 0$, so that $pb = n\pi$, $n = 1, 2, 3, \ldots$. Thus $Y(y) \propto \cos n\pi y/b$. Hence $X''(x) = p^2 X$ yields $X(x) = C_n e^{n\pi x/b} + D_n e^{-n\pi x/b}$. W.l.o.g. this gives

$$\phi_n(x,y) = \left(C_n e^{n\pi x/b} + D_n e^{-n\pi x/b}\right)\cos n\pi y/b\,, \qquad n = 1, 2, 3, \ldots .$$

4.2 Select separable solutions from Exercise 4.1 satisfying $\phi = 0$ at $x = a$ (i.e. $X(a) = 0$); thus $A_0 = -B_0 a$, $C_n e^{n\pi a/b} = -D_n e^{-n\pi a/b}$. For $n > 0$, write $\phi_n(x,y) = \sinh\{n\pi(x-a)/b\}\cos\{n\pi y/b\}$. Superposing gives

$$\phi = B_0(x-a) + \sum_{n=1}^{\infty} B_n \sinh\frac{n\pi(x-a)}{b}\cos\frac{n\pi y}{b}\,.$$

Set $x = 0$, so requiring that

$$ky^2(b-y) = -aB_0 - \sum_{n=1}^{\infty} B_n \sinh\frac{n\pi a}{b}\cos\frac{n\pi y}{b} \qquad \text{for} \quad 0 < y < b\,.$$

The half-range Fourier cosine series $\frac{1}{2}a_0 + \sum_{n=1}^{\infty} a_n \cos n\pi y/b$ represents $ky^2(b-y)$ if the coefficients are

$$a_n = \frac{2}{b}\int_0^b ky^2(b-y)\cos\frac{n\pi y}{b}\,dy\,, \qquad n = 0, 1, \ldots .$$

Thus, $a_0 = 2b^{-1}k[\frac{1}{3}by^3 - \frac{1}{4}y^4]_0^b = kb^3/6$ and (after integration by parts)

$$a_n = \frac{2k}{b}\int_0^b (by^2 - y^3)\cos\frac{n\pi y}{b}\,dy = \frac{-2kb^3}{n^4\pi^4}\{n^2\pi^2(-1)^n + 6[1-(-1)^n]\}\,.$$

Then, $B_0 = -\frac{1}{2}a_0/a = -kb^3/(12a)$ and $B_n = -a_n/(\sinh n\pi a/b)$ (comparing coefficients). Solution is

$$\phi = \frac{kb^3}{12a}(a-x) + \sum_{n=1}^{\infty}\frac{2kb^3\{6[1-(-1)^n] + n^2\pi^2(-1)^n\}}{n^4\pi^4\sinh(n\pi a/b)}\sinh\frac{n\pi(x-a)}{b}\cos\frac{n\pi y}{b}\,.$$

At $x = a$,

$$\text{total flux} = \int_0^b \frac{\partial\phi}{\partial x}\,dy = \int_0^b\left\{\frac{-kb^3}{12a} + \sum_{n=1}^{\infty}\frac{n\pi}{b}B_n\cos\frac{n\pi y}{b}\right\}dy = \frac{-kb^4}{12a}\,.$$

4.3 Consider $\nabla^2\Phi_2 = 0$; $\Phi_2(0,y) = 0$, $\partial\Phi_2/\partial y = 0$ over $y = 0, b$; $\Phi_2(a,y) = cy$. For $Y''(y) = 0$, use $\phi_0 = x$; for $Y''(y) = n^2\pi^2 b^{-2} Y$, separable solutions are $\phi_n(x,y) = \sinh\{n\pi x/b\}\cos\{n\pi y/b\}$. Superpose as $\Phi_2 = b_0 x + \sum_{n=1}^{\infty} b_n\phi_n(x,y)$, then set

$\Phi_2(a,y) = b_0 a + \sum_{n=1}^{\infty} b_n \sinh\{n\pi a/b\}\cos\{n\pi y/b\} = cy \; (0 < y < b)$.

Then $2ab_0 = 2b^{-1}\int_0^b cy\,dy = cb \;\Rightarrow\; b_0 = cb/2a$ and
$b_n\sinh\{n\pi a/b\} = 2b^{-1}\int_0^b cy\cos\{n\pi y/b\}\,dy = -2bc(n\pi)^{-2}[1-(-1)^n]$. Add

$$\Phi_2 = \frac{cbx}{2a} - 2cb\sum_{n=1}^{\infty}\frac{1-(-1)^n}{n^2\pi^2\sinh\{n\pi a/b\}}\sinh\frac{n\pi x}{b}\cos\frac{n\pi y}{b}$$

to expression for $\phi(x,y)$ in Exercise 4.2 (i.e. $\Phi_1(x,y)$), to yield new solution $\phi = \Phi_1 + \Phi_2$. At $x = a$, flux then is $\frac{-1}{12}kb^4/a + \frac{1}{2}cb^2/a$, which vanishes for $c = \frac{1}{6}b^2 k$.

4.4 For $\phi = k\sin\frac{\pi x}{a}\sin\frac{\pi y}{b}$, $\nabla^2\phi = -k\left(\frac{\pi^2}{a^2}+\frac{\pi^2}{b^2}\right)\sin\frac{\pi x}{a}\sin\frac{\pi y}{b}$.
This yields $k\pi^2(a^{-2}+b^{-2}) = -1$. Hence solution is

$$\phi(x,y) = -a^2b^2\pi^{-2}(a^2+b^2)^{-1}\sin\frac{\pi x}{a}\sin\frac{\pi y}{b}.$$

4.5 $0 = X''/X + Y''/Y + Z''/Z$ implies $Y''/Y =$ const. with $Y'(y) = 0 = Y'(b)$; hence $Y(y) \propto \cos m\pi y/b$ for $m = 0, 1, 2, \ldots$ (w.l.o.g.). Similarly, $Z''/Z =$ const. $\Rightarrow$ $Z(z) \propto \cos n\pi z/c$. Then $X''/X = (m\pi/b)^2 + (n\pi/c)^2 = (\alpha_{mn})^2$. Solving $X'' = (\alpha_{mn})^2 X$ with $X(0) = 0$ gives $X(x) \propto \sinh\alpha_{mn}x$ (for $(m,n) \neq (0,0)$), but $X(x) \propto x$ for $m = n = 0$. Superpose $\phi_{00}(x,y,z) \equiv x \times 1 \times 1 = x$ and $\phi_{mn}(x,y,z) \equiv \sinh\alpha_{mn}x\cos\dfrac{m\pi y}{b}\cos\dfrac{n\pi z}{c}$ to give required solution $\phi(x,y,z)$.
[N.B. $\sinh\alpha_{00}x \equiv 0$, so that the case $(m,n) = (0,0)$ disappears from the sum.]

4.6 (a) Use $\nabla z = \boldsymbol{k}$, $\nabla r = \boldsymbol{x}/r$, so that

$$\nabla(zr^{-3}) = \boldsymbol{k}r^{-3} - 3zr^{-4}\boldsymbol{x}/r = r^{-3}\boldsymbol{k} - 3zr^{-5}\boldsymbol{x}.$$

(b) $\dfrac{\partial^2}{\partial x^2}(zr^{-3}) = \dfrac{\partial}{\partial x}(-3zr^{-5}x) = 15zr^{-7}x^2 - 3zr^{-5}$;

$$\frac{\partial^2}{\partial y^2}(zr^{-3}) = 15zr^{-7}y^2 - 3zr^{-5}; \quad \frac{\partial}{\partial z}(zr^{-3}) = r^{-3} - 3zr^{-5}z$$

so $\dfrac{\partial^2}{\partial z^2}(zr^{-3}) = -3r^{-4}z/r - 6zr^{-5} + 15z^2r^{-7}z$.

Then adding gives $\nabla^2(zr^{-3})$

$$= \left(\frac{\partial^2}{\partial x^2}+\frac{\partial^2}{\partial y^2}+\frac{\partial^2}{\partial z^2}\right)(zr^{-3}) = 15zr^{-7}(x^2+y^2+z^2) - 15zr^{-5} = 0.$$

4.7 Let $\boldsymbol{p} = p_1\boldsymbol{i} + p_2\boldsymbol{j} + p_3\boldsymbol{k}$, so that

$$\boldsymbol{p}\cdot\nabla r^{-1} = \left(p_1\frac{\partial}{\partial x} + p_2\frac{\partial}{\partial y} + p_3\frac{\partial}{\partial z}\right)r^{-1} = -p_1\frac{x}{r^3} - p_2\frac{y}{r^3} - p_3\frac{z}{r^3}$$

which is a linear combination of x/r^3, y/r^3 and z/r^3.

4.8 In spherical polars, $x/r^3 = r^{-2}\sin\theta\cos\psi$, $y/r^3 = r^{-2}\sin\theta\sin\psi$ and $z/r^3 = r^{-2}\cos\theta$. For $\phi = r^{-2}\cos\theta\cos\psi$, the Laplacian is

$$\begin{aligned}\nabla^2\phi &= \{(-2)(-3)r^{-4} + 2(-2)r^{-4}\}\sin\theta\cos\psi \\ &\quad + \frac{1}{r^2\sin\theta}r^{-2}\frac{\partial}{\partial\theta}(\sin\theta\cos\theta\cos\psi) + \frac{1}{r^2\sin^2\theta}r^{-2}\sin\theta\,(-\cos\psi) \\ &= \frac{2\sin^2\theta + \cos^2\theta - \sin^2\theta - 1}{r^4\sin\theta}\cos\psi = 0.\end{aligned}$$

For $\phi = r^{-2}\sin\theta\sin\psi$, similar steps yield

$$\nabla^2\phi = \frac{2\sin^2\theta + \cos^2\theta - \sin^2\theta - 1}{r^4\sin\theta}\sin\psi = 0,$$

while for $\phi = r^{-2}\cos\theta$ it follows that

$$\nabla^2\phi = 2r^{-4}\cos\theta - r^{-4}(\sin\theta)^{-1}2\sin\theta\cos\theta = 0,$$

so that all three dipole potentials satisfy Laplace's equation.

4.9 In $r > a$, $\nabla^2\phi = 0$; try $\phi = -E_0 z + Az/r^3$.
At $r = a$, $\sigma \boldsymbol{E}\cdot\boldsymbol{n} = 0 \quad \Rightarrow \quad \partial\phi/\partial r = 0$.
But $\phi = -E_0 r\cos\theta + Ar^{-2}\cos\theta \quad \Rightarrow \quad \partial\phi/\partial r = (-E_0 - 2Ar^{-3})\cos\theta$.
Setting $\partial\phi/\partial r = 0$ at $r = a$ gives $E_0 + 2Aa^{-3} = 0$; hence $A = \frac{-1}{2}a^3 E_0$.
Thus, in $r \geq a$, $\quad \phi = -E_0 z - \dfrac{a^3 E_0}{2}\dfrac{z}{r^3}$, so that

$$\boldsymbol{E} = -\boldsymbol{\nabla}\phi = -E_0\boldsymbol{k} + \frac{a^3E_0}{2}\left(\frac{\boldsymbol{k}}{r^3} - \frac{3rz\boldsymbol{x}}{r^5}\right) \qquad \text{(by Exercise 4.6(a)).}$$

4.10 Choose

$$\phi = \begin{cases} Az + Bz/r^3 & a \leq r \leq b \\ -E_0 z + Cz/r^3 & r \geq b \end{cases}$$

with $\partial\phi/\partial r = 0$ at $r = a$ and with both ϕ and $\sigma_c\partial\phi/\partial r$ continuous across $r = b$. Thus

$$\begin{aligned} (A - 2a^{-3}B)\cos\theta &= 0 && (\partial\phi/\partial r = 0 \text{ at } r = a) \\ (Ab + Bb^{-2})\cos\theta &= (-E_0 b + Cb^{-2})\cos\theta && (\phi \text{ continuous at } r = b) \\ \sigma_1(A - 2b^{-3}B)\cos\theta &= \sigma_2(-E_0 - 2b^{-3}C)\cos\theta && (\sigma_c\partial\phi/\partial r \text{ continuous}). \end{aligned}$$

Using $B = \frac{1}{2}a^3 A$ gives

$$(b^3 + \tfrac{1}{2}a^3)A = -b^3E_0 + C\,, \qquad \sigma_1(b^3 - a^3)A = -\sigma_2 b^3 E_0 - 2\sigma_2 C\,.$$

Solving for A and C yields

$$A = \frac{-3\sigma_2 b^3 E_0}{(2\sigma_2 + \sigma_1)b^3 + (\sigma_2 - \sigma_1)a^3}, \; C = \frac{-(2\sigma_1 + \sigma_2)a^3 + 2(\sigma_1 - \sigma_2)b^3}{(2\sigma_2 + \sigma_1)b^3 + (\sigma_2 - \sigma_1)a^3}\frac{b^3E_0}{2}$$

so that

$$\phi = \begin{cases} -\left(1 + \dfrac{a^3}{2r^3}\right)\dfrac{3\sigma_2 b^3 E_0 z}{(2\sigma_2 + \sigma_1)b^3 + (\sigma_2 - \sigma_1)a^3} & a \leq r \leq b\,, \\ -E_0 z + \dfrac{2(\sigma_1 - \sigma_2)b^3 - (2\sigma_1 + \sigma_2)a^3}{(2\sigma_2 + \sigma_1)b^3 + (\sigma_2 - \sigma_1)a^3}\dfrac{b^3E_0}{2}\dfrac{z}{r^3} & b \leq r\,. \end{cases}$$

Chapter 5

5.1 For $v = \sin(n\pi x/a)Z_n(t)$, $\quad v_{xx} = -(n\pi/a)^2\sin(n\pi x/a)Z_n(t)$ and
$v_{tt} = \sin(n\pi x/a)\ddot{Z}_n(t)$, so that $\ddot{Z}_n(t) + (n\pi c/a)^2 Z_n = 0$.
Solutions are $Z_n = A_n\cos\omega_n t + B_n\sin\omega_n t$, with frequency $\omega_n = n\pi c/a$.
Superposing gives $v = \sum_n(A_n\cos\omega_n t + B_n\sin\omega_n t)\sin(n\pi x/a)$, hence
$v_t = \sum_n \omega_n(B_n\cos\omega_n t - A_n\sin\omega_n t)\sin(n\pi x/a)$.
This satisfies $v_t(x,0) = 0 \;\forall x \in (0,a)$ only if $B_n = 0 \;\forall n = 1, 2, \ldots$.
At $t = 0$, $v(x,0) = \sum_n A_n\sin(n\pi x/a) = kx(a - x)$, so (for $n \neq 0$)

$$\begin{aligned} A_n &= \frac{2}{a}\int_0^a kx(a-x)\sin(n\pi x/a)\,\mathrm{d}x \\ &= 2ka^{-1}\left\{0 + (a/n\pi)\int_0^a (a - 2x)\cos(n\pi x/a)\mathrm{d}x\right\} \\ &= 2(k/n\pi)(a/n\pi)(-2a/n\pi)[\cos n\pi x/a]_0^a = 4ka^2[1 - (-1)^n]/(n\pi)^3 \end{aligned}$$

(integrating by parts twice). Also $A_0 = 2ka^{-1}\int_0^a x(a-x)\mathrm{d}x = ka^2/3$. Since $Z_n(t)$ has fundamental period $2\pi/\omega_n = 2a/(nc)$, all functions $Z_n(t)$ have period $2a/c$. As $A_1 \neq 0$, the fundamental period of $v(x,t)$ is $2a/c$.

5.2 $v = V(x)\cos(\omega t-\alpha)$ gives $v_{tt} = -\omega^2 V(x)\cos(\omega t-\alpha)$, $v_{xx} = V''(x)\cos(\omega t - \alpha)$. Substituting gives $\{-\omega^2 V(x) - [T_0/m(x)]V''(x)\}\cos(\omega t-\alpha) = 0$ which is satisfied provided that $V(x)$ satisfies the ODE

$$V''(x) + \omega^2 m(x)(T_0)^{-1}V(x) = 0\,.$$

Boundary conditions $V(0) = 0 = V(a)$ ensure that $v(0,t) = 0 = v(a,t)$.

5.3 Separable solutions of $v_{tt} = c^2 v_{xx}$ satisfying $v(0,t) = 0 = v(a,t)$ are of the form $Z_n(t)\sin n\pi x/a$ (with $Z_n(t)$ as in Exercise 5.1). Superposing as $v = \sum_n Z_n(t)\sin n\pi x/a$ and setting $v(x,0) = 0$ gives $A_n = 0 \quad \forall n$.
Thus, $v = \sum_n B_n \sin(n\pi ct/a)\sin(n\pi x/a)$, so
$v_t(x,0) = \sum_n (n\pi c/a)B_n \cos 0\, \sin(n\pi x/a)$, a Fourier sine series.
Its coefficients are given by

$$(n\pi c/a)B_n = \frac{2}{a}\int_{a/3}^{2a/3} V\sin(n\pi x/a)\mathrm{d}x = (2V/a)\left[\frac{-a}{n\pi}\cos\frac{n\pi x}{a}\right]_{a/3}^{2a/3}.$$

Hence $B_n = 2Va/(n^2\pi^2 c)\,(\cos(n\pi/3) - \cos(2n\pi/3))$.
(N.B. $\cos(n\pi/3) - \cos(2n\pi/3)$ takes successive values $1, 0, -2, 0, 1, 0, 1, 0, -2, \ldots$ for $n = 1, 2, 3, \ldots$ repeating cyclically with period 6.)

5.4 Setting $\boldsymbol{r} = \boldsymbol{i}x$ gives $\boldsymbol{r}_{tt} = \boldsymbol{i}x_{tt} = \boldsymbol{i}U_t$ and $\boldsymbol{r}_X = \boldsymbol{i}x_X = \lambda\boldsymbol{i}$, where $U \equiv \partial x/\partial t$ and $\lambda \equiv \partial x/\partial X$. Then, in (5.6), $\hat{m}U_t\boldsymbol{i} = (T(\lambda)\boldsymbol{i})_X$; also, compatibility requires that $U_X = \partial^2 x/\partial X\partial t = \lambda_t$. Hence,

$$\hat{m}U_t = T_X\,, \quad \lambda_t = U_X\,, \quad \text{where } T = T(\lambda) \quad [\text{i.e. } \hat{m}\partial^2\lambda/\partial t^2 = \partial^2\{T(\lambda)\}/\partial X^2].$$

5.5 The time-derivatives of the unit tangent $\boldsymbol{t}$ and unit normal $\boldsymbol{n}$ are

$$\boldsymbol{t}_t = -\boldsymbol{i}\sin\psi\,\psi_t + \boldsymbol{j}\cos\psi\,\psi_t = \boldsymbol{n}\psi_t\,, \quad \boldsymbol{n}_t = -\boldsymbol{i}\cos\psi\,\psi_t - \boldsymbol{j}\sin\psi\,\psi_t = -\boldsymbol{t}\psi_t\,.$$

Using the velocity in the form $\boldsymbol{v} = U\boldsymbol{t} + V\boldsymbol{n}$ gives $\boldsymbol{v}_t = U_t\boldsymbol{t} + V_t\boldsymbol{n} + U\boldsymbol{n}\psi_t - V\boldsymbol{t}\psi_t$, as required. Similarly, since $\boldsymbol{t}_X = \boldsymbol{n}\psi_X$, $(T\boldsymbol{t})_X = \boldsymbol{t}T_X + T\boldsymbol{n}\psi_X$. Inserting into (5.6) and resolving into components along $\boldsymbol{t}$ and $\boldsymbol{n}$ yields

$$\hat{m}(U_t - V\psi_t) = T_X = T'(\lambda)\lambda_X \quad , \qquad \hat{m}(V_t + U\psi_t) = T(\lambda)\psi_X\,.$$

From the identity for $\boldsymbol{v}$, it follows that $U_X\boldsymbol{t} + U\boldsymbol{n}\psi_X + V_X\boldsymbol{n} - V\boldsymbol{t}\psi_X = \boldsymbol{v}_X = (\lambda\boldsymbol{t})_t = \lambda_t\boldsymbol{t} + \lambda\boldsymbol{n}\psi_t$. Taking components along $\boldsymbol{t}$ and $\boldsymbol{n}$ gives

$$\partial U/\partial X - V\partial\psi/\partial X = \partial\lambda/\partial t \quad , \qquad \partial V/\partial X + U\partial\psi/\partial X = \lambda\partial\psi/\partial t\,.$$

5.6 In (5.14), $\cos\omega_n t\sin(n\pi x/a) = \frac{1}{2}[\sin(n\pi x/a - \omega_n t) + \sin(n\pi x/a + \omega_n t)]$
$= \frac{1}{2}\sin[n\pi a^{-1}(x-ct)] + \frac{1}{2}\sin[n\pi a^{-1}(x+ct)]$
and $\sin\omega_n t\sin(n\pi x/a) = \frac{1}{2}[\cos(n\pi x/a - \omega_n t) - \cos(n\pi x/a + \omega_n t)]$
$= \frac{1}{2}\cos[n\pi a^{-1}(x-ct)] - \frac{1}{2}\cos[n\pi a^{-1}(x+ct)]$,
so $v = \frac{1}{2}\sum_n\{A_n\sin[n\pi(x-ct)/a] + A_n\sin[n\pi(x+ct)/a]$
$+B_n\cos[n\pi(x-ct)/a] - B_n\cos[n\pi(x+ct)/a]\}$
$= f(x-ct) + g(x+ct)$,
where

$$f(\xi) = \tfrac{1}{2}\sum_n\left(A_n\sin\frac{n\pi\xi}{a} + B_n\cos\frac{n\pi\xi}{a}\right),$$

$$g(\eta) = \tfrac{1}{2}\sum_n\left(A_n\sin\frac{n\pi\eta}{a} - B_n\cos\frac{n\pi\eta}{a}\right).$$

5.7 The lines ξ = constant are parallel; so are the lines η = constant, which are equally inclined to the Ox axis. Writing $v(x,t) \equiv u(x-ct, x+ct) = u(\xi,\eta)$ and using $\xi_x = \eta_x = 1$, $\xi_t = -c$, $\eta_t = c$ gives $v_t = -cu_\xi + cu_\eta$, $v_x = u_\xi + u_\eta$, $v_{tt} = (-c)^2 u_{\xi\xi} - 2c^2 u_{\xi\eta} + c^2 u_{\eta\eta}$, $v_{xx} = u_{\xi\xi} + 2u_{\xi\eta} + u_{\eta\eta}$.
Inserting into $v_{tt} - c^2 v_{xx} = 0$ gives $-4c^2 u_{\xi\eta} = 0$, or $\partial^2 v/\partial\xi\partial\eta = 0$.
Now, if $F = F(\xi,\eta)$, the equation $\partial F/\partial\xi = 0$ has *general* solution $F = \bar{F}(\eta)$, an arbitrary function of η. Hence, since $\partial(\partial v/\partial\eta)/\partial\xi = 0$, it is possible to write $\partial v/\partial\eta = g'(\eta)$ for some differentiable function $g(\eta)$. Then, from $\partial[v-g(\eta)]/\partial\eta = 0$ it follows that $v - g(\eta) = f(\xi)$, for some function $f(\xi)$. Thus, $v = f(\xi) + g(\eta)$.

5.8 Write $F(s) \equiv f(-cs)$, $G(s) \equiv g(cs)$, so $v = f(x-ct) + g(x+ct) = F(t - x/c) + G(t + x/c)$. Then, $v_t = F'(t - x/c) + G'(t + x/c)$. Put $x = 0$, so $0 = v(0,t) = F(t) + G(t) \Rightarrow G(t) = -F(t)$. Put $x = a$, so $0 = F(t - a/c) + G(t + a/c)$. Thus, $G(t + 2a/c) = -F(t) = G(t)$.

5.9 Use $v = f(x-ct) + g(x+ct)$; $0 = v_t(x,0) \Rightarrow f'(x) = g'(x)\ \forall\ x \in [0,a]$. Choose $f(x) = g(x) = \tilde{f}(x)$, where $\tilde{f}(x) \equiv bx/a$ in $[0, a/2]$ and $\tilde{f}(x) \equiv b(a-x)/a$ in $[a/2, a]$. Also, $0 = f(-ct) + g(ct)\ \ \forall\ \ t \geq 0$ gives $f(\xi) = -g(-\xi) = -f(-\xi)$ for $\xi < 0$; while $0 = f(a-ct) + g(a+ct)\ \ \forall\ \ t > 0$ leads to $f(\xi) = -g(2a-\xi) = f(\xi - 2a)\ \ \forall\ \ \xi \leq a$. Thus, in $\xi \leq a$, $f(\xi)$ is the odd, periodic extension of $\tilde{f}(\xi)$ of period $2a$.
Evaluating at $t = 0, a/4c, a/2c, a/c, 3a/2c$ and $2a/c$ gives:

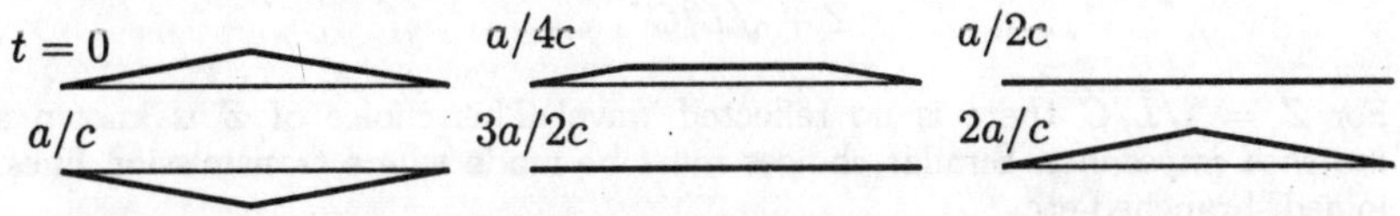

5.10 Use $v = f(x-ct) + g(x+ct)$, so

$$P(x) = f(x) + g(x) \text{ and } Q(x) = -cf'(x) + cg'(x) \text{ in } 0 \leq x \leq a.$$

Define $R(x) = \int Q(x)\,\mathrm{d}x$; then $f(x) - g(x) = -c^{-1}R(x)$.
Hence, $f(x) = \frac{1}{2}[P(x) - c^{-1}R(x)]$, $g(x) = \frac{1}{2}[P(x) + c^{-1}R(x)]$ in $0 \leq x \leq a$. Condition $v(0,t) = 0$ gives $f(\xi) = -g(-\xi)$ for $\xi < 0$; $v(a,t) = 0$ gives $g(\eta) = -f(2a-\eta)$ for $\eta \geq a$. Therefore, $f(\xi) = -\frac{1}{2}[P(-\xi) + c^{-1}R(-\xi)]$ in $-a \leq \xi \leq 0$, also $g(\eta) = \frac{1}{2}[c^{-1}R(2a-\eta) - P(2a-\eta)]$ in $a \leq \eta \leq 2a$. So $f(\xi) = -g(-\xi) = \frac{1}{2}[P(\xi+2a) - c^{-1}R(\xi+2a)] = f(\xi+2a)$ in $-2a \leq \xi \leq -a$, while $g(\eta) = -f(2a-\eta) = \frac{1}{2}[P(\eta-2a) + c^{-1}R(\eta-2a)] = g(\eta-2a)$ for $2a \leq \eta \leq 3a$. Thus, $v(x, t+2a) = v(x,t)$. For the given functions P, Q in $0 < x < a$, choose $R(x) = Vx$ in $0 \leq x \leq a$, so that $f(\xi) = -(V/2c)|\xi|$ in $|\xi| \leq a$, $g(\eta) = (V/2c)|\eta|$ in $|\eta| \leq a$, with $2a$-periodic extensions. Hence, the configurations are as shown:

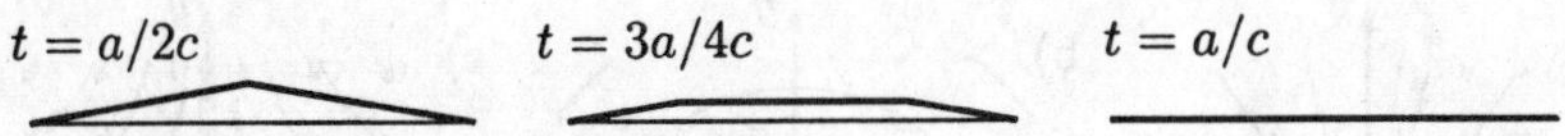

5.11 In $0 < x < a$, $c^2\rho_0 v_{xx} = -c^2\tilde{\rho}_{xt} = \rho_0 v_{tt}$, so $v_{tt} = c^2 v_{xx}$. At $x = a$, $\rho_0 v_x(a,t) = -\tilde{\rho}_t(a,t) = 0$. Seek $v = X(x)\sin\omega_0 t$, so $c^2 X''(x)\sin\omega_0 t = -\omega_0^2 X(x)\sin\omega_0 t$. Thus, solving $X''(x) + (\omega_0/c)^2 X(x) = 0$ with $X(0) = A$ gives $X = A\cos(\omega_0 x/c) + B\sin(\omega_0 x/c)$. Set $0 = X'(a)$, so $0 = -\omega_0 A c^{-1}\sin(\omega_0 a/c) + \omega_0 B c^{-1}\cos(\omega_0 a/c)$. Thus, $B = A\tan(\omega_0 a/c)$ so that

$$X(x) = A[\cos(\omega_0 x/c) + \sin(\omega_0 x/c)\ \tan(\omega_0 a/c)] = A\frac{\cos\omega_0(x-a)/c}{\cos(\omega_0 a/c)}.$$

If $\omega_0 a/c \approx \frac{1}{2}\pi + n\pi$, the amplitude becomes large. The forcing frequency ω_0 is close to one of the natural frequencies $(2n+1)\pi c/(2a)$ for an open-ended tube – a *resonance phenomenon.*

5.12 Write $v = X(x)\sin\omega_0 t + \tilde{v}(x,t)$ so that $\tilde{v}_{tt} = c^2\tilde{v}_{xx}$ follows from

$$v_{tt} = -\omega_0^2 X(x)\sin\omega_0 t + \tilde{v}_{tt} = c^2 v_{xx} = c^2 X''(x)\sin\omega_0 t + c^2\tilde{v}_{xx}\,,$$

so $\tilde{v}_{tt} = c^2\tilde{v}_{xx}$, since $c^2X''(x) = -\omega_0^2 X$ from Exercise 5.11.
Also, $\tilde{v}(0,t) = v(0,t) - X(0)\sin\omega_0 t = 0$, $\tilde{v}_x(a,t) = v_x(0,t) - X'(a)\sin\omega_0 t = \rho_0^{-1}\tilde{\rho}_t(a,t) + 0 = 0$. In $0 < x < a$, setting $\tilde{\rho}(x,0) = 0$ implies that $v_t(x,0) = 0$. With $v(x,0) = 0$ this gives the initial data
$\tilde{v}(x,0) = 0$, $\tilde{v}_t(x,0) = -\omega_0 X(x) = -\omega_0 A\{\cos\omega_0(x-a)/c\}/\cos(\omega_0 a/c)$.

5.13 From $V_x = -LI_t$, it follows that $ZV_x(a,t) = -LI_t(a,t) = -LV_t(a,t)$. Assuming the representation $V = f(x - t/\sqrt{LC}) + g(x + t/\sqrt{LC})$ gives
$V_x(a,t) = f'(a - t/\sqrt{LC}) + g'(a + t/\sqrt{LC})$
and $\quad V_t(a,t) = -(LC)^{-1/2}\left\{f'(a - t/\sqrt{LC}) - g'(a + t/\sqrt{LC})\right\}$;
thus $[Z - (L/C)^{1/2}]f'(a - t/\sqrt{LC}) + [Z + (L/C)^{1/2}]g'(a + t/\sqrt{LC}) = 0$. Hence, the reflected wave is given by

$$g(x + t/\sqrt{LC}) = \frac{Z - \sqrt{L/C}}{Z + \sqrt{L/C}}\,f(2a - x - t/\sqrt{L/C})\,.$$

For $Z = \sqrt{L/C}$ there is no reflected wave! This choice of Z is known as a *matched impedance.* Similar choices must be made where transmission lines are joined, branched etc.

Chapter 6

6.1 Streamlines satisfy $\mathrm{d}\boldsymbol{x}/\mathrm{d}\alpha = 2U\boldsymbol{j} + U\boldsymbol{k}$, or $\mathrm{d}x : \mathrm{d}y : \mathrm{d}z = 0 : 2 : 1$. Integrating gives $x = x_0$, $y = y_0 + 2U\alpha$ and $z = z_0 + U\alpha$, or $\boldsymbol{x} = \boldsymbol{x}_0 + (2\boldsymbol{j} + \boldsymbol{k})U\alpha$. These are the parallel straight lines $x =$ constant, $y - 2z =$ constant. The flow speed is $\sqrt{4U^2 + U^2} = \sqrt{5}U$.
Flow is *uniform*, at speed $\sqrt{5}U$ parallel to $2\boldsymbol{j} + \boldsymbol{k}$.

6.2 (a) Here $\boldsymbol{u} = \boldsymbol{i}u + \boldsymbol{j}v = q(x\boldsymbol{i} + y\boldsymbol{j})/(x^2 + y^2) = q\boldsymbol{x}/|\boldsymbol{x}|^2$, where $\boldsymbol{x} \equiv x\boldsymbol{i} + y\boldsymbol{j}$ points radially away from the Oz axis. Streamlines lie in planes $z =$ constant with $\mathrm{d}y/\mathrm{d}x = v/u = y/x$; i.e. with $y/x =$ constant.
(b) Since $\boldsymbol{u} = \gamma(-y\boldsymbol{i} + x\boldsymbol{j})/(x^2 + y^2)$, streamlines again lie in planes $z =$ constant,

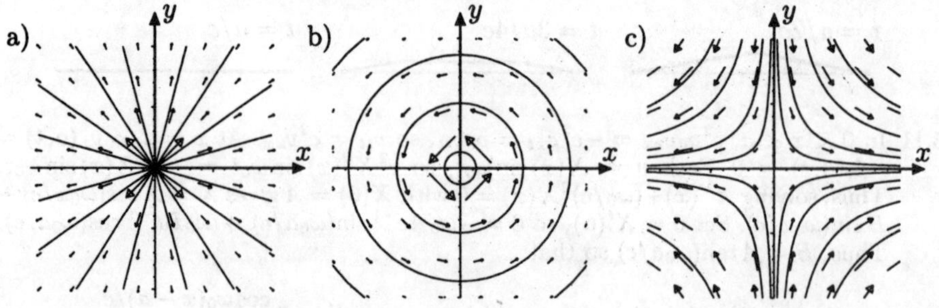

but with $\mathrm{d}y/\mathrm{d}x = -x/y$. Hence, along each streamline, $y^2 + x^2 =$ constant. The streamlines are the set of circles centred on the Oz axis and lying in the planes $z =$ constant. (Note that $\boldsymbol{u} \cdot \boldsymbol{x} = 0$.)
(c) Again $z =$ constant. Also $\mathrm{d}y/\mathrm{d}x = -y/x$, so that $\ln|y| = -\ln|x| +$ constant. Hence $|yx| =$ constant. The streamlines are rectangular hyperbolae in planes $z =$ constant. When $A > 0$, the flow approaches parallel to the Oy axis and departs parallel to the Ox axis. Each half-axis is also a streamline.

6.3 Using $x^2 + y^2 = r^2$ gives $\mathrm{d}x/\mathrm{d}\alpha = qx/r^2$, $\mathrm{d}y/\mathrm{d}\alpha = qy/r^2$ and $\mathrm{d}z/\mathrm{d}\alpha = W$. Hence, as in Exercise 6.2(a), $y/x =$ constant along each streamline.
Write $x = r\cos\theta$ and $y = r\sin\theta$ (confirming that $\mathrm{d}\theta/\mathrm{d}\alpha = 0$ along each streamline). Also,

$$r\frac{\mathrm{d}r}{\mathrm{d}\alpha} = x\frac{\mathrm{d}x}{\mathrm{d}\alpha} + y\frac{\mathrm{d}y}{\mathrm{d}\alpha} = q\frac{x^2+y^2}{x^2+y^2} = q \qquad \text{so that} \quad r^2 = q\alpha + r_0^2 .$$

Since $z = W\alpha + z_0$, elimination of α gives $r^2 - (q/W)z = r_0^2 - (q/W)z_0$. Each streamline is parabolic, lying in a plane $\theta =$ constant; it is related to a streamline in the Oxz plane by rotation about the Oz axis.
The flow is a superposition of the two-dimensional radial flow in Exercise 6.2(a) and a uniform axial flow.

6.4 Velocity is perpendicular to Oz. Flow field is independent of z. (a) On plane $y = b$ outward unit normal is $\boldsymbol{n} = \boldsymbol{j}$, so that $\boldsymbol{u}\cdot\boldsymbol{n} = -Ay = -Ab =$ constant, giving total volume outflow rate $Q_1 = -Ab \times 2a \times 1 = -2Aab$. (b) Over $x = a$ the outward unit normal is $\boldsymbol{n} = \boldsymbol{i}$, so that $\boldsymbol{u}\cdot\boldsymbol{n} = Ax = Aa =$ constant. The total volume outward flow rate is $Q_2 = Aa \times b \times 1 = Aab$.
Similarly, the outflow rate through the face at $x = -a$ is $Q_3 = Aab$. On the face $y = 0$, $\boldsymbol{n}\cdot\boldsymbol{u} = -\boldsymbol{j}\cdot\boldsymbol{u} = 0$ giving zero outflow through that face. Also, $\boldsymbol{u}\cdot\boldsymbol{n} = 0$ on each end $z = 0, 1$. Thus, adding the outflow rates through all six faces gives $-2Aab + Aab + Aab = 0$.
Result follows also from setting $\operatorname{div}\boldsymbol{u} = 0$ in the divergence theorem.

6.5 At general $\boldsymbol{x}$, velocity is $\boldsymbol{u} = q\boldsymbol{x}/r$, where $q = q(r,t)$ is the speed. At bubble surface, $q = \mathrm{d}R/\mathrm{d}t = \dot{R}(t)$.
Consider region $\mathcal{R}$ between spheres of radius $R(t)$ (variable) and r (fixed, but exceeding $R(t)$). It contains fluid volume $V = \frac{4}{3}\pi(r^3 - R^3)$. Then $\mathrm{d}V/\mathrm{d}t =$ rate of fluid volume inflow across $\partial\mathcal{R}$. Hence, $\dot{V}(t) = 0 - 4\pi r^2 q(r,t)$. Thus,

$$q(r,t) = \tfrac{1}{3}r^{-2}\mathrm{d}R^3/\mathrm{d}t = R^2\dot{R}(t)/r^2 .$$

Then, $\boldsymbol{u} = R^2\dot{R}(t)\boldsymbol{x}/r^3 = R^2\dot{R}(t)\boldsymbol{\nabla}(-r^{-1}) = \boldsymbol{\nabla}(-R^2\dot{R}(t)r^{-1}) \equiv \boldsymbol{\nabla}\phi$.

6.6 Check $\phi_{xx} + \phi_{yy} = 0$; for $\phi = Ux$ then $\phi_{xx} = 0$, $\phi_{yy} = 0$, so $\nabla^2\phi = 0 + 0 = 0$; similarly, for $\phi = Vy$, $\nabla^2\phi = 0 + 0 = 0$; for $\phi = A(x^2 - y^2)$, $\nabla^2\phi = 2A - 2A = 0$. For $\phi = \log r$, $\nabla^2\phi = \partial(xr^{-2})/\partial x + \partial(yr^{-2})/\partial y$

$$= r^{-2} - 2xr^{-3}x/r + r^{-2} - 2yr^{-3}y/r = 2r^{-2} - 2(x^2+y^2)/r^4 = 0.$$

Finally, use $\tan\theta = y/x$, so $(\sec^2\theta)\theta_y = \partial(\tan\theta)/\partial y = x^{-1}$.
Since $1 + \tan^2\theta \equiv \sec^2\theta$, this gives $\theta_y = x/(x^2+y^2)$ and $\theta_{yy} = -2xy/(x^2+y^2)^2$. Also, from $(\sec^2\theta)\theta_x = \partial(\tan\theta)/\partial x = -yx^{-2}$ it follows that $\theta_{xx} = (-y/(x^2+y^2))_x = 2yx/(x^2+y^2)^2$, so that $\nabla^2\theta = 0$.
$\phi = Ux\ :\ \Rightarrow\ \boldsymbol{u} = U\boldsymbol{i}\ ;\quad \psi_x = -v = 0$ so $\psi = \Psi(y)$. Then $U = \psi_y = \Psi'(y)$.
Stream function is $\psi = Uy +$ constant.
$\phi = Vy\ :\ \Rightarrow\ \boldsymbol{u} = V\boldsymbol{j}\ ;\quad \psi_y = u = 0$ so $\psi = \Psi(x)$. Then $-V = \psi_x = \Psi'(x)$.
Stream function is $\psi = -Vx +$ constant.
$\phi = A(x^2 - y^2)\ :\ \Rightarrow\ \boldsymbol{u} = 2A(x\boldsymbol{i} - y\boldsymbol{j})$ (cf. Exercise 6.2(c)). Then $\psi_y = 2Ax$
$\Rightarrow\ \psi = 2Axy + f(x)$. Hence $-2Ay = -\psi_x = -2Ay - f'(x)\ \Rightarrow\ f'(x) = 0$.

Finally $\psi = 2Axy+$ constant.
$\phi = \log r\ :\ \Rightarrow \boldsymbol{u} = (x\boldsymbol{i} + y\boldsymbol{j})/r^2$ (cf. Exercise 6.2(a)). Then $\psi_y = x/(x^2+y^2)$ so $\psi = \tan^{-1}(y/x) + f(x)$. From $(-y/x^2)\{1+(y/x)^2\}^{-1} + f'(x) = \psi_x = -v = -y/(x^2+y^2)$, $f(x) =$ constant; hence $\psi = \tan^{-1}(y/x) = \theta$ (plus a constant).
$\phi = \theta :\ \Rightarrow \boldsymbol{u} = (-y\boldsymbol{i} + x\boldsymbol{j})/r^2$ (cf. Exercise 6.2(b)). Then $\psi_y = -y/(x^2+y^2)$ so $\psi = -\frac{1}{2}\log(x^2+y^2) + f(x)$. From $-x\{x^2+y^2\}^{-1} + f'(x) = \psi_x = -v = -x/(x^2+y^2)$, $f(x) =$ constant; hence $\psi = -\log r$.

6.7 (a) Streamline at $\boldsymbol{x}$ is tangential to $\boldsymbol{u} = \boldsymbol{\nabla}\phi$, i.e. normal to equipotential $\phi =$ constant passing through $\boldsymbol{x}$. (b) Using $\boldsymbol{u} = \boldsymbol{i}\phi_x + \boldsymbol{j}\phi_y = \boldsymbol{i}\psi_y - \boldsymbol{j}\psi_x$ gives flowspeed $= q = \sqrt{\phi_x^2+\phi_y^2} = \sqrt{\psi_y^2+\psi_x^2} = |\boldsymbol{\nabla}\phi| = |\boldsymbol{\nabla}\psi|$
(c) Write $\hat{u} = \phi_y = -\psi_x$ and $\hat{v} = -\phi_x = -\psi_y$ (i.e. treating ϕ as a stream function). Then, $\hat{\boldsymbol{u}} \equiv \hat{u}\boldsymbol{i} + \hat{v}\boldsymbol{j} = \boldsymbol{\nabla}(-\psi)$ is the velocity field for a flow having $-\psi(x,y)$ as velocity potential. (cf. the pair $\log r$ and θ in Exercise 6.6).

6.8 $\boldsymbol{\nabla}p = -\rho g\boldsymbol{k} \quad \Longrightarrow\ p_x = 0 = p_y$ and $p_z = -\rho g$. Hence, $p = p_0 - \rho g z$. The bath cross-section is $\frac{1}{2}a(4a+2a) = 3a^2$, so mass $= M = \rho 3a^2 L$.
At $z = -a$, $p = p_0 + \rho g a$; outward unit normal is $-\boldsymbol{k}$. Thus, force *on* fluid across base is $2aL(-p)(-\boldsymbol{k})$; so reaction on the bath is $-2aL(p_0+\rho g a)\boldsymbol{k} \equiv \boldsymbol{F}_1$ (say).
On $x = z + 2a$, outward normal is $\boldsymbol{n} = (\boldsymbol{i} - \boldsymbol{k})/\sqrt{2}$ so force *on* the wall is

$$\boldsymbol{F}_2 = L\int_{-a}^{0} p\frac{1}{\sqrt{2}}(\boldsymbol{i}-\boldsymbol{k})\frac{\mathrm{d}s}{\mathrm{d}z}\mathrm{d}z = (\boldsymbol{i}-\boldsymbol{k})L\int_{-a}^{0}(p_0 - \rho g z)\mathrm{d}z$$
$$= (\boldsymbol{i}-\boldsymbol{k})L[p_0 a - \tfrac{1}{2}\rho g(0-a^2)] = (\boldsymbol{i}-\boldsymbol{k})aL(p_0 + \tfrac{1}{2}\rho g a)\,.$$

Similarly, on $x = -(z+2a)$; $\boldsymbol{F}_3 = -(\boldsymbol{i}+\boldsymbol{k})aL(p_0 + \frac{1}{2}\rho g a)$. On the ends, the resultant forces are parallel to $\boldsymbol{j}$, but equal and opposite. Hence, the *total force* on the walls of the bath is
$\boldsymbol{F}_1 + \boldsymbol{F}_2 + \boldsymbol{F}_3 = aL\{(-2p_0 - 2\rho g a)\boldsymbol{k} + (p_0 + \frac{1}{2}\rho g a)(\boldsymbol{i}-\boldsymbol{k}) - (p_0+\frac{1}{2}\rho g a)(\boldsymbol{i}+\boldsymbol{k})\}$ $= aL(-4p_0 - 3\rho g a)\boldsymbol{k}$. This is the sum of the force $-4aLp_0\boldsymbol{k}$ due to the pressure acting over the surface $z = 0$ and the force $-Mg\boldsymbol{k}$ due to the weight Mg.

6.9 On $r = a$, the stream function is $\psi = -\gamma\log a =$ constant. Hence $\boldsymbol{\nabla}\psi$ is radial $\Longrightarrow\ |\boldsymbol{\nabla}\psi| = |\psi_r(a,\theta)|$. Since $\partial\psi/\partial r = -\gamma r^{-1} - U\sin\theta\,(1-(-1)a^2r^{-2})$, on the surface of cylinder the (clockwise) flow speed is $\partial\psi/\partial r = -\gamma a^{-1} - 2U\sin\theta$. Thus, stagnation points occur where $\sin\theta = -\gamma/(2aU)$, when $|\gamma| \le 2a|U|$.
If $|\gamma/U| > 2a$, no stagnation points occur on $r = a$. From $u = \psi_y = -\gamma y r^{-2} - U + Ua^2(r^2 - 2y^2)r^{-4}$, $v = -\psi_x = -x\{\gamma + 2Ua^2 y r^{-2}\}r^{-2}$ $= -x\{\gamma + 2Ua^2 r^{-1}\sin\theta\}r^{-2}$, stagnation points occur only $x = 0$. Thus, $y^2 + a^2 + \gamma U^{-1}y = 0$, so $y = -\{\gamma + \sqrt{\gamma^2 - 4a^2U^2}\}/(2U)$ (the root with $|y| \ge a$).

6.10 Put $u = \phi_x$, $v = \phi_y$ in $u_x + v_y = 0$, giving $\phi_{xx} + \phi_{yy} = 0$. Substituting into the Euler equations gives $\phi_{xt} + \phi_x\phi_{xx} + \phi_y\phi_{xy} = -\rho^{-1}p_x$ and
$\phi_{yt} + \phi_x\phi_{yx} + \phi_y\phi_{yy} = -\rho^{-1}p_y$. Rearrange as

$$\partial\{\phi_t + \tfrac{1}{2}(\phi_x^2+\phi_y^2) + \rho^{-1}p\}/\partial x = 0 = \partial\{\phi_t + \tfrac{1}{2}(\phi_x^2+\phi_y^2) + \rho^{-1}p\}/\partial y\,,$$

so (a) $\phi_t + \frac{1}{2}(\phi_x)^2 + \frac{1}{2}(\phi_y)^2 + \rho^{-1}p$ cannot depend on either x or y. It can depend (possibly) on t. Denote this function (for convenience) as $f'(t)$. Write $\phi(x,y,t) - f(t) \equiv \hat{\phi}(t)$, so that $u = \hat{\phi}_x$ and $v = \hat{\phi}_y$, with

$$\hat{\phi}_t + \tfrac{1}{2}(u^2+v^2) + \rho^{-1}p = \hat{\phi}_t + \tfrac{1}{2}[(\hat{\phi}_x)^2 + (\hat{\phi}_y)^2] + \rho^{-1}p = 0$$

(this is a special case of the required result – varying the constant of integration within $f(t)$ replaces the 0 on the right-hand side by any constant).

Chapter 7

7.1 Using e.g. $\partial x_1/\partial X_1 = A + DX_3$, $\partial x_1/\partial X_2 = 0$, $\partial x_1/\partial X_3 = DX_1$ gives

(a) $\mathbf{F} = \begin{pmatrix} A+DX_3 & 0 & DX_1 \\ 0 & B+DX_3 & DX_2 \\ 0 & 0 & C-2DX_3 \end{pmatrix}$

(b) $\mathbf{F} = \begin{pmatrix} a\cos\kappa X_3 & -a\sin\kappa X_3 & -a\kappa(X_1\sin\kappa X_3 + X_2\cos\kappa X_3) \\ a\sin\kappa X_3 & a\cos\kappa X_3 & a\kappa(X_1\cos\kappa X_3 - X_2\sin\kappa X_3) \\ 0 & 0 & b \end{pmatrix}$

7.2

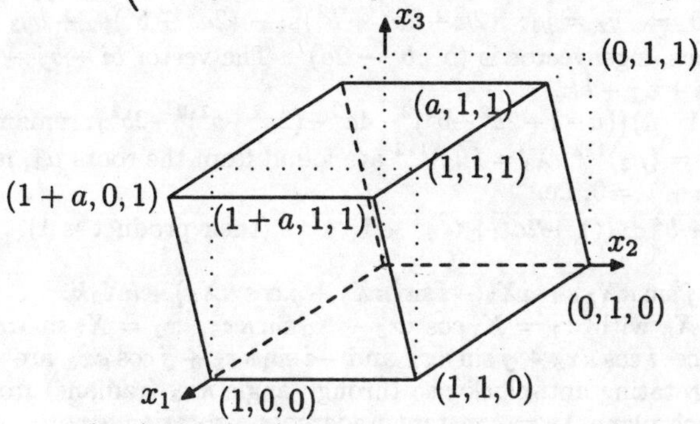

$\mathbf{F} = \begin{pmatrix} 1 & 0 & a \\ 0 & 1 & 0 \\ 0 & 0 & 1 \end{pmatrix}$

$\mathbf{G} = \begin{pmatrix} 1 & 0 & a \\ 0 & 1 & 0 \\ a & 0 & 1+a^2 \end{pmatrix}$

so $\mathbf{G}\begin{pmatrix} 0 \\ 1 \\ 0 \end{pmatrix} = \begin{pmatrix} 0 \\ 1 \\ 0 \end{pmatrix}$ with eigenvalue 1 (zero stretching).

7.3 (a) Since $\boldsymbol{x} = \boldsymbol{X} + (a\boldsymbol{i} + b\boldsymbol{j})X_3$, all points of *each* plane X_3 = constant undergo the same displacement $(a\boldsymbol{i}+b\boldsymbol{j})X_3$ within that plane. Displacement is parallel to $a\boldsymbol{i}+b\boldsymbol{j}$ and linear in X_3, so deformation is a simple shear.
(b) $\boldsymbol{x} = \boldsymbol{X} + (a\boldsymbol{i}+a\boldsymbol{j}+b\boldsymbol{k})X_1 - (a\boldsymbol{i}+a\boldsymbol{j}+b\boldsymbol{k})X_2 = \boldsymbol{X} + (a\boldsymbol{i}+a\boldsymbol{j}+b\boldsymbol{k})(X_1 - X_2)$. Displacements are parallel to $a\boldsymbol{i}+a\boldsymbol{j}+b\boldsymbol{k}$, of magnitude proportional to $X_1 - X_2$. Each plane $X_1 - X_2$ = constant is moved in direction $a\boldsymbol{i}+a\boldsymbol{j}+b\boldsymbol{k}$, which is perpendicular to $\boldsymbol{i}-\boldsymbol{j}$. Hence, deformation is a simple shear.

7.4 (a) $\mathbf{G} = \begin{pmatrix} 1 & 0 & a \\ 0 & 1 & b \\ a & b & 1+a^2+b^2 \end{pmatrix}$,

$\det(\mathbf{G}-\mu\mathbf{I}) = (1-\mu)\{(1-\mu)(1+a^2+b^2-\mu)-b^2\} - a^2(1-\mu) = (1-\mu)\{\mu^2-\mu(2+a^2+b^2)+1\}$. Thus $\mu = 1$ is a root of $\det(\mathbf{G}-\mu\mathbf{I}) = 0$, corresponding eigenvector is $(-b\ a\ 0)^T$. The vector $-b\boldsymbol{i}+a\boldsymbol{j}$ is perpendicular to the displacement direction $a\boldsymbol{i}+b\boldsymbol{j}$. Remaining eigenvalues are $\mu_{2,3} = 1+\frac{1}{2}(a^2+b^2) \pm \frac{1}{2}\{(2+a^2+b^2)^2-4\}^{1/2}$. Principal stretches are $\lambda_2 = \sqrt{\mu_2}$ and $\lambda_3 = \sqrt{\mu_3}$. Since $\mu_2\mu_3 = 1$, then $\lambda_2\lambda_3 = 1$.

(b)

$$\mathbf{G} = \begin{pmatrix} 1+a & a & b \\ -a & 1-a & -b \\ 0 & 0 & 1 \end{pmatrix}\begin{pmatrix} 1+a & -a & 0 \\ a & 1-a & 0 \\ b & -b & 1 \end{pmatrix}$$
$$= \begin{pmatrix} 1+2a+2a^2+b^2 & -2a^2-b^2 & b \\ -2a^2-b^2 & 1-2a+2a^2+b^2 & -b \\ b & -b & 1 \end{pmatrix}.$$

Since

$$\det(\mathbf{G}-\mathbf{I}) = \begin{vmatrix} 2a+2a^2+b^2 & -(2a^2+b^2) & b \\ -(2a^2+b^2) & -2a+2a^2+b^2 & -b \\ b & -b & 0 \end{vmatrix}$$

$$= \begin{vmatrix} 2a & -(2a^2+b^2) & b \\ -2a & -2a+2a^2+b^2 & -b \\ 0 & -b & 0 \end{vmatrix} = 0,$$

$\mu = 1$ is one root of $\det(\mathbf{G}-\mu\mathbf{I}) = 0$. Corresponding eigenvector $(y_1 \;\; y_2 \;\; y_3)^T$ satisfies $by_1 - by_2 = 0 \Rightarrow y_2 = y_1$; $(2a+2a^2+b^2)y_1 - (2a^2+b^2)y_2 + by_3 = 0$ $\Rightarrow 2a_1 + by_3 = 0$. Thus, eigenvector is $(b \;\; b \;\; -2a)^T$. The vector $b\boldsymbol{i}+b\boldsymbol{j}-2a\boldsymbol{k}$ is perpendicular to $a\boldsymbol{i}+a\boldsymbol{j}+b\boldsymbol{k}$.
Since $\det(\mathbf{G}-\mu\mathbf{I}) = (1-\mu)\{(\mu-1-2a^2-b^2)^2-4a^2-(2a^2+b^2)^2-2b^2\}$, remaining principal stretches $\lambda_2 = (\mu_2)^{1/2}$, $\lambda_3 = (\mu_3)^{1/2}$ are found from the roots μ_2, μ_3 of $\mu^2 - 2(1+2a^2+b^2)\mu + 1 = 0$, i.e.

$$\mu_2, \mu_3 = 1+2a^2+b^2 \pm \{(1+2a^2+b^2)^2-1\}^{1/2} \quad \text{(their product is 1).}$$

7.5 In Exercise 7.1(b),
$\boldsymbol{x} = aX_1(\boldsymbol{i}\cos\kappa X_3 + \boldsymbol{j}\sin\kappa X_3) + aX_2(-\boldsymbol{i}\sin\kappa X_3 + \boldsymbol{j}\cos\kappa X_3) + bX_3\boldsymbol{k}$.
For $a = b = 1$, $x_3 = X_3$ with $x_1 = X_1\cos\kappa x_3 - X_2\sin\kappa x_3$, $x_2 = X_1\sin\kappa x_3 + X_2\cos\kappa x_3$. Thus, since $\boldsymbol{i}\cos\kappa x_3 + \boldsymbol{j}\sin\kappa x_3$ and $-\boldsymbol{i}\sin\kappa x_3 + \boldsymbol{j}\cos\kappa x_3$ are unit vectors, obtained by rotating anti-clockwise through angle κx_3 (radians) from $\boldsymbol{i}$ and $\boldsymbol{j}$ respectively, each plane X_3 = constant undergoes a rotation κx_3.

$$\mathbf{F} = \begin{pmatrix} \cos\kappa X_3 & -\sin\kappa X_3 & 0 \\ \sin\kappa X_3 & \cos\kappa X_3 & 0 \\ 0 & 0 & 1 \end{pmatrix}\begin{pmatrix} 1 & 0 & -\kappa X_2 \\ 0 & 1 & \kappa X_1 \\ 0 & 0 & 1 \end{pmatrix} = \mathbf{RS}.$$

Treat (X_1, X_2) as parameters. $\mathbf{S}$ would arise from shearing of planes X_3 = const. in direction $-\kappa X_2\boldsymbol{i} + \kappa X_1\boldsymbol{j}$. Shear magnitude is $\kappa\sqrt{X_1^2+X_2^2}$, proportional to distance from OX_3. Direction of shear is perpendicular to $X_1\boldsymbol{i}+X_2\boldsymbol{j}$ (the normal to $X_1^2+X_2^2$ = constant).

7.6 (a)

$$\left(\frac{\partial u_i}{\partial X_j}\right) = \begin{pmatrix} 0 & 0 & a \\ 0 & 0 & b \\ 0 & 0 & 0 \end{pmatrix},$$

(b)

$$\left(\frac{\partial u_i}{\partial X_j}\right) = \begin{pmatrix} 0 & -a & 0 \\ a & -a & 0 \\ b & -b & 0 \end{pmatrix}.$$

(a) $a\boldsymbol{i}+b\boldsymbol{j}$ (b) $a\boldsymbol{i}+a\boldsymbol{j}+b\boldsymbol{k}$ are directions of shear for planes X_3 = constant and $X_1 - X_2$ = constant, respectively.

7.7 (a)

$$\mathbf{G} = \mathbf{I} + \begin{pmatrix} 0 & 0 & a \\ 0 & 0 & b \\ a & b & 0 \end{pmatrix} + \begin{pmatrix} 0 & 0 & 0 \\ 0 & a^2+b^2 & 0 \\ 0 & 0 & a^2+b^2 \end{pmatrix}, \quad \mathbf{e} = \tfrac{1}{2}\begin{pmatrix} 0 & 0 & a \\ 0 & 0 & b \\ a & b & 0 \end{pmatrix}.$$

(b)

$$\mathbf{G} - \mathbf{I} = \begin{pmatrix} 2a & 0 & b \\ 0 & -2a & -b \\ b & -b & 0 \end{pmatrix} + \begin{pmatrix} 2a^2+b^2 & -2a^2-b^2 & 0 \\ -2a^2-b^2 & 2a^2+b^2 & 0 \\ 0 & 0 & 0 \end{pmatrix},$$

$$(e_{ij}) = \frac{1}{2}\begin{pmatrix} 2a & 0 & b \\ 0 & -2a & -b \\ b & -b & 0 \end{pmatrix}.$$

7.8 (a) $\partial u_1/\partial X_1 = a,\ \partial u_2/\partial X_2 = b,\ \partial u_i/\partial X_j = 0$ otherwise; $\Delta = a + b$,

$$\begin{pmatrix} \sigma_{11} & \sigma_{12} & \sigma_{13} \\ \sigma_{21} & \sigma_{22} & \sigma_{23} \\ \sigma_{31} & \sigma_{32} & \sigma_{33} \end{pmatrix} = \begin{pmatrix} (\lambda+2\mu)a+\lambda b & 0 & 0 \\ 0 & \lambda a+(\lambda+2\mu)b & 0 \\ 0 & 0 & \lambda a+\mu b \end{pmatrix}.$$

(b) $\partial u_2/\partial X_1 = \gamma,\ \partial u_i/\partial X_j = 0$ otherwise; so $\Delta = 0$,

$$\boldsymbol{\sigma} = \begin{pmatrix} 0 & \mu\gamma & 0 \\ \mu\gamma & 0 & 0 \\ 0 & 0 & 0 \end{pmatrix}.$$

(c) $\partial u_1/\partial X_3 = A,\ \partial u_2/\partial X_3 = B;\ \partial u_i/\partial X_j = 0$ otherwise; $\Delta = 0$ so

$$\boldsymbol{\sigma} = \begin{pmatrix} 0 & 0 & \mu A \\ 0 & 0 & \mu B \\ \mu A & \mu B & 0 \end{pmatrix}.$$

(d) For

$$\mathbf{e} = \begin{pmatrix} a & 0 & \frac{1}{2}b \\ 0 & -a & \frac{-1}{2}b \\ \frac{1}{2}b & \frac{-1}{2}b & 0 \end{pmatrix},\ \boldsymbol{\sigma} = \begin{pmatrix} 2\mu a & 0 & \mu b \\ 0 & -2\mu a & -\mu b \\ \mu b & -\mu b & 0 \end{pmatrix}\ (\text{since } \Delta = 0).$$

7.9 $v_1 = \omega A \sin(X_2 - \omega t)\ ,\ v_2 = 0\ ,\ v_3 = \omega A\cos(X_2 - \omega t)$.

$$\left(\frac{\partial u_i}{\partial X_j}\right) = \begin{pmatrix} a & -A\sin(X_2-\omega t) & 0 \\ 0 & b & c \\ 0 & -A\cos(X_2-\omega t) & d \end{pmatrix},\quad \Delta = a+b+d\,,$$

$$\mathbf{e} = \tfrac{1}{2}\begin{pmatrix} 2a & -A\sin(X_2-\omega t) & 0 \\ -A\sin(X_2-\omega t) & 2b & c-A\cos(X_2-\omega t) \\ 0 & c-A\cos(X_2-\omega t) & 2d \end{pmatrix},$$

$$\boldsymbol{\sigma} = \begin{bmatrix} (\lambda+2\mu)a+\mu b+\mu d & -\mu A\sin(X_2-\omega t) & 0 \\ -\mu A\sin(X_2-\omega t) & \mu a+(\lambda+2\mu)b+\mu d & \mu c-\mu A\cos(X_2-\omega t) \\ 0 & \mu c-\mu A\cos(X_2-\omega t) & \mu a+\mu b+(\lambda+2\mu)d \end{bmatrix}.$$

$i = 1;\ \partial v_1/\partial t = -\omega^2 A\cos(X_2-\omega t),\ \partial\sigma_{11}/\partial X_1 = 0,$
$\partial\sigma_{12}/\partial X_2 = -\mu A\cos(X_2-\omega t),\ \partial\sigma_{13}/\partial X_3 = 0$, so $\rho\partial v_1/\partial t = \rho\omega^2\mu^{-1}\partial\sigma_{1j}/\partial X_j$.
$i = 2;\ \partial v_2/\partial t = 0,\ \partial\sigma_{2j}/\partial X_j = 0.$
$i = 3;\ \rho\partial v_3/\partial t = \rho\omega^2 A\sin(X_2-\omega t) = \rho\omega^2\mu^{-1}\partial\sigma_{32}/\partial X_2 = \partial\sigma_{3j}/\partial X_j$.

7.10 For any $\boldsymbol{x}(\boldsymbol{X},t)$, deformation gradient may be written as $\mathbf{F} = \mathbf{RU}$, where $\mathbf{R}^T\mathbf{R} = \mathbf{I}$, $\det\mathbf{R} = +1$, $\mathbf{U} = \mathbf{U}^T$. Then

$$\mathbf{F} = \mathbf{RUR}^T\mathbf{R} = \mathbf{VR},\ \text{where}\ \mathbf{V}^T = (\mathbf{RUR}^T)^T = (\mathbf{R}^T)^T\mathbf{U}^T\mathbf{R}^T = \mathbf{RUR}^T$$

i.e. $\mathbf{V}$ is symmetric. Eigenvalues λ and eigenvectors $\mathbf{Y}$ satisfy $\mathbf{VY} = \lambda\mathbf{Y}$, then $\mathbf{RUR}^T\mathbf{Y} = \lambda\mathbf{Y}$ so $\mathbf{U}(\mathbf{R}^T\mathbf{Y}) = \mathbf{R}^T(\lambda\mathbf{Y}) = \lambda\mathbf{R}^T\mathbf{Y}$; i.e. λ is also an eigenvalue of $\mathbf{U}$.

7.11 Since $\mathbf{U} = \mathbf{I} + \mathbf{e} +$ higher order terms; $(\mathbf{I} - \mathbf{e})\mathbf{U} = \mathbf{I} - \mathbf{e} + \mathbf{e} +$ higher order terms, so $\mathbf{U}^{-1} = \mathbf{I} - \mathbf{e}$ correct to linear terms. Then $\mathbf{R} = \mathbf{F}\mathbf{U}^{-1}$

$$\approx \begin{pmatrix} 1 + \frac{\partial u_1}{\partial X_1} & \frac{\partial u_1}{\partial X_2} & \frac{\partial u_1}{\partial X_3} \\ \frac{\partial u_2}{\partial X_1} & 1 + \frac{\partial u_2}{\partial X_2} & \frac{\partial u_2}{\partial X_3} \\ \frac{\partial u_3}{\partial X_1} & \frac{\partial u_3}{\partial X_2} & 1 + \frac{\partial u_3}{\partial X_3} \end{pmatrix} \begin{pmatrix} 1 - e_{11} & -e_{12} & -e_{13} \\ -e_{21} & 1 - e_{22} & -e_{23} \\ -e_{31} & -e_{32} & 1 - e_{33} \end{pmatrix}$$

gives (e.g.) $r_{11} = 1 + \partial u_1/\partial X_1 - \partial u_1/\partial X_1 +$ quadratic terms, $r_{12} = \partial u_1/\partial X_2 - e_{12} = \frac{1}{2}(\partial u_1/\partial X_2 - \partial u_2/\partial X_1)$ (neglecting terms quadratic in elements of $\partial u_i/\partial X_j$).

Chapter 8

8.1 In (8.10), $\phi = \mathrm{Re}\{\mathcal{A}(e^{ik_3z} + e^{-ik_3z})e^{i(k_1x-\omega t)}\}$; so choose $\Phi(z) = 2\mathcal{A}\cos k_3z$. If $\phi = \mathrm{Re}\{\Phi(z)\exp i(k_1x - \omega t)\}$, then $\nabla^2\phi = \mathrm{Re}\{(-k_1^2\Phi + \Phi'')\exp i(k_1x - \omega t)\}$, $\phi_{tt} = \mathrm{Re}\{-\omega^2\Phi\exp i(k_1x - \omega t)\}$, $\partial\phi/\partial z = \mathrm{Re}\{\Phi'(z)\exp i(k_1x - \omega t)\}$. The wave equation gives $\mathrm{Re}\{(c^2\Phi''(z) - c^2k_1^2\Phi + \omega^2\Phi)\exp i(k_1x - \omega t)\} = 0$, so that

$$\Phi''(z) + (\omega^2/c^2 - k_1^2)\Phi(z) = 0\,, \quad z > 0\,.$$

At $z = 0$, the boundary condition $\partial\phi/\partial z = 0$ gives $\Phi'(z) = 0$. Thus, for $k_1^2 < \omega^2/c^2$, the solution $\Phi(z) = \mathcal{A}\cos k_3z + \mathcal{B}\sin k_3z$ with $k_3 = -\sqrt{\omega^2c^{-2} - k_1^2}$ must have $\mathcal{B} = 0$. Hence, $\Phi(z) = \mathcal{A}\cos k_3z$, so $\phi = \mathrm{Re}\{\mathcal{A}\cos k_3z\ \exp i(k_1x - \omega t)\}$ (which differs from (8.10) only by a factor 2 in the choice of the *complex* arbitrary constant $\mathcal{A}$).

8.2 For $\phi = e^{i(k_1x+k_2y-\omega t)}\Phi(z)$, $\phi_{tt} = -\omega^2 e^{i(k_1x+k_2y-\omega t)}\Phi(z)$ and $\nabla^2\phi = \{-(k_1^2 + k_2^2)\Phi(z) + \Phi''(z)\}e^{i(k_1x+k_2y-\omega t)}$. Then

$$\phi_{tt} - c^2\nabla^2\phi = \{[c^2(k_1^2 + k_2^2) - \omega^2]\Phi(z) - c^2\Phi''(z)\}e^{i(k_1x+k_2y-\omega t)} = 0$$

must hold as x, y and t are varied. Thus, $\Phi''(z) = (k_1^2 + k_2^2 - \omega^2/c^2)\Phi(z)$. If $c = c_1$, with $k_1^2 + k_2^2 < \omega^2/c_1^2$, then $\Phi''(z) + k_3^2\Phi(z) = 0$ with $k_1^2 + k_2^2 + k_3^2 = \omega^2/c_1^2$, so that $\Phi(z) = \mathcal{A}_1\cos k_3z + \mathcal{A}_2\sin k_3z$.
In $z < 0$, where $k_1^2 + k_2^2 > \omega^2/c_2^2$, then $\Phi''(z) = \gamma^2\Phi(z)$ where $\gamma^2 = k_1^2 + k_2^2 - \omega^2/c_2^2$. Choose $\gamma = \sqrt{k_1^2 + k_2^2 - \omega^2/c_2^2} > 0$, so general solution is $\Phi(z) = \mathcal{C}e^{\gamma z} + \mathcal{D}e^{-\gamma z}$. Unless $\mathcal{D} = 0$, this is unbounded as $z \to -\infty$. Hence $\Phi(z) = \mathcal{C}e^{\gamma z}$.
If $c = c_1$ in $z > 0$, then $\phi = e^{i(k_1x+k_2y-\omega t)}\Phi(z)$, with $\Phi(z) = \mathcal{A}_1\cos k_3z + \mathcal{A}_2\sin k_3z$ in $z \geq 0$, $\Phi(z) = \mathcal{C}e^{\gamma z}$ in $z \geq 0$, with $\Phi'(z)$ and $i\omega\rho\Phi(z)$ both continuous at $z = 0$. Thus $\rho_1\mathcal{A}_1 = \rho_2\mathcal{C}$, $k_3\mathcal{A}_2 = \gamma\mathcal{C}$ gives

$$\phi = \Phi(z)e^{i(k_1x+k_2y-\omega t)} = \begin{cases} \left(\frac{\rho_2}{\rho_1}\cos k_3z + \frac{\gamma}{k_3}\sin k_3z\right)\mathcal{C}e^{i(k_1x+k_2y-\omega t)} & z \geq 0 \\ \mathcal{C}e^{\gamma z}e^{i(k_1x+k_2y-\omega t)} & z \leq 0 \end{cases}$$

$$= \begin{cases} \mathcal{C}\frac{1}{2}(\rho_2/\rho_1 - i\gamma/k_3)e^{i(k_1x+k_2y+k_3z-\omega t)} \\ \qquad +\mathcal{C}\frac{1}{2}(\rho_2/\rho_1 + i\gamma/k_3)e^{i(k_1x+k_2y-k_3z-\omega t)} & z \geq 0 \\ \mathcal{C}e^{\gamma z}e^{i(k_1x+k_2y-\omega t)} & z \leq 0\,. \end{cases}$$

Then identifying $\mathcal{A} = \frac{1}{2}\mathcal{C}(\rho_2/\rho_1 - i\gamma/k_3)$ as the amplitude of the incident wave gives $\mathcal{C} = 2(\rho_2/\rho_1 - i\gamma/k_3)^{-1}$, while the reflected amplitude is $\mathcal{R} = (\rho_2/\rho_1 + i\gamma/k_3)\mathcal{C} = (\rho_2/\rho_1 + i\gamma/k_3)/(\rho_2/\rho_1 - i\gamma/k_3)$ (so generalizing (8.17)).

8.3 For $\phi = \phi(z,t)$, $\phi_{tt} = c^2\phi_{zz}$ with incident wave $\phi_I = \mathrm{Re}\{\mathcal{A}_1 e^{\imath(kz-\omega t)}\}$ and $\omega^2 = c_1^2k^2$, reflected wave $\phi = \mathrm{Re}\{\mathcal{A}_R e^{\imath(-kz-\omega t)}\}$ and transmitted wave $\phi = \mathrm{Re}\{\mathcal{A}_2 e^{\imath(k_2 z-\omega t)}\}$ and $\omega^2 = c_2^2k_2^2$. Thus $k_2 = (c_1/c_2)k < 0$. At $z = 0$, continuity of $\partial\phi/\partial z$ gives $\imath k(\mathcal{A}_1 - \mathcal{A}_R) = \imath k_2\mathcal{A}_2$ and continuity of $\rho\partial\phi/\partial t$ gives $-\imath\omega\rho_1(\mathcal{A}_1 + \mathcal{A}_R) = -\imath\omega\rho_2\mathcal{A}_2$. Solving gives $c_2\rho_2(\mathcal{A}_1 - \mathcal{A}_R) = c_1\rho_2\mathcal{A}_2 = c_1\rho_1(\mathcal{A}_1 + \mathcal{A}_R)$, so that $\mathcal{A}_R = 0$ if $\rho_2c_2 = \rho_1c_1$ (impedance matching). Then, $\mathcal{A}_2/\mathcal{A}_1 = \rho_1/\rho_2$.
If $\rho = \rho_n$ and $c = c_n$ in $-h_{n-1} > z > -h_n$ (with $h_1 = 0$), then $c = c_2$ and $\rho = \rho_2$ in $0 > z > -h_2$, etc. At the interface $z = -h_{n-1}$, the amplitudes $\mathcal{A}_{n-1}$ of the incident wave and $\mathcal{A}_n$ of the reflected wave satisfy $\mathcal{A}_n e^{-\imath k_n h_{n-1}}/[\mathcal{A}_{n-1}e^{-\imath k_{n-1}h_{n-1}}] = \rho_{n-1}/\rho_n$, when there is no reflection. Hence, for a succession of non-reflecting interfaces,

$$\left|\frac{\mathcal{A}_n}{\mathcal{A}_1}\right| = \left|\frac{\mathcal{A}_n}{\mathcal{A}_{n-1}}\right|\left|\frac{\mathcal{A}_{n-1}}{\mathcal{A}_{n-2}}\right|\cdots\left|\frac{\mathcal{A}_2}{\mathcal{A}_1}\right| = \frac{\rho_{n-1}}{\rho_n}\frac{\rho_{n-2}}{\rho_{n-1}}\cdots\frac{\rho_1}{\rho_2} = \frac{\rho_1}{\rho_n} = \frac{c_n}{c_1}.$$

8.4 Since, when $r^2 = x^2 + y^2 + z^2$, the Laplacian of $\Phi(r,t)$ is (cf. Section 2.3) $\nabla^2\Phi = \Phi_{rr} + 2r^{-1}\Phi_r$, the wave equation becomes

$$\frac{\partial^2\Phi}{\partial t^2} = c^2\left(\frac{\partial^2\Phi}{\partial r^2} + \frac{2}{r}\frac{\partial\Phi}{\partial r}\right) = \frac{c^2}{r}\frac{\partial^2(r\Phi)}{\partial r^2}.$$

Hence $(r\Phi)_{tt} = c^2(r\Phi)_{rr}$. Solutions with frequency $\omega = kc$ have $r\Phi = A\sin k(r - ct - \alpha) + B\cos k(r + ct - \beta)$ for some parameters A, B, α and β. Outgoing waves have $B = 0$, so giving

$$\phi = \Phi(r,t) = Ar^{-1}\sin k(r - ct - \alpha).$$

The pressure disturbance is $\bar{p} = -\rho_0\phi_t = kcAr^{-1}\cos k(r - ct - \alpha)$. The acoustic velocity is
$\boldsymbol{\nabla}\phi = \boldsymbol{r}r^{-1}\partial\Phi/\partial r = A\boldsymbol{r}r^{-2}\left[k\cos k(r - ct - \alpha) - r^{-1}\sin k(r - ct - \alpha)\right]$.

8.5 In Exercise 8.4 replace r by $|\boldsymbol{x} - a\boldsymbol{e}_3| \equiv r_1$ (say), so giving acoustic velocity $\boldsymbol{\nabla}\phi = A(\boldsymbol{x} - a\boldsymbol{e}_3)r_1^{-2}[k\cos k(r_1 - ct - \alpha) - r_1^{-1}\sin k(r_1 - ct - \alpha)]$.
Let $\boldsymbol{r}_2 \equiv \boldsymbol{x} + a\boldsymbol{e}_3$, $r_2 = |\boldsymbol{x} + a\boldsymbol{e}_3|$. On $z = 0$, this gives $r_1 = r_2 = (x^2 + y^2 + a^2)^{1/2}$ and $\boldsymbol{x} - a\boldsymbol{e}_3 = x\boldsymbol{e}_1 + y\boldsymbol{e}_2 - a\boldsymbol{e}_3$, $\boldsymbol{x} + a\boldsymbol{e}_3 = x\boldsymbol{e}_1 + y\boldsymbol{e}_2 + a\boldsymbol{e}_3$. Superposing potentials to give $\phi = Ar_1^{-1}\sin k(r_1 - ct - \alpha) + Ar_2^{-1}\sin k(r_2 - ct - \alpha)$
then yields

$$\begin{aligned} u_3 =& A(z - a)r_1^{-1}[k\cos k(r_1 - ct - \alpha) - r_1^{-1}\sin k(r_1 - ct - \alpha)] \\ &+ A(z + a)r_2^{-1}[k\cos k(r_2 - ct - \alpha) - r_2^{-1}\sin k(r_2 - ct - \alpha)] \end{aligned}$$

i.e. $u_3 = 0$ on $z = 0$.

8.6 For $\phi = \mathrm{Re}\{e^{-\imath\omega t}X(x)Z(z)\}$, wave equation gives $c^2(X''Z + XZ'') + \omega^2XZ = 0$. Hence $X''(x)/X(x)$ and $Z''(z)/Z(z)$ each are constant. Also $Z'(0) = 0 = Z'(a)$, so Z''/Z must be a negative constant. Relevant solutions are $Z = \cos(n\pi z/a)$, $n = 0, 1, \ldots$. Thus $Z''/Z = -n^2\pi^2/a^2$, so $X''/X = n^2\pi^2a^{-2} - \omega^2/c^2$.
Wavelike solutions require that $\omega^2/c^2 > n^2\pi^2/a^2$. For $\omega > n\pi c/a$, write $\beta_n = +\sqrt{c^{-2}\omega^2 - n^2\pi^2/a^2}$, so $X'' + \beta_n^2X = 0$ gives $X = Ae^{\imath\beta_n x} + Be^{-\imath\beta_n x}$. The factor $e^{\imath(\beta_n x - \omega t)}$ is associated with sinusoidal waves travelling towards *increasing* x, i.e. *away from* $x = 0$ into $x > 0$.
For $0 < \omega < n\pi c/a$, write $\gamma_n = \sqrt{a^{-2}n^2\pi^2 - \omega^2/c^2}$, so $X(x) = Ae^{-\gamma_n x} + Be^{\gamma_n x}$. This is unbounded as $x \to \infty$ unless $B = 0$. Hence, $\phi \propto e^{-\imath\omega t}e^{-\gamma_n x}\sin(n\pi z/a)$ is the required solution.

8.7 Suppose $N\pi c/a < \omega < (N+1)\pi c/a$. For $n = 0, 1, \ldots, N$, the potential $e^{-\imath\omega t}e^{\imath\beta_n x}\cos(n\pi z/a)$ describes a mode travelling *away from* $x = 0$. For $n = N+1, \ldots$, $e^{-\imath\omega t}e^{-\gamma_n x}\cos(n\pi z/a)$ describes an oscillation in $x \geq 0$, decaying as $x \to +\infty$. Superpose all these, so showing that the specified function is the required potential $\phi(x,z,t)$ in $x \geq 0$, $0 \leq z \leq a$.
Differentiate to give $\phi_x(x,z,t)$, then set $x = 0$;

$$U(z)\cos\omega t = \text{Re}\left\{e^{-\imath\omega t}\left[\sum_{n=0}^{N}\imath\beta_n A_n\cos\frac{n\pi z}{a} + \sum_{n=N+1}^{\infty}(-\gamma_n B_n)\cos\frac{n\pi z}{a}\right]\right\}$$
$$= \tfrac{1}{2}a_0 + \textstyle\sum_{n=1}^{\infty}a_n\cos(n\pi z/a),$$

where $\imath\beta_0 = \frac{1}{2}a_0$; $\imath\beta_n A_n = a_n$, $n = 1, 2, \ldots, N$; $-\gamma_n B_n = a_n$, $n = N+1, N+2, \ldots$.
Euler formulae give $a_0 = (2/a)\int_0^a U(z)\mathrm{d}z$, $a_n = 2a^{-1}\int_0^a U(z)\cos(n\pi z/a)\mathrm{d}z$.
Thus, $A_0 = (-\imath c/a\omega)\int_0^a U(z)\mathrm{d}z$;
$A_n = (-2\imath/a\beta_n)\int_0^a U(z)\cos(n\pi z/a)\,\mathrm{d}z$, $n = 1, 2, \ldots, N$;
$B_n = (-2/a\gamma_n)\int_0^a U(z)\cos(n\pi z/a)\,\mathrm{d}z$, $n > N$.

8.8 From $\gamma\rho_0\cos k_3 a = k_3\rho_1\sin k_3 a$, it follows that $\rho_0 V\cos U = \rho_1 U\sin U$, so that $V = (\rho_1/\rho_0)U\tan U$. Also $U^2 = k_3^2a^2 = -k^2a^2 + \omega^2a^2/c_0^2$,
$V^2 = \gamma^2a^2 = k^2a^2 - \omega^2a^2/c_1^2$ gives $U^2 + V^2 = a^2\omega^2(c_0^{-2} - c_1^{-2})$.
For fixed ω, this describes a circle in the U, V plane, centred at the origin.
Function $V(U) \equiv (\rho_1/\rho_0)U\tan U$ is *even*; its graph passes through the origin and has asymptotes at $U = \pm\pi/2, \pm 3\pi/2, \ldots$. Branch through $(0,0)$ must intersect the circle (at two points, where $U = \pm U_1(\omega)$). [Other modes exist, corresponding to $\pm U_2(\omega), \pm U_3(\omega)$ etc. as ω increases; e.g. the circle may intersect (twice) each branch having $\frac{1}{2}\pi < |U| < \frac{3}{2}\pi$.]

8.9 $X^2 = a^2\omega^2c_0^{-2} - a^2k^2$, $Y^2 = a^2k^2 - a^2\omega^2/c_1^2$ so $X^2 + Y^2 = a^2\omega^2(c_0^{-2} - c_1^{-2})$.
Also $\mu_0 X\sin X = \mu_1 Y\sin X$ for Love waves, so $Y = (\mu_0/\mu_1)X\tan X$. As in Exercise 8.8, two intersections of the circle with the (even) branch in $-\frac{1}{2}\pi < X < \frac{1}{2}\pi$ exist for *all* ω. [Since this root has $|ak_3| = |X| < \frac{1}{2}\pi$, $U(z) > 0$ throughout $-a \leq z \leq 0$, also $U(z) > 0$ in $z \leq -a$, so $U(z) > 0$ in all $z < 0$.]
For $\pi(c_0^{-2} - c_1^{-2})^{-1/2} < a\omega < \frac{3}{2}\pi(c_0^{-2} - c_1^{-2})^{-1/2}$, circle is $X^2 + Y^2 = R^2$ with $\pi < R < \frac{3}{2}\pi$. Circle crosses branch lying in $\frac{1}{2}\pi < X < \frac{3}{2}\pi$ once in $\pi < X < \frac{3}{2}\pi$ and also somewhere in $\frac{1}{2}\pi < X < \pi$. Each intersection defines k_3 and γ for a mode (corresponding negative values for X yield the same function $U(z)$, for a mode travelling in the direction of *decreasing* x).

8.10 $u_1 = 0$, $u_2 = \text{Re}\{\mathcal{A}_2e^{\imath(kx-\omega t)}\}$, $u_3 = \text{Re}\{\mathcal{A}_3e^{\imath(kx-\omega t)}\}$, so (putting $x = x_1$) $\partial u_i/\partial x_1 = \text{Re}\{\imath k\mathcal{A}_ie^{\imath(kx-\omega t)}\}$ for $i = 2, 3$, $\partial u_i/\partial x_j = 0$ otherwise. Each material plane $X =$ constant moves within $x =$ constant as a rigid body.
Since $\sigma_{12} = \sigma_{21} = \mu\partial u_2/\partial x_1$, $\sigma_{13} = \sigma_{31} = \mu\partial u_3/\partial x_1$, $\sigma_{ij} = 0$ otherwise, elastodynamic equations become
$0 = 0$, $\quad\rho\text{Re}\{-\omega^2\mathcal{A}_2e^{\imath(kx_1-\omega t)}\} = \partial\sigma_{21}/\partial x_1 = \mu\text{Re}\{-k^2\mathcal{A}_2e^{\imath(kx_1-\omega t)}\}$;
$\rho\text{Re}\{-\omega^2\mathcal{A}_3e^{\imath(kx_1-\omega t)}\} = \mu\text{Re}\{-k^2\mathcal{A}_3e^{\imath(kx_1-\omega t)}\}$.
Hence, all are satisfied if $\rho\omega^2 = \mu k^2$.
(a) $\mathcal{A}_3/\mathcal{A}_2 = a$ (real) gives $\boldsymbol{u} = (\boldsymbol{e}_2 + a\boldsymbol{e}_3)\text{Re}\{a_2e^{\imath(kx-\omega t)}\}$, polarized parallel to $\boldsymbol{e}_2 + a\boldsymbol{e}_3$.
(b) Write $\mathcal{A}_2 = a_2e^{\imath\alpha}$, a_2 real, so $u_2 = a_2\cos(kx_1 - \omega t + \alpha)$, $u_3 = \mp a_2\sin(kx_1 - \omega t + \alpha)$; right-handed for $\mathcal{A}_3 = \imath\mathcal{A}_2$ and left-handed circularly polarized for $\mathcal{A}_3 = -\imath\mathcal{A}_2$.

Since $\boldsymbol{u} = \mathrm{Re}\{b(\boldsymbol{e}_2 + \imath\boldsymbol{e}_3)\mathrm{e}^{\imath(kx-\omega t)}\} + \mathrm{Re}\{c(\boldsymbol{e}_2 - \imath\boldsymbol{e}_3)\mathrm{e}^{\imath(kx-\omega t)}\}$, the wave is a superposition of a right-handed and a left-handed circularly polarized shear wave, of respective amplitudes $|b|$ and $|c|$. Here, $|b| = \frac{1}{2}|\mathcal{A}_2 - \imath\mathcal{A}_3|$, $|c| = \frac{1}{2}|\mathcal{A}_2 + \imath\mathcal{A}_3|$.

8.11 In Exercise 8.10, put $x = x_1$ and $\mathcal{A}_2\boldsymbol{e}_2 + \mathcal{A}_3\boldsymbol{e}_3 = \boldsymbol{a}$, then $\boldsymbol{a}\cdot\boldsymbol{e}_1 = 0$ with $\mathcal{A}_2$, $\mathcal{A}_3$ as arbitrary complex components of $\boldsymbol{a}$. Write $\boldsymbol{a} = \boldsymbol{b} + \imath\boldsymbol{c}$ with $\boldsymbol{b}$ and $\boldsymbol{c}$ real. Then, $\boldsymbol{u} = \boldsymbol{b}\,\mathrm{Re}\{\mathrm{e}^{\imath(kx_1-\omega t)}\} + \boldsymbol{c}\,\mathrm{Re}\{\imath\mathrm{e}^{\imath(kx_1-\omega t)}\} = \boldsymbol{b}\cos\alpha + \boldsymbol{c}\sin\alpha$, where $\alpha = \omega t - kx_1$. Write $\boldsymbol{b} = b_2\boldsymbol{e}_2 + b_3\boldsymbol{e}_3$, $\boldsymbol{c} = c_2\boldsymbol{e}_2 + c_3\boldsymbol{e}_3$ so $u_2 = b_2\cos\alpha + c_2\sin\alpha$, $u_3 = b_3\cos\alpha + c_3\sin\alpha$. Eliminating α gives
$(c_3u_2 - c_2u_3)^2 + (b_3u_2 - b_2u_3)^2 = (b_2c_3 - b_3c_2)^2 =$ constant. Displacements (u_2, u_3) lie on a conic section, and are everywhere finite; hence conic is an ellipse. [You can check that the conic has no real asymptotes.]

8.12 $\boldsymbol{u} = u_1\boldsymbol{e}_1 + u_2\boldsymbol{e}_2 \Rightarrow \Delta = \partial u_1/\partial x + \partial u_2/\partial y$;
$\sigma_{11} = \lambda\Delta + 2\mu\partial u_1/\partial x = (\lambda + 2\mu)\partial u_1/\partial x + \mu\partial u_2/\partial y$,
$\sigma_{12} = \sigma_{21} = \mu(\partial u_1/\partial y + \partial u_2/\partial x)$;
$\sigma_{22} = \mu\partial u_1/\partial x + (\lambda + 2\mu)\partial u_2/\partial y$. Also $\sigma_{33} = \lambda\Delta$, so $\partial\sigma_{33}/\partial z = 0$ and the elastodynamic equations are
$\rho\partial^2u_1/\partial t^2 = \partial\sigma_{11}/\partial x + \partial\sigma_{12}/\partial y$, $\rho\partial^2u_2/\partial t^2 = \partial\sigma_{21}/\partial x + \partial\sigma_{22}/\partial y$ so yielding
$-\rho k^2c^2A = -(\lambda + 2\mu)k^2A + \mu\gamma^2A + \imath(\lambda + \mu)\gamma kB$,
$-\rho k^2c^2B = (\lambda + 2\mu)\gamma^2B - \mu k^2B + \imath(\lambda + \mu)\gamma kA$.
Writing $\lambda + 2\mu = \rho c_L^2$, $\mu = \rho c_S^2$, $\gamma/k \equiv \Gamma$ gives
$(c^2 - c_L^2 + \Gamma^2c_S^2)A + \imath\Gamma(c_L^2 - c_S^2)B = 0 = \imath\Gamma(c_L^2 - c_S^2)A + (c^2 + \Gamma^2c_L^2 - c_S^2)B$.
Hence, the compatibility condition gives
$(\Gamma^2c_S^2 + c^2 - c_L^2)(\Gamma^2c_L^2 - c_S^2 + c^2) + \Gamma^2(c_L^2 - c_S^2)^2 = 0$, or
$\Gamma^4 + [(c/c_L)^2 + (c/c_S)^2 - 2]\Gamma^2 + [(c/c_L)^2 - 1][(c/c_S)^2 - 1] = 0$,
with roots $\Gamma^2 = 1 - c^2/c_S^2,\ 1 - c^2/c_L^2$. Thus $\Gamma = \pm\Gamma_1$, $\Gamma = \pm\Gamma_2$, as required.
Then, for $\Gamma = \Gamma_1$, $\imath B_1/A_1 = -(c^2 - c_L^2 + \Gamma_1^2c_S^2)/(c_L^2 - c_S^2)\Gamma_1 = \Gamma_1^{-1}$;
for $\Gamma = \Gamma_2$; $\imath B_2/A_2 = (c_L^2 - c_S^2)\Gamma_2/(c^2 + \Gamma_2^2c_L^2 - c_S^2) = \Gamma_2$.

8.13 For A_1, A_2 real, $u_1 = \mathrm{Re}\{(A_1\mathrm{e}^{k\Gamma_1y} + A_2\mathrm{e}^{k\Gamma_2y})\mathrm{e}^{\imath k(x-ct)}\}$,
$u_2 = (\Gamma_1^{-1}A_1\mathrm{e}^{k\Gamma_1y} + \Gamma_2A_2\mathrm{e}^{k\Gamma_2y})\sin k(x - ct)$. Then, on $y = 0$,
$\sigma_{12} = \mu(\partial u_1/\partial y + \partial u_2/\partial x) = k\mu[(\Gamma_1 + \Gamma_1^{-1})A_1 + 2\Gamma_2A_2]\cos k(x - ct)$, $\sigma_{22} = \lambda\partial u_1/\partial x + (\lambda + 2\mu)\partial u_2/\partial y = k[2\mu A_1 + ((\lambda + 2\mu)\Gamma_2^2 - \lambda)A_2] \times \sin k(x - ct)$.
Vanishing traction yields $(\Gamma_1^2 + 1)A_1 = -2\Gamma_1\Gamma_2A_2$ and
$2c_S^2A_1 = -(c_L^2\Gamma_2^2 - c_L^2 + 2c_S^2)A_2 = -c_S^2(1 + \Gamma_1^2)A_2$. Eliminating A_1/A_2 gives $(\Gamma_1^2 + 1)^2 = 4\Gamma_1\Gamma_2$. Squaring both sides, then substituting for Γ_i^2 gives (after removing a common factor c^2) the required cubic equation $f(c^2) = 0$. Since $f(0) = -16c_S^6(1 - c_S^2/c_L^2) < 0$ and $f(c_S^2) = c_S^6 > 0$, the cubic equation has a root c_R^2 lying in $0 < c_R^2 < c_S^2$. [Actually, for all physically relevant ratios c_S/c_L, there is a *unique* Rayleigh wave speed c_R.]

Chapter 9

9.1 Take divergence of (9.10) as $0 = \mathrm{div}(\mathrm{curl}\,\boldsymbol{H}) = \mathrm{div}(\partial\boldsymbol{D}/\partial t) + \mathrm{div}\,\boldsymbol{J}$ $= \partial(\mathrm{div}\,\boldsymbol{D})/\partial t + \mathrm{div}\,\boldsymbol{J}$. Thus, inserting (9.7) gives $\partial\rho_e/\partial t + \mathrm{div}\,\boldsymbol{J} = 0$.

9.2 (a) From (9.10), $\partial^2\boldsymbol{E}/\partial t^2 = \epsilon^{-1}\partial(\partial\boldsymbol{D}/\partial t)/\partial t = \epsilon^{-1}\partial(\boldsymbol{\nabla}\times\boldsymbol{H})\partial t$ $= \epsilon^{-1}\boldsymbol{\nabla}\times(\partial\boldsymbol{H}/\partial t) = (\epsilon\mu)^{-1}\boldsymbol{\nabla}\times(\partial\boldsymbol{B}/\partial t) = -(\epsilon\mu)^{-1}\boldsymbol{\nabla}\times(\boldsymbol{\nabla}\times\boldsymbol{E}) = (\epsilon\mu)^{-1}(\nabla^2\boldsymbol{E} - \boldsymbol{\nabla}(\boldsymbol{\nabla}\cdot\boldsymbol{E}))$. Since $\boldsymbol{\nabla}\cdot\boldsymbol{E} = \epsilon^{-1}\boldsymbol{\nabla}\cdot\boldsymbol{D} = 0$, this gives
$\partial^2\boldsymbol{E}/\partial t^2 = (\epsilon\mu)^{-1}\nabla^2\boldsymbol{E}\ \ (= c^2\nabla^2\boldsymbol{E})$.
(b) From (9.9), $\partial^2\boldsymbol{H}/\partial t^2 = -\mu^{-1}\partial(\boldsymbol{\nabla}\times\boldsymbol{E})/\partial t = -\mu^{-1}\boldsymbol{\nabla}\times(\partial\boldsymbol{E}/\partial t) = -(\epsilon\mu)^{-1}\boldsymbol{\nabla}\times(\boldsymbol{\nabla}\times\boldsymbol{H}) = -(\epsilon\mu)^{-1}\{\boldsymbol{\nabla}(\boldsymbol{\nabla}\cdot\boldsymbol{H}) - \nabla^2\boldsymbol{H}\}$. Using $\boldsymbol{H} = \mu^{-1}\boldsymbol{B}$ and $\boldsymbol{\nabla}\cdot\boldsymbol{B} = 0$, this gives $\partial^2\boldsymbol{H}/\partial t^2 = (\epsilon\mu)^{-1}\nabla^2\boldsymbol{H}\ \ (= c^2\nabla^2\boldsymbol{H})$.

9.3 Let $\boldsymbol{a} = A_2\boldsymbol{e}_2 + A_3\boldsymbol{e}_3$; components of $\boldsymbol{D}$ are $D_1 = 0$, $D_2 = \mathrm{Re}\{\epsilon A_2 e^{\imath(kx-\omega t)}\}$, $D_3 = \mathrm{Re}\{\epsilon A_3 e^{\imath(kx-\omega t)}\}$. Since $\partial D_2/\partial y = 0 = \partial D_3/\partial z$, then $\operatorname{div}\boldsymbol{D} = 0$ satisfying (9.7).
Also, $\epsilon\mu\partial^2\boldsymbol{E}/\partial t^2 = -\omega^2\epsilon\mu\boldsymbol{E}$, while $\nabla^2\boldsymbol{E} \equiv \boldsymbol{e}_1\nabla^2 E_1 + \boldsymbol{e}_2\nabla^2 E_2 + \boldsymbol{e}_3\nabla^2 E_3 = 0 + (-k^2)\boldsymbol{e}_2 E_2 + (-k^2)\boldsymbol{e}_3 E_3 = -k^2\boldsymbol{E}$. Hence $\boldsymbol{E}$ satisfies the wave equation when $k^2 = \epsilon\mu\omega^2$. Since A_2, A_3 are real, then $\boldsymbol{E} = (A_2\boldsymbol{e}_2 + A_3\boldsymbol{e}_3)\cos(kx - \omega t)$ so $\boldsymbol{E}$ has constant direction, perpendicular to $\boldsymbol{e}_1$ (a linearly polarized wave). Use $\mu\partial\boldsymbol{H}/\partial t = -\nabla\times\boldsymbol{E} = -\boldsymbol{e}_3\partial E_2/\partial x + \boldsymbol{e}_2\partial E_3/\partial x = \mathrm{Re}\{\imath k(A_3\boldsymbol{e}_2 - A_2\boldsymbol{e}_3)e^{\imath(kx-\omega t)}\}$, then $\mu\boldsymbol{H}(\boldsymbol{x},t) = k\omega^{-1}(A_2\boldsymbol{e}_3 - A_2\boldsymbol{e}_3)\mathrm{Re}\{e^{\imath(kx-\omega t)}\} + \mu\hat{\boldsymbol{H}}(\boldsymbol{x})$. Also, since $\nabla\times\boldsymbol{H} = \epsilon\partial\boldsymbol{E}/\partial t = \epsilon\mathrm{Re}\{-\imath\omega(A_2\boldsymbol{e}_2 + A_3\boldsymbol{e}_3)e^{\imath(kx-\omega t)}\}$ and $\nabla\times\boldsymbol{H} = k^2/(\omega\mu)\mathrm{Re}\{-\imath k(A_3\boldsymbol{e}_3 + A_2\boldsymbol{e}_2)e^{\imath(kx-\omega t)}\} + \nabla\times\hat{\boldsymbol{H}}$. Using $k^2 = \epsilon\mu\omega^2$, this is consistent if $\nabla\times\hat{\boldsymbol{H}} = \boldsymbol{0}$. Actually, $\hat{\boldsymbol{H}}$ is a static magnetic field (which must be irrotational, i.e. $\hat{\boldsymbol{H}} = -\mu^{-1}\nabla\phi m$, see Section 4.4). The contribution $\hat{\boldsymbol{H}}$ is steady and so is not part of the electromagnetic wave. The required magnetic field is $\boldsymbol{H} = k/(\omega\mu)(A_2\boldsymbol{e}_3 - A_3\boldsymbol{e}_2)\cos(kx - \omega t) = k/(\omega\mu)\,\boldsymbol{e}_1\times\boldsymbol{E}$.
Since $\boldsymbol{E} = \mathrm{Re}\{(\boldsymbol{a} + \imath\boldsymbol{a}\times\boldsymbol{e}_1)e^{\imath(kx-\omega t)}\} = \boldsymbol{a}\cos(kx - \omega t) + \boldsymbol{e}_1\times\boldsymbol{a}\sin(kx - \omega t)$ is the sum of two plane waves of equal amplitude, but one quarter-period out of phase (so that $|\boldsymbol{E}| = $ constant), this describes a right-handed circularly polarized wave. Similarly, $\boldsymbol{E} = \mathrm{Re}\{(\boldsymbol{a} - \imath\boldsymbol{a}\times\boldsymbol{e}_1)e^{\imath(kx-\omega t)}\} = \boldsymbol{a}\cos(kx - \omega t) + \boldsymbol{a}\times\boldsymbol{e}_1\sin(kx - \omega t)$ describes a left-handed circularly polarized wave.

9.4 For $\boldsymbol{k} = k(\boldsymbol{e}_1\sin\beta - \boldsymbol{e}_3\cos\beta)$ with $\omega/k = (\epsilon\mu)^{-1/2}$, a wave with
$\boldsymbol{E} = E_0(\boldsymbol{e}_1\cos\beta + \boldsymbol{e}_3\sin\beta)\cos\alpha\cos(\boldsymbol{k}\cdot\boldsymbol{x} - \omega t)$
corresponds to a TM incident wave, with magnetic field
$\boldsymbol{H} = (\omega\mu)^{-1}\boldsymbol{k}\times\boldsymbol{E} = -(\epsilon/\mu)^{1/2}E_0\boldsymbol{e}_2\cos\alpha\,\cos(\boldsymbol{k}\cdot\boldsymbol{x} - \omega t)$.
The corresponding reflected wave has (from Section 9.1.3)
$\boldsymbol{H} = -(\epsilon/\mu)^{1/2}E_0\boldsymbol{e}_2\cos\alpha\,\cos(\boldsymbol{k}_R\cdot\boldsymbol{x} - \omega t)$, where $\boldsymbol{k}_R = k(\boldsymbol{e}_1\sin\beta + \boldsymbol{e}_3\cos\beta)$ and $\boldsymbol{E} = E_0(-\boldsymbol{e}_1\cos\beta + \boldsymbol{e}_3\sin\beta)\cos\alpha\,\cos(\boldsymbol{k}_R\cdot\boldsymbol{x} - \omega t)$.
Somewhat similarly, the linearly polarized TE wave with
$\boldsymbol{E} = E_0\boldsymbol{e}_2\sin\alpha\,\cos(\boldsymbol{k}\cdot\boldsymbol{x} - \omega t)$ gives rise to a reflected wave with
$\boldsymbol{E} = -E_0\boldsymbol{e}_2\sin\alpha\cos(\boldsymbol{k}_R\cdot\boldsymbol{x} - \omega t)$. The total reflected wave is
$\boldsymbol{E}_R = E_0\{(-\boldsymbol{e}_1\cos\beta + \boldsymbol{e}_3\sin\beta)\cos\alpha - \boldsymbol{e}_2\sin\alpha\}\cos(\boldsymbol{k}_R\cdot\boldsymbol{x} - \omega t)$.
This is linearly polarized. Its polarization direction is given by the unit vector $(\boldsymbol{e}_3\sin\beta - \boldsymbol{e}_1\cos\beta)\cos\alpha - \boldsymbol{e}_2\sin\alpha$, obtained from that of $\boldsymbol{E}_I$ by changing the sign of the $\boldsymbol{e}_1$ and $\boldsymbol{e}_2$ components, but leaving the $\boldsymbol{e}_3$ component unchanged (i.e. obtained from $\boldsymbol{k}$ by rotating through 180° around the normal to the conducting surface).

9.5 From chain rule,

$$\frac{\partial\phi}{\partial x} = \frac{\partial\phi}{\partial r}\frac{\partial r}{\partial x} + \frac{\partial\phi}{\partial\theta}\frac{\partial\theta}{\partial x} = xr^{-1}\phi_r - yr^{-2}\phi_\theta\,, \quad \frac{\partial\phi}{\partial y} = \frac{\partial\phi}{\partial r}\frac{y}{r} + \frac{\partial\phi}{\partial\theta}\frac{x}{r^2}\,.$$

Differentiating again gives $\phi_{xx} = r^{-1}\phi_r + x^2r^{-1}[r^{-1}\phi_{rr} - r^{-2}\phi_r] - xyr^{-3}\phi_{r\theta}$ $-yxr^{-1}(-2r^{-3}\phi_\theta + r^{-2}\phi_{r\theta}) + y^2r^{-4}\phi_{\theta\theta}$ and $\phi_{yy} = r^{-1}\phi_r + y^2r^{-1}[r^{-1}\phi_{rr} - r^{-2}\phi_r] + xyr^{-3}\phi_{r\theta}$ $+xyr^{-1}(-2r^{-3}\phi_\theta + r^{-2}\phi_{r\theta}) + x^2r^{-4}\phi_{\theta\theta}$, so that adding gives the required result $\nabla^2\phi = \phi_{xx} + \phi_{yy} = \phi_{rr} + r^{-1}\phi_r + r^{-2}\phi_{\theta\theta}$.

9.6 Since $\phi_r = F'(r)\cos\theta$, $\phi_{rr} = F''(r)\cos\theta$, $\phi_{\theta\theta} = -F(r)\cos\theta$; substituting
$\Rightarrow\ \{F''(r) + r^{-1}F'(r) + (K - r^{-2})F(r)\}\cos\theta = 0$
$\Rightarrow\ r^2F''(r) + rF'(r) + (\kappa^2r^2 - 1)F(r) = 0$ (when $K = \kappa^2$).
In $F(r) = CJ_0'(\kappa r) + DY_0'(\kappa r)$, write $\kappa r = s$, so $F = CJ_0'(s) + DY_0'(s)$,
$F'(r) = \kappa[CJ_0''(s) + DY_0''(s)]$, $\quad F''(r) = \kappa^2[CJ_0'''(s) + DY_0'''(s)]$.

Differentiate $s^2 J_0''(s) + sJ_0'(s) + s^2 J_0(s) = 0$
giving $s^2 J_0'''(s) + sJ_0''(s) + s^2 J_0'(s) + 2sJ_0''(s) + J_0'(s) + 2sJ_0(s) = 0$
$= s^2 J_0'''(s) + sJ_0''(s) + (s^2-1)J_0'(s)$.
Similarly $s^2 Y_0'''(s) + sY_0''(s) + (s^2-1)Y_0'(s) = 0$,
so $r^2 F''(r) + rF'(r) + (\kappa^2 r^2 - 1)F(r) = 0 \quad \forall C, D$ (see footnote on p.203).

Chapter 10

10.1 $c_x = 2\pi^{-1/2}e^{-(x^2/4Dt)}(4Dt)^{-1/2} = (\pi Dt)^{-1/2}e^{-(x^2/4Dt)}$ which satisfies $(c_x)_t = D(c_x)_{xx}$. Rename this as $c(x,t) = (\pi Dt)^{-1/2}e^{-(x^2/4Dt)}$.
(a) $\max e^{-(x^2/4Dt)} = 1$, so c has maximum value $(\pi Dt)^{-1/2}$.
(b) It occurs at $x = 0$.
(c) Write $\sigma = x(4Dt)^{-1/2}$, so $dx = 2(Dt)^{1/2}d\sigma$;
$\int_{-\infty}^{\infty} c(x,t)\,dx = \pi^{-1/2}2\int_{-\infty}^{\infty} e^{-\sigma^2}\,d\sigma = 1-(-1) = 2$.
Thus, at each t, graph of $c(x,t)$ is bell-shaped, centred at $x=0$; height diminishes $\propto t^{-1/2}$, width expands like $t^{1/2}$; area under *each* curve = 2.

10.2 Let $xt^{-1/4} \equiv \zeta$, assume $c(x,t) = t^{-1/4}F(\zeta)$; so
$c_x = t^{-1/2}F'$, $c_t = -\frac{1}{4}t^{-5/4}F - \frac{1}{4}t^{-5/4}xt^{-1/4}F' = -\frac{1}{4}t^{-5/4}[F(\zeta) + \zeta F'(\zeta)]$.
Also $c^2 c_x = t^{-1}F^2F'(\zeta)$, so $(c^2c_x)_x = t^{-5/4}[F^2F'(\zeta)]'$.
Substituting gives $B[F^2F']' = -\frac{1}{4}[F+\zeta F'] = -\frac{1}{4}(\zeta F)'$,
so $4BF^2F'(\zeta) + \zeta F = C$, a constant. If $F \to 0$ within solution, then $C = 0$.
Hence $(4BFF'(\zeta)+\zeta)F = 0$. Thus, $F^2 + \zeta^2/(4B) = a^2$ (a constant).
Hence $F(\zeta) = \sqrt{a^2 - \zeta^2/4B} \quad \Rightarrow \quad c(x,t) = t^{-1/4}\{a^2 - x^2/(4Bt^{1/2})\}^{1/2}$ for $x^2 \le 4Bt^{1/2}a^2$, i.e. $|x| < 2aB^{1/2}t^{1/4}$.

10.3 For $u = e^{\alpha t}c$; $u_t = e^{\alpha t}(c_t + \alpha c)$, $u_{xx} = e^{\alpha t}c_{xx}$.
Substitute, requiring $e^{\alpha t}(c_t+\alpha c) = e^{\alpha t}[Dc_{xx} + \alpha c]$, so $c_t = Dc_{xx}$.
Solution $c = (\pi Dt)^{-1/2}\exp(-x^2/4Dt)$ has $\int_{-\infty}^{\infty} c(x,t)\,dx = 2$. Required solution to $u_t = Du_{xx} + u$ is $u(x,t) = [\frac{1}{2}M(\pi Dt)^{-1/2}\exp(-x^2/4Dt)]e^{\alpha t}$.

10.4 $\dot{X} = X(-\alpha X - bY + a)$, $\dot{Y} = Y(dX + \beta Y - c)$ so, for $X>0$, $Y>0$, regions are
(a) $\alpha X + bY < a$, (b) $\alpha X + bY > a$, (c) $dX + \beta Y > c$, (d) $dX + \beta Y < c$.
Equilibrium state lies at intersection of lines $\alpha X + bY = a$, $dX + \beta Y = c$.
If $a/\alpha > c/d$ with $a/b < c/\beta$, or $a/\alpha < c/d$ with $a/b > c/\beta$,
they intersect in $X>0$, $Y>0$ at (X_0, Y_0),
with $X_0 = (cb - a\beta)/(bd - \alpha\beta)$, $Y_0 = (ad - \alpha c)/(bd - \alpha\beta)$.

10.5 For $\hat{u} = J_0(kr)U(t)$, $\hat{u}_t = J_0(kr)\dot{U}(t)$,
$\hat{u}_{xx} + \hat{u}_{yy} = [k^2J_0''(kr) + r^{-1}kJ_0'(kr)]U(t)$, etc.
Substitute, using $k^2[J_0''(kr) + (kr)^{-1}J_0'(kr)] = -k^2J_0(kr)$, to obtain (10.25).
[N.B. System (10.25) governs evolution of solutions of the form $\hat{u} = U(t)F(x,y)$, $\hat{v} = V(t)F(x,y)$ with F *any* solution $F_{xx} + F_{yy} = -k^2F$.]

10.6 Use $f(u_0+\hat{u}, v_0+\hat{v}) \approx f(u_0,v_0) + \hat{u}f_u(u_0,v_0) + \hat{v}f_v(u_0,v_0)$
+ higher order terms.
Since $f(u_0,v_0) = 0$, linearization gives $a_{11} = f_u(u_0,v_0)$, $a_{12} = f_v(u_0,v_0)$, with a_{21}, a_{22} found similarly. For $f(u,v) = au - \alpha u^2 - buv$, $g(u,v) = -cv + duv$; u_0 and v_0 are found from $a - \alpha u_0 - bv_0 = 0$, $-c + du_0 = 0$, so $u_0 = c/d$, $v_0 = (ad - \alpha c)/bd$.
Then $a_{11} = f_u(u_0,v_0) = a - 2\alpha u_0 - bv_0 = -\alpha u_0$; $a_{12} = f_v(u_0,v_0) = -bu_0$, $a_{21} = g_u(u_0,v_0) = dv_0$, $a_{22} = -c + du_0 = 0$.

[Or use $f = a(u_0 + \hat{u}) - \alpha(u_0 + \hat{u})^2 - b(u_0 + \hat{u})(v_0 + \hat{v})$
$= au_0 - \alpha u_0^2 - bu_0v_0 + \hat{u}(a - 2\alpha u_0 - bv_0) - \hat{v}bu_0 +$ quadratic terms, etc.]
Equation defining p becomes
$p^2 + [\alpha u_0 + k^2(D_1 + D_2)]p + (\alpha u_0 + k^2 D_1)k^2 D_2 + bdu_0v_0 = 0.$
Roots p_1, p_2 satisfy $p_1 + p_2 = -[\alpha u_0 + k^2(D_1 + D_2)] < 0$ and
$p_1p_2 = (\alpha u_0 + k^2 D_1)k^2 D_2 + bdu_0v_0 > 0$. If p_1, p_2 are real, then both are negative. If complex conjugates, each has real part $-\frac{1}{2}[\alpha u_0 + k^2(D_1 + D_2)] < 0$. Thus, no roots with positive real part can arise.

10.7 Given data lead to $p^2 + (1+9k^2)p + 8k^4 - 6k^2 + 1 = 0$. Also $k_*^2 = (-2+8)/16 = \frac{3}{8}$. Since roots p_1, p_2 have $p_1p_2 = 8(k^2 - \frac{1}{4})(k^2 - \frac{1}{2})$ then, for $\frac{1}{4} < k^2 < \frac{1}{2}$, both are real, one being negative, the other positive (so giving *diffusive instability*).
At $k = k_* = \frac{3}{8}$, positive root is $p = \frac{1}{16}[-35 + \sqrt{35^2 + 32}] = 0.02839$.

10.8 Solutions of (10.24) may be superposed. Use $(k_1, k_2) = (k, 0)$ and $(k_1, k_2) = (0, k)$, giving $\hat{u} = U(\cos kx + \cos ky)e^{\imath\omega t}$, etc. Since $\cos kx + \cos ky$ has maximum value 2 at $(x, y) = k^{-1}(2m\pi, 2n\pi)$ and minimum value -2 at midpoints $(x, y) = k^{-1}((2m + 1)\pi, (2n + 1)\pi)$ of that square array (m, n integers) while $\cos kx + \cos ky = 2\cos\frac{k}{2}(x + y)\cos\frac{k}{2}(x - y) = 0$ on lines $x + y = (2m + 1)\pi/k$ and on $x - y = (2n + 1)\pi/k$, then $\hat{u} = 2\cos\frac{k}{2}(x + y)\cos\frac{k}{2}(x - y)\,e^{\imath\omega t}$ is an oscillating chequerboard pattern. Pattern $\hat{v} = (V/U)\hat{u}(x, y, t)$ is similar.

10.9 $(10.33)_3$ may be rewritten as $d(e^{-\gamma\theta/c}W)/d\theta = -(b\gamma/c)e^{-\gamma\theta/c}U(\theta)$, so integration and use of $u \to 0$ as $\theta \to \infty$ gives (10.34).
Taking $U(s) = 0$ for $s > 0 \Rightarrow W(\theta) = 0$ for $\theta > 0$. For $-(2D)^{-1/2} \le \theta \le 0$, evaluate integral as $\int_\theta^\infty e^{-\gamma s/c}U(s)ds = -\int_\theta^0 e^{-\gamma s/c}(-\sqrt{2D})s\,ds$, which simplifies to $\Gamma^{-1}\left\{\frac{c}{\gamma} - \frac{c}{\gamma}e^{-\gamma\theta/c} - \theta e^{-\gamma\theta/c}\right\}$ for $\Gamma \equiv (\gamma/c)(2D)^{-1/2}$.
Thus, $W(\theta) = b\Gamma^{-1}[e^{\gamma\theta/c} - 1 - \gamma\theta/c]$ in $-(2D)^{-1/2} \le \theta \le 0$.
Put $\theta = -(2D)^{-1/2} \Rightarrow W(-(2D)^{-1/2}) = b\Gamma^{-1}(e^{-\Gamma} - 1 + \Gamma) = W_1$.
For $\theta < -(2D)^{-1/2}$, $W'(\theta) = (\gamma/c)(W - b)$
$\Rightarrow W(\theta) - b = (W_1 - b)\exp\{(\gamma/c)[\theta + (2D)^{-1/2}]\}$.
Thus, $W(\theta) = b + e^{\Gamma}(W_1 - b)e^{\gamma\theta/c} = b + b(1 - e^{\Gamma})e^{\gamma\theta/c}$ for $\theta < -(2D)^{-1/2}$.
If $\Gamma \ll 1$, then $W_1/b \approx \frac{1}{2}\Gamma^2/\Gamma = \frac{1}{2}\Gamma \ll 1$, while $W \to b$ as $\theta \to -\infty$.

Bibliography

Arrowsmith DK and Place CM (1992) Dynamical systems: Differential equations, maps and chaotic behaviour. Chapman and Hall, London.

Billingham J and King AC (2000) Wave motion. Cambridge University Press, Cambridge.

Brown JW and Churchill RV (1996) Complex variables and applications (6th edn). McGraw-Hill, New York.

Evans GA, Blackledge JM and Yardley PD (1999) Analytical methods for partial differential equations, Springer, London.

Fitzhugh R (1969) Mathematical models of excitation and propagation in nerve fibres, in: *Biological Engineering* (ed. HP Schwan), McGraw-Hill, New York.

Fourier J (1822) The analytical theory of heat (transl. A Freeman) Dover, New York (1955).

Hodgkin AL and Huxley AF (1952) A qualitative description of membrane current and its application to conduction and excitation of nerves. *Journal of Physiology* **117**, 500–544.

Hoppenstaedt FC (1982) Mathematical methods of population biology. Cambridge University Press, Cambridge.

Hunter SC (1976) Mechanics of continuous media. Ellis Horwood, Chichester.

Jones DS and Sleeman BD (2002) Differential equations and mathematical biology. Chapman and Hall/CRC, London.

Lorrain P, Corson DR and Lorrain F (2000) Fundamentals of electromagnetic phenomena. Freeman, New York.

Love AEH (1911) Some problems of geodynamics. Cambridge University Press, Cambridge.

MacCallum WG, Hughes-Hallett D, Gleason AM *et al.* (1997) Multivariable calculus. John Wiley, New York.

Matthews PC (1998) Vector calculus. Springer, London.

Midwinter JE and Guo YL (1992) Optoelectronics and lightwave technology. Wiley, Chichester.

Milne-Thomson LM (1968) Theoretical hydrodynamics (6th edn). Macmillan, London.

Osborne AD (1999) Complex variables and their application. Addison-Wesley, Harlow.

Pain HJ (1998) The physics of vibrations and waves. Wiley, Chichester.

Paterson AR (1983) A first course in fluid dynamics. Cambridge University Press, Cambridge.

Simmons GF (1972) Differential equations with applications and historical notes. McGraw-Hill, New York.

Snyder AW and Love JD (1983) Optical waveguide theory. Chapman and Hall, London.

Spencer AJM (1979) Continuum mechanics. Longman, London.

Tolstoy I and Clay CS (1987) Ocean acoustics. American Institute of Physics, New York.

Volterra V (1931) Leçons sur la théorie mathématique de la lutte pour la vie. Gauthier-Villars, Paris.

Index